Implementation of Smart Healthcare Systems using AI, IoT, and Blockchain

Intelligent Data-Centric Systems

Implementation of Smart Healthcare Systems using AI, IoT, and Blockchain

Edited by

Chinmay Chakraborty

Department of Electronics and Communication Engineering, Birla Institute of Technology, Mesra, Jharkhand, India

Subhendu Kumar Pani

Krupajal Engineering College, Bhubaneswar, Khordha, Odisha, India

Mohd Abdul Ahad

Department of Computer Science and Engineering, School of Engineering Sciences and Technology, Jamia Hamdard, New Delhi, Delhi, India

Qin Xin

Faculty of Science and Technology, University of the Faroe Islands, Torshavn, Faroe Islands, Denmark

Series Editor: Fatos Xhafa

UPC-BarcelonaTech, Barcelona, Spain

Academic Press is an imprint of Elsevier
125 London Wall, London EC2Y 5AS, United Kingdom
525 B Street, Suite 1650, San Diego, CA 92101, United States
50 Hampshire Street, 5th Floor, Cambridge, MA 02139, United States
The Boulevard, Langford Lane, Kidlington, Oxford OX5 1GB, United Kingdom

ISBN: 978-0-323-91916-6

For Information on all Academic Press publications
visit our website at https://www.elsevier.com/books-and-journals

Publisher: Mara Conner
Editorial Project Manager: Emily Thomson
Production Project Manager: Prasanna Kalyanaraman
Cover Designer: Victoria Pearson

Typeset by MPS Limited, Chennai, India

Contents

List of contributors

Muyideen AbdulRaheem
Department of Computer Sciences, University of Ilorin, Ilorin, Kwara State, Nigeria

Emmanuel Abidemi Adeniyi
Department of Computer Sciences, Landmark University, Omu-Aran, Kwara State, Nigeria

Rishabh Anand
Service Deliver Manager, HCL Technologies Limited, New Delhi, India

F. Antony Xavier Bronson
Department of Biotechnology, Dr. MGR Educational and Research Institute, Chennai, Tamil Nadu, India

Joseph Bamdele Awotunde
Department of Computer Sciences, University of Ilorin, Ilorin, Kwara State, Nigeria

K. Bhaskar
Department of Automobile Engineering, Rajalakshmi Engineering College, Chennai, Tamil Nadu, India

Akash Kumar Bhoi
KIET Group of Institutions, Delhi-NCR, Ghaziabad, Uttar Pradesh, India; Directorate of Research, Sikkim Manipal University, Gangtok, Sikkim, India

Korhan Cengiz
Department of Telecommunication, Trakya University, Edirne, Turkey

Chinmay Chakraborty
Department of Electronics and Communication Engineering, Birla Institute of Technology, Mesra, Jharkhand, India

Anubha Dubey
Independent Researcher and Analyst, Noida, Uttar Pradesh, India

Pijush Dutta
Department of Electronics and Communication Engineering, Greater Kolkata College of Engineerting & Management, West Bengal, India

A. Ganesan
Department of EEE, RRASE College of Engineering, Chennai, Tamil Nadu, India

Mostafa Ghobaei-Arani
Department of Computer Engineering, Qom Branch, Islamic Azad University, Qom, Iran

Rajeswary Hari
Department of Biotechnology, Dr. MGR Educational and Research Institute, Chennai, Tamil Nadu, India

R. Jayagowri
Department of Electronics and Communication Engineering, BMS College of Engineering, Bengaluru, Karnataka, India

Roohum Jegan
Department of Electronics and Communication Engineering, BMS College of Engineering, Bengaluru, Karnataka, India

Rasheed Gbenga Jimoh
Department of Computer Sciences, University of Ilorin, Ilorin, Kwara State, Nigeria

A. Kalaivani
Department of CSE, Saveetha School of Engineering, SIMATS, Chennai, Tamil Nadu, India

Madhurima Majumder
Department of Computer Science Engineering, Global Institute of Management and Technology, Krishnagar, West Bengal, India

G. Nalinashini
Department of EIE, RMD Engineering College, Chennai, Tamil Nadu, India

Idowu Dauda Oladipo
Department of Computer Sciences, University of Ilorin, Ilorin, Kwara State, Nigeria

Shobhandeb Paul
Guru Nanak Institute of Technology, West Bengal, India

G.Boopathi Raja
Department of Electronics and Communication Engineering, Velalar College of Engineering and Technology, Erode, Tamil Nadu, India

K. Sethil
Department of Mechanical Engineering, Dr. MGR Educational and Research Institute, Chennai, Tamil Nadu, India

Ali Shahidinejad
Department of Computing and IT, Global College of Engineering and Technology, Muscat, Oman

Devendra Kumar Sharma
Faculty of Engineering and Technology, Department of Electronics and Communication Engineering, SRM Institute of Science & Technology, Delhi NCR Campus, Ghaziabad, Uttar Pradesh, India

Abhishek Singhal
Faculty of Engineering and Technology, Department of Electronics and Communication Engineering, SRM Institute of Science & Technology, Delhi NCR Campus, Ghaziabad, Uttar Pradesh, India

Sanjay Kumar Sinha
Department of Physics, Birla Institute of Technology, Mesra, Patna Campus, Patna, Bihar, India

K. Sujatha
Department of EEE, Dr. MGR Educational and Research Institute, Chennai, Tamil Nadu, India

Aumnat Tongkaw
Faculty of Science and Technology, Songkhla Rajabhat University, Songkhla, Thailand

Apurva Saxena Verma
Researcher Computer Science, Bhopal, Madhya Pradesh, India

Mohsen Hosseini Yekta
Department of Engineering, Garmsar Branch, Islamic Azad University, Garmsar, Iran

Preface

The novel applications of data fusion and analytics for healthcare can be regarded as an emerging field in computer science, information technology, and biomedical engineering. Data fusion is a fertile area of research that is rapidly growing since it presents a means for combining pieces of information from various sources/sensors, resulting in ameliorated overall system performance (better decision making, improved detection capabilities, reduced number of false alarms, improved reliability in various situations) using separate sensors/sources. Various data fusion techniques have been developed to optimize the overall system output in a variety of applications for which data fusion might be helpful: security (military, humanitarian), medical diagnosis, environmental monitoring, remote sensing, robotics, etc. Data fusion and analytics discover the data-driven detection model in science and the need to manage large amounts of varied data. Drivers of this transformation include the enhanced availability and accessibility of hyphenated analytical stand, imaging techniques, and the expansion of information technology. As big data-driven analytical research deals with an inductive approach that intends to extract information and build models capable of gathering the fundamental phenomena from the data itself. The internet of things (IoT) helps in the creation of smart spaces by changing existing environments into sensor-enabled data-centric cyber-physical systems with a rising degree of automation, leading to Industry 4.0. When implemented in commercial/industrial contexts, this trend is transforming many features of our day-to-day life, considering the way people access and get healthcare services. As we progress towards Healthcare Industry 4.0, the underlying data-rich IoT systems of Smart Healthcare spaces are growing in size and complexity, making it significant to make sure that tremendous amounts of collected data are correctly processed to give helpful insights and decisions according to necessities in place.

This book will be a pivotal reference source that provides imperative research on the development of data fusion and analytics for healthcare and their implementation into current issues in a real-time environment. While highlighting topics such as IoT, bio-inspired computing, big data, and evolutionary programming, this publication will explore various concepts and theories of data fusion, IoT, and Big Data Analytics. This book will be ideally designed for IT specialists, researchers, academicians, engineers, developers, practitioners, and students seeking current research on data fusion and analytics for healthcare. This book also investigates the challenges and methodologies required to integrate data from heterogeneous multiple sources, and analytical platforms in healthcare sectors. With the recent COVID-19 pandemic, nations are embracing the new normal that has completely transformed the healthcare sector. Technology-driven and innovative healthcare facilities are the need of the hour. This book is unique in the way that it will provide a useful insight into the implementation of a “Smart and Intelligent Healthcare System” in a post-pandemic world using enabling technologies like Artificial Intelligence, the IoT, and blockchain. This book would focus on recent advances and different research areas in multi-modal data fusion under smart healthcare and would also seek out theoretical, methodological, well-established, and validated empirical work dealing with these different topics.

At last, we would like to extend our sincere thanks to authors from industry, academia, and policy expertise to complete this work for aspiring researchers in this domain. We are confident that this book would play a key role in providing readers a comprehensive view of medical sensor data and developments around it and can be used as a learning resource for various examinations, which deal with cutting-edge technologies.

Chinmay Chakraborty
Subhendu Kumar Pani
Mohd Abdul Ahad
Qin Xin

CHAPTER

Internet of medical things for enhanced smart healthcare systems

1

Joseph Bamdele Awotunde[1], Chinmay Chakraborty[2], Muyideen AbdulRaheem[1], Rasheed Gbenga Jimoh[1], Idowu Dauda Oladipo[1] and Akash Kumar Bhoi[3,4]

[1]Department of Computer Sciences, University of Ilorin, Ilorin, Kwara State, Nigeria [2]Department of Electronics and Communication Engineering, Birla Institute of Technology, Mesra, Jharkhand, India [3]KIET Group of Institutions, Delhi-NCR, Ghaziabad, Uttar Pradesh, India [4]Directorate of Research, Sikkim Manipal University, Gangtok, Sikkim, India

1.1 Introduction

The healthcare system has been on the frontline in recent years, researchers have tried to find solutions to different diseases by applying various modern methods. But the major difference among them is that in recent years' other powerful new tools have emerged, which could be used as an instrument in the healthcare system and keeping it within reasonable limits. One of those technological tools is the Internet of medial things (IoMT) and Artificial intelligence (AI). Recently, AI-enabled with IoMT-based systems is causing a paradigm shift in the healthcare zone and the applicability might yield profit, especially in diagnosis, prediction, and treatment of different diseases outbreak. The application of AI-enabled with IoMT-based systems in the healthcare system can be expediting the diagnoses and monitoring of disease and minimizes the burden of medical processes.

To have high-level abstractions with multiple nonlinear transformations, DL is based on a series of ML techniques use to the model data (Folorunso, Awotunde, Ayo, & Abdullah, 2021). The artificial neural net (ANNs) work system runs on deep learning technology. The algorithms follow learning efficacy and are enhanced by a continuous increase in the amount of data. Efficiency depends on large amounts of data. The training process is referred to as intense as the number of neural network layers grows with time (Amit, Chinmay, & Wilson, 2021; Brown, Abbasi, & Lau, 2015; Chakraborty & Abougreen, 2021; Jayanthi & Valluvan, 2017). The efficacy of machine-learning algorithms has historically depended heavily on the consistency of input data representation. As opposed to a good data representation, poor data interpretation may also result in worse outcomes. As a result, for a long time, feature engineering has been a significant study direction in machine learning, concentrating on creating features from raw data and contributing to a vast number of research studies (Pouyanfar et al., 2018). The execution of the deep learning experience is strictly dependent on two stages, known as the training phase and the assumption phase. The training phase entails labeling and evaluating the matching features of massive amounts of data, while the assumption phase entails making conclusions and using prior knowledge to mark new unexposed data (Ahmed, Boudhir, Santos, El Aroussi, & Karas, 2020; Rodgers, 2020; Tokmurzina, 2020).

Implementation of Smart Healthcare Systems using AI, IoT, and Blockchain. DOI: https://doi.org/10.1016/B978-0-323-91916-6.00009-6

Health professionals are in desperate need of technology for decision-making to tackle any diseases and allow them to get timely feedback in real-time to prevent their transmission. AI works to simulate the human intellect competently. This may also play a crucial role in interpreting and recommending the creation of a vaccine for any infectious diseases. This result-driven engineering is used to better scan, evaluate, forecast, and monitor current clinicians and patients expected to be future. The relevant technologies relate to the monitoring of verified, recovered, and death cases. The data science analysis using AI is newly evolving, intending to empower health care systems and organizations to connect to harness information and convert it to usable knowledge and preferably personalized clinical decision-making. Utilizing deep learning, the implementation of AI in the field of infectious diseases has implemented a range of improvements in the modeling of knowledge generation. Big data can be interpreted, stored, and collected in healthcare through the constantly emerging field of AI models, thereby allowing the understanding, rationalization, and use of data for various reasons.

To control the resources of megacity populations, any device that is part of a smart healthcare system must collaborate with others. If the hospital is to be genuinely "intelligent," these machines must interact with each other. This is where the IoT comes in, offering the ideal model of a body of communicating devices that provide daily challenges with smart solutions (Woodhead, Stephenson, & Morrey, 2018). For learning purposes, applications of IoT can stand to benefit from the decision process. Even in the case of location-aware services, for example, location estimation may be described as a decision-making procedure wherein the exact or nearest value to a given goal is determined by a software agent. In this respect, to formulate and solve the issue, reinforcement learning can be used. A virtual machine communicates with its surroundings in a reinforcement learning solution and alters by carrying out certain operations, to improve the condition of the world (Bhadoria, Saha, Biswas, & Chowdhury, 2020; Mohammadi, Al-Fuqaha, Guizani, & Oh, 2017). Fig. 1.1 shows the applications and services in IoMT-based systems.

The IoMT-based system creates a huge amount of data named Big Data and thereby Influences the creation and growth of better-customized healthcare systems. Wearable medical devices can have active surveillance functionality that can gather a vast amount of medical data, resulting in Big Data, from which physicians can foresee the future condition of the patient (Marques & Pitarma, 2018). This observational study and the extraction of information is a dynamic process that must ensure enhanced security methods (Manogaran et al., 2018). The use of AI on generated Big Data from IoMT-based systems offers several opportunities for healthcare systems (Özdemir & Hekim, 2018). The application of AI in the process of generating Big Data can significantly improve global healthcare systems (Allam & Dhunny, 2019; Marques & Pitarma, 2016a,b; Marques, Roque Ferreira, & Pitarma, 2018). The IoMT-based system has been used to reduce the global cost of infectious disease prevention. The IoMT-based system can be used in real-time data capture to help patients during self-administration treatments. The integration of mobile apps is common in IoMT-based sensors data capture for telemedicine and mHealth systems (Adeniyi, Ogundokun, & Awotunde, 2021; Marques, Pitarma, 2016a,b).

The data interpretation becomes easier with AI-based data analytics and decreases the time needed for data performance analysis (Dimitrov, 2016). Besides, a new system has been created, "Personalized Preventative Health Coaches." It retains relationships and can be used to clarify and understand data on health and well-being (Marques, Ferreira, & Pitarma, 2019). For efficient health monitoring, the networked sensors enable people without direct access to medical facilities to be

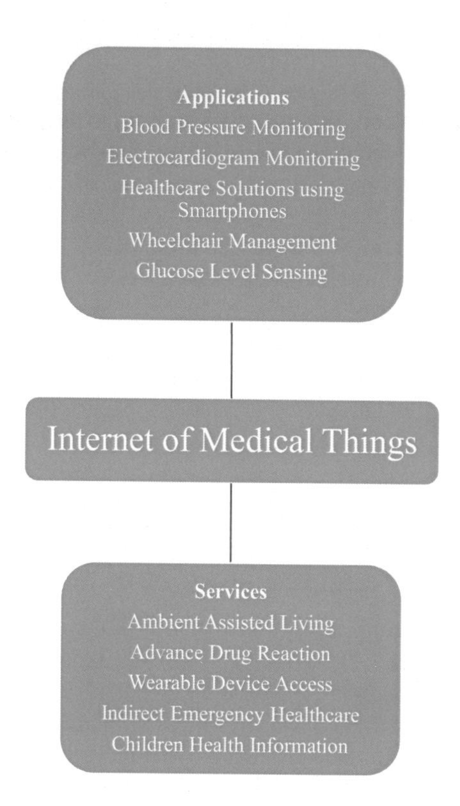

FIGURE 1.1

The applications and services on the Internet of medical things.

appropriately monitored (Kaur, Kumar, & Kumar, 2019; Solanki, & Nayyar, 2019). The use of an AI-based system with wireless communication has helped physicians to make appropriate recommendations to patients. A thorough analysis of the IoMT framework in the medical areas has greatly helped in reducing the cost of diagnosing a patient in the healthcare system. Furthermore, the IoMT-based helps in several healthcare systems like the generation of big data through the use of sensors and devices for vital physiological and biophysical parameters supervision, and big data analytics can be performed on them in other medical decision-making support methods. Hence, the combination of AI and IoMT will greatly improve the healthcare system and significantly help in disease diagnosis, monitoring, predicting, and patient treatment.

The contributions of the chapter are:

- A general overview of the IoT system applications was presented.
- The applications of ML-enabled IoT-based systems in the healthcare system were discussed.
- We propose a framework for machine learning-enabled with IoT-based systems.
- A practical case was used to implement the proposed system diagnosis the Diabetes Mellitus using Fuzzy Logic.

Therefore this chapter discusses the significance of AI-enabled with IoMT-based systems for better expansion and research in the healthcare system. The hope of using AI in an IoMT-based system will have a revolutionize the smart healthcare systems in the areas of disease diagnosis, prediction, and treatment and can be delivered quality care to patients across socioeconomic and geographic boundaries. The chapter offered a review of the AI and IoMT-based systems in the healthcare sectors, and the practice of AI-enabled and IoMT in the medical fields was presented. The extraordinary opportunities brought by AI-enabled and IoMT-based healthcare and the research challenges in deploying them in healthcare are also discussed. The chapter is organized as follows: Section 1.2 explains the AI-enabled Internet of Things Medical in smart healthcare systems, and Section 1.3 presents applications of AI in enabled IoMT-based systems in the healthcare systems. Section 1.4 discusses the challenges of the AI-enabled IoMT-based system in healthcare industries. Finally, the chapter was concluded in Section 1.5 with recommended future directions.

1.2 Artificial intelligence-enabled Internet of medical things

The word "smart" refers to a computerized process that is adopted within a domain to conducting the desired operation. For example, all devices and sensors like heart rate, blood pressure, body temperature, and smart wristwatch among others are embedded in smart healthcare with miniature sensing devices. The devices can use to collect information, monitor, predict, treat, and move information to process hubs for making a decision using dynamic rules and regulations. AI is a rapidly growing technology, in recent years AI methods have been used in smart healthcare to achieve remarkable breakthroughs in developing various models, and are currently implemented in many fields. It is a computational intelligence paradigm, has attracted substantial interest from the academic community, and has shown greater promise over traditional techniques (Nguyen et al., 2019).

AI is not a specific approach to knowledge, but it conforms to different methods and topographies that could be useful for a wide-ranging of complicated problems. The method learns the illustrative and differential characteristics in a rather heterogeneous method (Aggour et al., 2019; Khan, Fan, Lu, & Lau, 2020). Robotics is the branch of AI that automatically, without being explicitly programmed, provides the system with the benefits of learning from concepts and ideas. These produce better results and make future decisions using observations like direct experiences to prepare for information characteristics and patterns (Arrieta et al., 2020).

AI-based approaches have the potential to help with resource management through a large range of devices. As a result, AI and big data analytics are well-suited to extracting information from high-dimensional data and making complex decisions (Mahdavinejad et al., 2018). The patients will have new forms of data that can help doctors achieve their goals. The use of self-tracking devices, social media, and health help patients to get real-time information about various diseases, thus reducing the spread of diseases and greatly aiding in healthcare control. The embedded AI devices can be used for medical diagnosis, monitoring, and treatment by the patient, and medical experts can give real-time advice to caregivers and patients (Nathani & Vijayvergia, 2017; Shen et al., 2020; Zheng, Sun, Mukkamala, Vatrapu, & Ordieres-Meré, 2019). This field not only presents exciting obstacles for any patients, but it also has the potential to increase data collection and treatment efficacy. The world of "things" in an IoMT-based system is made of various subsystems serving the important part of the IoMT system mainly for the capturing of data from the users (Mehta, 2018).

The AI-based enabled robots can be used in healthcare for teleconferencing with patients in real-time, thus both medical experts and the patients are saved especially during the COVID-19 outbreak where there is no physical contact. During hazardous environmental problems, teleoperated robots can perform nursing tasks with high precision and productivity. These robots can be used for collecting specimens without physical contact, delivery of drugs and meals, and transporting waste products within the hospital (Panzirsch et al., 2017). The most benefit of using the robots is the ability to monitor multiple robots by a single operator when moving between quarantine areas for monitoring and delivering various tasks. Also, the use of a virtual telepresence system in real-time to communicate with patients is another benefit of using robots. The use of the TRINA robot a Tele-Robotic Intelligent Nursing Assistant to perform nursing tasks was an example of a promising robot in healthcare systems (Li, Hu, & Zhang, 2017; Marques & Pitarma, 2018).

The application of AI-enabled IoMT has transformed the healthcare system. Health professionals are in desperate need of technology for decision-making to tackle the outbreak of infectious diseases, and a system that allows them to get timely feedback in real-time to prevent transmission of such diseases. AI works to simulate the human intellect competently, and using the methods to enhance IoMT-based systems will be of great benefit. Also, AI with an IoMT-based system plays a crucial role in interpreting and recommending the creation of a vaccine for any pandemic outbreak. This result-driven engineering is used to better scan, evaluate, forecast, and monitor current clinicians and patients expected to be future. The application of AI-enabled IoMT in any disease outbreak can expedite the diagnoses and monitoring of such illness and minimizes the burden on physicians during these processes. Therefore this section discusses the areas of applicability of AI-enabled IoMT in enhanced smart healthcare systems.

The application of AI-enabled IoMT in the smart healthcare system has increased tremendously. This has been used to achieve precise diagnosis accuracy, and reduce the burden on healthcare experts. Also, the system reduces the time of evaluation and diagnosis associated with the conventional approach in the detection procedure. The AI-enabled IoMT techniques are seen as a major aspect in identifying the risk of infectious diseases in enhancing the forecasting and identification of potential world health threats. The continued expansion of AI-enabled IoMT for infectious disease has dramatically improved monitoring, diagnosis, analysis, forecasting, touch trailing, and medications/vaccine production process and minimized human involvement in nursing treatment (Bravo et al., 2012; Kishor, Chakraborty, & Jeberson, 2021).

Methods of artificial intelligence have produced an increased focus level within the research community. As defined in many recent findings, machine learning approaches offer valuable detection accuracy in comparison with different data classification techniques (Dey, Bajpai, Gandhi, & Dey, 2008; Liberti, Lavor, Maculan, & Mucherino, 2014).

The AI with IoMT-based application is still emerging in the smart framework, smart healthcare system, and other areas of human activity, and thus many aspects of conceptualizing and leveraging it remains a work in progress. From industrial applications to emergency services, treatment and mobility, public safety, diagnosis, and other Healthcare applications, the Internet of Things is in every business and government field today. Cities are becoming increasingly connected as IoT technology advances, in an effort to improve the performance of infrastructure installations. As a result, the emergency services' dependability and awareness are improved at a lower cost. And there is continuous innovation. In the years to come, we expect to see many smarter healthcare ideas using IoMT technologies for this market. Nevertheless, there are many reasons for municipalities to switch to the methods of wireless communication given by IoMT technologies.

In previous research works, AI has been used in medical and the biomedical field (Ayo, Awotunde, Ogundokun, Folorunso, & Adekunle, 2020; Kurd, Kelly, & Austin, 2007; Oladele, Ogundokun, Awotunde, Adebiyi, & Adeniyi, 2020), for the involvement of heart disease and diabetes (Lingaraj, Devadass, Gopi, & Palanisamy, 2015; Oladipo, Babatunde, Aro, & Awotunde, 2020; Oladipo, Babatunde, Awotunde, & Abdulraheem, 2021), Among other things, Berglund et al. (2018) investigated diabetes proteins. Academics have used ANN, Support Vector Machine (SVM), Fuzzy Logic Systems, K-means classifier, and many other AI methods (Awotunde et al., 2020; Awotunde, Matiluko, & Fatai, 2014; Ayo, Ogundokun, Awotunde, Adebiyi, & Adeniyi, 2020).

Hitherto, the implementations of AI methods have shown positive results in numerous commercial and tourism environments to discover new behavior and identify the potential for the future. The recent modern systems, like machine learning, and ANN techniques, have shown positive outcomes in the extraction of a massive dataset of nonlinear dynamic structures. Recent findings using AI techniques to monitor rodent reservoirs of future zoonotic diseases within a disease-related framework (Han, Schmidt, Bowden, & Drake, 2015), Predict Extended-spectrum β-lactamase (ESBL) generating species (Goodman et al., 2016; and regulation of outbreaks of tuberculosis (TB) and gonorrhea (Wong, Zhou, & Zhang, 2019). It can be hard to forecast public response to infectious diseases. However, we are progressively able to compare population behavior with deadly diseases with the accessibility of Big Data and the emergence of AI methods. For example, dedicated studies employed a psychosocial informatics and data science method (Goodman et al., 2016; Mahroum et al., 2018). During the outbreak era, examined infection control outbreaks correlated with digital activity trends through search engine trends (such as Google Trend). It is envisaged that the development of AI technologies for infection control data analytics will boost our opportunity to actively monitor social interaction with infectious diseases and effectively forecast infectious spread, which can help policymakers take timely steps to respond to any pandemic. Applications of AI can be useful tools in assisting diagnoses and decision-making in disease treatment.

AI will create a smart framework to automatically track and forecast the spreading of any disease outbreak (Agbehadji, Awuzie, Ngowi, & Millham, 2020; Ganasegeran, & Abdulrahman, 2020; Londhe & Bhasin, 2019). A genetic algorithm may also be built to remove the visual characteristics of this infection. It has the potential to provide patients with daily alerts and also to offer better options for disease follow-up.

AI can easily determine this virus' level of transmission by recognizing the fragments and "hot spots" and can effectively track the individuals' contacts and even monitor them. It can foresee this spread of the disease's future path, and possibly reoccurrence. These technologies can also monitor and predict the existence of the epidemic from the data, social media, and broadcasting channels present, about the threats of the outbreak and its probable spreading. It can also forecast the number of use cases and deaths in any area. AI will help recognize the areas, citizens, and communities most affected and take effective measures.

1.3 Applications of artificial intelligence in enabled Internet of medical things

With the rapid development of the IoMT technologies, researchers have been motivated to develop smart services that extract knowledge from big data generated from IoMT-based devices/sensors (Awotunde, Adeniyi, Ogundokun, Ajamu, & Adebayo, 2021). The development of various models like forecast, diagnosing, predicting, monitoring, and ambiguity exploration in the healthcare system has been enhanced by the applications of AI techniques, and for healthcare development. There has also yielded greater results in the process of the huge data and input variables coming from the IoMT-based cognitive healthcare system. The AI methods that are mostly applicable in the development of various models in the healthcare system, and smart healthcare system, the most commonly used AI are Decision Trees, SVM, Artificial Neural Network, Bayesians, Neuro-Fuzzy, ensembles, and their hybridizations.

Fig. 1.2 displayed the general architecture of an AI-enabled IoMT-based system for disease diagnosis and monitoring. In Big Data Analytics AI models can be used for intelligent decision-making due to the huge data generation by the IoMT-based devices. The method of implementing data processing techniques for particular fields requires specifying the data involves like the velocity, variety, and volume of such data. Normal data analysis modeling involves the model of the neural network, the model of classification and the process of clustering, and the implementation of efficient algorithms as well. The IoMT devices can be used to generate various data formats from several sources, hence, it is very important to describe the features of the data generated for proper data handling. These help in handling the various characteristics of capture data for scalability, and velocity, thus helping in finding the best model that can provide the best results in real-time globally without any challenges. These are all known to be one of the IoMT's big problems (Abikoye, Ojo, Awotunde, & Ogundokun, 2020; Hoofnagle, van der Sloot, & Borgesius, 2019; Shaban-Nejad, Michalowski, & Buckeridge, 2018). However, in the latest technologies, these are all problems that generate a large number of possibilities. Such information can be accessed using the latest healthcare applications, and the data is securely stored on the cloud server.

1.3.1 Disease diagnosis

Accurate and quick diagnosis of any disease can be useful using IoT-based devices in generating data to train AI models. The information is also imperative in limiting the spread of the disease and saving lives. AI may provide valuable input in making a diagnosis based on images of chest radiography. AI can be as accurate as human beings in the diagnosis of various diseases that are peculiar to a human being. It means that it can save the time that physicians deploy in the diagnosis of the disease. It also performs the diagnosis at a cheaper standard than a physician or radiologist, and it is quicker than a human. Technologies like computed tomography (CT) and chest x-rays (CXR) can be coupled with AI to ensure the detection of the disease. Most disease test kits are very expensive and in short supply, but all hospitals have CXR machines (Zheng et al., 2020). The technology can be used in smartphones to scan CT images. Many initiatives have been deployed to help understand the conditions, such as the deep Convolutional Neural Network (CNN) that uses CXR images to detect infectious diseases (Folorunso et al., 2021).

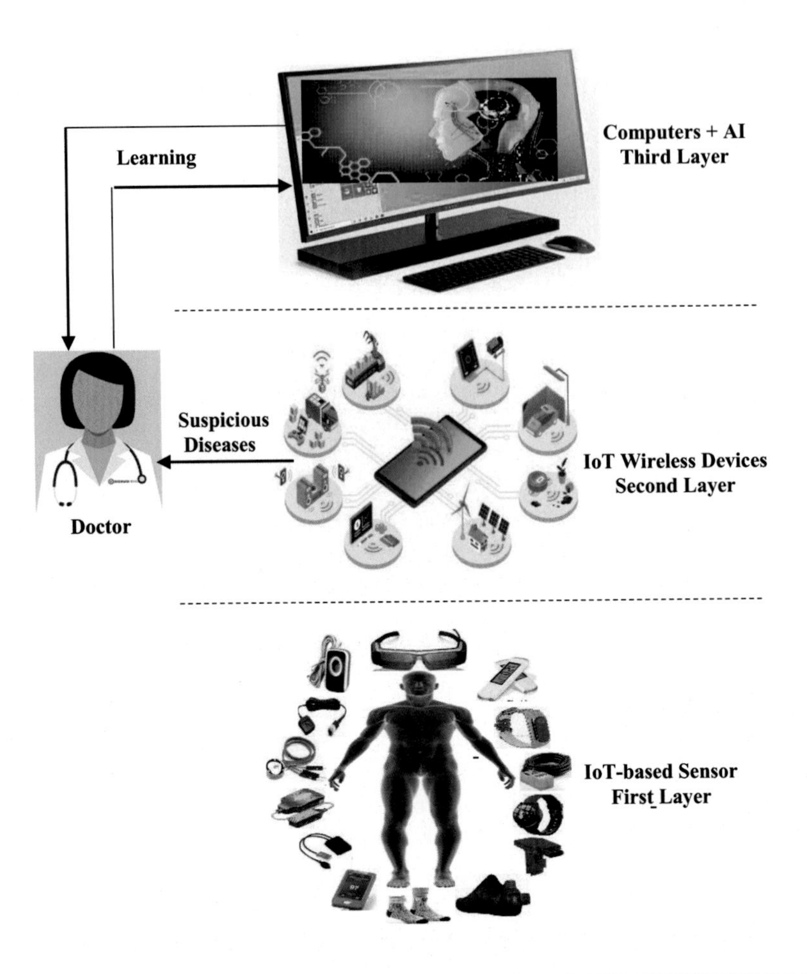

FIGURE 1.2

The architecture of an artificial intelligence-enabled Internet of medical of things-based system for disease diagnosis and monitoring.

The main aim of IoT is to make the environment smarter by providing the requisite historical or real-world data and automatically applying AI to make smart decisions. Several forms of research have been documented in current contributions and are capable of enabling early detection and prognosis based on different techniques (Arunkumar & Ramakrishnan, 2018; Pati, 2018), and the evaluation of the information is varied by each method (Zhang et al., 2019). IoT devices use lung cancer knowledge to understand and control complex environments, facilitating great automation, greater performance, accuracy, wealth creation, productivity, and better decision-making (Das et al., 2019). The timely processing of huge quantities of data to produce highly steady and reliable analyses and recommendations so that IoT can fulfill its promise is a serious obstacle in these conditions.

Various illnesses need real-time monitoring like pulmonary disease, glaucoma, and hypertension in healthcare fields. This has created a huge burden on medical doctors because the patients need to book an appointment for regular check-ups and daily medical evaluations of their health status. Hence, this creates a burden for many patients because going to visit their doctor daily can disturb their daily route too. Introducing IoT-based systems in the healthcare industry will help this situation. Hypertension, or elevated blood pressure, is a condition in which there is a persistent increase in pressure in the arteries' vessels. The high pressure makes pumping more difficult for the heart. This has caused over 7.5 million mortalities, and affects 1.13 billion people globally, with about 12.8% of all deaths. The management of high blood pressure can lead to a reduction in the complications of heart disease. In many applications, IoT has increasingly begun to emerge and improve. The use of IoT in the healthcare system, nevertheless, is still restricted. The research was based on real-time monitoring of diseases, emergencies, and operational services (Kumar, Sangwan, & Nayyar, 2020).

Emergency services provide patients with emergency responders, ambulance services, and vehicles. The system use sensors to detect and transmit patient data during emergency services using the capture physiological health information to medical doctors and health caregivers. The capture data was used for emergency and health surveillance purposes and emergency reaction purposes, but no particular disease is listed. Neyja et al. (2017) proposed a system for surveillance of cardiovascular illness using an electrocardiogram monitor system. The embedded algorithm was used for enabled abnormal cases, thus the medical experts can respond in real-time.

Gia et al. (2019) suggested a framework for glucose monitoring based on an IoT-based system composed of a sensor interface, gateway, and a cloud back-end system. The application was used by the medical doctor to track patients in real-time using a web browser application. The system alerts medical experts and patients in case there is any emergency using mobile apps. In addition, their proposed method limits its measurements to glucose monitoring only. Li et al. (2017) suggested a tracking device for heart disease using IoT. The architecture of the device is broken down into the sensing layer, the transport layer, and the application layer. The proposed system can be used to track the physiological healthcare of any patient in real-time using the devices and sensors.

Besides, for atrial fibrillation identification, mobile cardiac telemetry systems are important. For the prolonged duration of heart rhythm, these systems conduct real-time monitoring and lead to the detection of arrhythmia. Also, medical wearables and AI can be used through remote patient monitoring to establish effective methods for diagnosing heart disease (Kumar & Gandhi, 2018). Symptoms of heart disease can be detected by a machine learning system (MLS) by reviewing scan data from recent patients and by analyzing previous data. Besides, MLS will estimate the probability of a potential occurrence of a heart attack. Researchers investigated the use of AI and ECG at the Mayo Clinic. MLS has been used to measure electrocardiography and classify pulmonary vascular immunocompromised dysfunction. A co-evolutionary neural network was run by these researchers (CNN). CNN uses ECG data to detect ejection fraction in ventricular dysfunction patients by 35%, according to the study (Attia et al., 2019). The findings show that AI applications based on ECG data are a low-cost method for detecting ALVD in asymptomatic people.

To facilitate sensor data in disease diagnosis, AI technology is used with the IoMT-based system devices. The MapReduce framework with cluster analysis and effective differential private method (EDPDCS) based can be used to diagnose patients from the captured data using IoMT-based devices (Guan, Lv, Du, Wu, & Guizani, 2019). Via analysis of the normalized intra-cluster variance, the k-means algorithm (NICV) was used to forecast using capture data. Even, with grasshopper optimization, there is an effective hybrid NN method. The grasshopper optimization algorithm for NN has a major influence on optimization problems, particularly for versatile and adaptive results for ML search mechanisms (Guan et al., 2019).

Another initiative from March 2020 that uses an AI model to diagnose COVID-19 deploys CXR images and has been used in the past to diagnose pulmonary Tuberculosis (TB). The potential has not yet been implemented in clinical practice settings, but several hospitals in China have used radiology technologies based on AI. Other radiologists have expressed concerns over the amount of data available to train AI models. Many hospitals in China are bound to suffer from selection bias because using CXR and CT scans can contaminate equipment and result in the increased spread of the disease (Alwashmi, 2020). The use of such scans in European hospitals reduced significantly after the pandemic broke to reflect the concern. Once an individual has been diagnosed with the condition, the fear intensifies and it may affect the patronage of the hospital. However, the point remains that ML algorithms can develop prognostic prediction models that predict the mortality risks of infected individuals. They can provide more than 80% accuracy concerning an individual who has been infected and determine if he or she can develop acute respiratory distress syndrome.

1.3.2 Prediction and forecasting

One of the most fascinating parts of humanity's quest for knowledge is the future. The future is all about estimating the values and probabilities of favorable events. The concept of forecasting and prediction is born from the inculcation of the thought of prediction. The two words appear to be nearly identical. They aren't overly complicated, and most people mistake the two terms for one another. But, in reality, they are vastly different. This is because they are used in a variety of situations. Prediction is the process of predicting the outcomes of unknown facts. Fitting a model to a training dataset does this, yielding an estimator capable of making predictions for new samples. Forecasting is a sub-discipline of prediction in which we use time-series data to make predictions. The sole distinction between prediction and forecasting is that the temporal component is taken into account. The form of a forecasting estimator is [Math Processing Error], which denotes historic measurements at time points, and the estimate denotes a time point or another time in the future. This is known as an autoregressive model because it is based on past observations.

AI can be used in managing any disease especially infectious diseases through the deployment of thermal imaging. The technology is useful in scanning public spaces for specific individuals who have the potential to be infected (Alehegn, Joshi, & Mulay, 2018; Ayo et al., 2020; Lingaraj et al., 2015; Neri, Miele, Coppola, & Grassi, 2020). It is then valuable to enforce lockdown and social distancing measures. Infrared cameras are used in train stations and airports across China to scan individuals for high temperatures. They were also deployed in facial recognition to pinpoint individuals with high temperatures. Baidu is one of the producers of infrared cameras that use computer vision in scanning different individuals in the crowds (Awotunde et al., 2014; Barstugan, Ozkaya, & Ozturk, 2020; Oladele et al., 2020). The camera can scan 200 individuals every minute and recognize body temperature that exceeds 370C. However, thermal imaging has been inadequate in identifying a fever in individuals wearing glasses because the most reliable indication was the deployment of scanning of the inner tear duct. It was also hard to determine if a person's temperature was increased due to any disease/outbreak or other sources.

As noted earlier, AI is useful in predicting and diagnosing diseases but it is hampered due to the lack of historical data. However, robots and computer vision cannot be hindered by such limitations. This type of AI will be useful for social control. Related technologies like mobile phones with wearables and applications that have AI can help locate and control whole populations. In line with this, Rodrigues et al. (2020) did research to determine the effectiveness of social distancing in Europe which was conducted by the application of the AI to determine the impact of social distancing on the spread of COVID-19. The very first and basic action that was taken against the coronavirus was social distancing so that the virus may not spread more rapidly or maybe to control the virus. Although social distance measures may be awkward to normal social norms and some medic-patient interactions, it was still adopted because the condition is becoming a matter of life and death.

Various diseases like diabetes, blood pressure, and hypertension among others has being regulate by monitoring the dosage during diseases' progress using wearable devices for the clinical monitoring. For all procedures, the physician serves as the central access. To determine the patient's decision, the system is connected to the cloud server to access the records and for a future forecast. This remote-tracking helps reduce the cost of management of infrastructure, human resources, etc. the encryption system was used to protect the transfer of sensitive information from the proposed system.

IoMT-based systems are very useful in tracking illnesses and fitness programs because they are effective and detailed with powerful miniature handheld sensors. In the cloud, significant medical data must be stored and transmitted. To deal with the privacy budget and decide the centroids for initialization, k-Means clustering was suggested under various privacy-related techniques. The number of iterations can be determined using improved k-means algorithms to represent fixed and unfixed iterations. The mean square error between the noisy and true centroids is determined during the recruitment analysis to identify the budgetary allocations and formulations. Random collection, and fair division of datasets is often done to compensate for the uncertainty in the production of the datasets instead of selecting a method for allocations. By dividing each subset by the original dataset, DPLK optimizations were proposed to increase the initial centroid selection for the k-means algorithm. To reduce the time complexity of cluster analysis computational speed, a distributed environment was implemented. By evolving the parallel k-means algorithm, which is more efficient than sequential programming, MapReduce can be used to perform distributed operations. The closest centroids are also updated regularly.

Through sickness assessment and evaluation, a greater degree of IoMT-based system innovation for predictive modeling has been established. The expertise provided by IoMT devices is growing in lockstep with the growth of the technologies. To circumvent this, big data analysis techniques have been employed to process and analyze cloud data. There is a strong probability of leakage of the data, which includes privacy information, in case of problems. While some privacy-preserving strategies exist, there are still several private information loopholes that need to be filled with an algorithm like k-anonymity, heterogeneity, and empirical privacy. To obtain the desired outcome in this situation, the trade-off between providing good accuracy and maintaining privacy must be digested (Ma & Pang, 2019; Miller & Brown, 2018; Xiuqin, Zhang, Zhang, & Li, 2019).

There are major ongoing attempts to enhance human health programs, where IoT technologies have achieved considerable success as part of AI techniques. However, it is still important to further explore the awareness and application of these fundamental processes, the IoMT application, concerning healthcare practices. The science of behavioral analysis aimed at avoiding health issues of people such as mild cognitive impairment and frailty is an important endeavor attributed to the AI paradigm (Ayo et al., 2020; Ma & Pang, 2019). The use of advanced AI-inclined technologies, therefore, allows discrete collection of personal data for automated identification of behavioral changes for accurate and efficient prediction of health care. This improves behavioral healthcare processes and improves patients' customized health care, especially for the prevention of heart disease. As described in the preceding section, several AI techniques have been investigated. Success stories have been developed from the implementation of AI methods to enhance the accuracy of classification models and ultimately improve the efficiency of healthcare systems in disease prediction. More needs to be done, however, to test disease prediction systems based on multiple evaluation metrics, because the use of a single evaluation metric, such as the accuracy of model classification, does not guarantee optimum system efficiency.

The technique of computational intelligence requires historical data for computing and healthcare systems to produce such data that can be used to build intelligent healthcare systems. These frameworks may be in the form of customized models of healthcare that use the IoMT for their operations as a basis. Various sources of medical data collection and the potential for data to expand exponentially require resources and systems with high processing capacity and performance. To process and handle medical data, the implementation of big data analytics and cloud-based frameworks as artificial intelligence methods is therefore suitable. Another significant explanation for the adoption of computer intelligence is that it is capable of replicating and mimicking expert expertise, thus complementing human experts' efforts and thereby reducing the chances of medical diagnosis and disease management errors.

1.3.3 Monitoring system

New emerging technologies, especially the IoMT, are increasingly being used in remote health monitoring, care, and therapy in today's telemedicine. With a focus on developing smart applications, this has gained rapid acceptance in the healthcare sector. In addition, the IoMT offers a platform for devices and sensors to communicate seamlessly in a smart world, allowing for simple data and knowledge sharing over the Internet. The Internet of Things (IoMT) is increasingly being used in remote health monitoring, and as a result, it has gained rapid adoption in the healthcare field, with a focus on developing smart applications. The IoMT offers a platform for devices and sensors to communicate seamlessly in a smart world, allowing for simple data and knowledge sharing over the Internet. Modern IoMT-based implementations of many wireless equipment positions have advanced technology by delivering the full potential of digital technology.

The IoMT-based system has gained ground, especially in the healthcare sector in recent years. The IoMT reshapes contemporary healthcare systems, changing the traditional ways of the medical system to a smart healthcare system. The conventional healthcare systems have gradually moved to smart healthcare systems where patients' diagnosis, monitoring, and treatments are becoming easier and effortless. Wearable body sensor networks have enhanced the chance to transform our lifestyle with numerous innovations in areas such as healthcare, entertainment, transportation, retail, business, and emergency services control. The creation of a multidisciplinary concept of ambient intelligence to answer the challenges we face in our everyday lives has resulted from the integration of wireless sensors and sensor networks with simulation and intelligent systems research (Özdemir & Hekim, 2018).

The IoMT-based system with AI could be used to help patients get proper medical care at home when applied during any epidemic, and the healthcare policymakers and government can make use of the robust database created for infectious disease outbreak management (Awotunde, Folorunso, Jimoh, et al., 2021). Monitoring and healthcare devices such as thermometers, smart helmets, smart wristwatches, medications, protective masks, and monitoring infection kits may be purchased for people with moderate symptoms. Periodically, the health status of patients can be uploaded over the Internet-based IoMT network to the clinical cloud storage, and their data could be forwarded to the nearest clinics or health center hospitals, and the center for disease control.

Subsequently, a medical expert will provide online health consultations based on the health status of each patient and if necessary, the policymakers and healthcare experts assign facilities and designate quarantine stations to the affected person. People may dynamically monitor their clinical diagnosis and obtain adequate medical needs using the IoMT platform with AI without virus transmission to others. Hence, minimize the costs, alleviate the shortages of medical equipment, and provide a systemic database that could be used by physicians to track the spread of infectious diseases effectively, the supplies of relative tools become easier, and enforce emergency strategies.

For example, the current COVID-19 pandemic we are fighting is increasingly suffocating the healthcare sector toward its ending levels; the hospitals and clinics are filled with reported suspected cases pending the evidence of the diagnosis. As the demands are rising, the shortage of medical diagnostic equipment and supplies is increasing, and with increases in patients that need care without adequate tools for complements the upward increase in admitted of patients in the hospital, and the place of mission-critical healthcare staff at higher risk on the front lines. To mitigate this crisis, a powerful and supportive medical system is needed to save the lives of the populace.

A more robust medical framework for the fight against the COVI-19 pandemic is needed. An IoMT-based system is needed to alleviate the diagnostic and monitoring problems, and this will help in enforcing stay-at-home protocols and limiting the clinical resources required. This will support the appropriate distribution of equipment and supplies by government and private donors to clinics and various hospitals, and the approach would provide information on healthcare facilities to establish effective patient care. To save countless lives, this combined strategy can be very helpful, and also safeguard strained economies and build a blueprint to tackle future threats more effectively.

IoMT is an innovative way of combining healthcare gadgets and their applications to interact with human resources. The adoption of an IoMT-based system with AI during the infectious outbreak provided equal rights for both rich and the poor populace in having equal access to healthcare facilities without any form of preference (Awotunde, Folorunso, Bhoi, Adebayo, & Ijaz, 2021). Various cloud-based AI-enabled IoMT techniques managements are the exchange of knowledge, report verification, investigation, diagnosis, treatment, and patient monitoring among other services provided by the system. This creates rewarding various patients with a prevailing treatment and diagnosis that are more fulfilling and creates a new working system of medical services, particularly during this outbreak. The use of an IoMT-based system gives health workers full concentration on the patient by easily identifying an infected person and those that have contact with them, and moving them to an isolation center. The tools provided by IoMT devices can be used to curb the spread of the outbreak, such tools could be an early warning system like the geographic information system, and wearable sensors embedded within the human body. Sensors like temperature and other signs might be used at airports around the world to detect people infected with any diseases.

For instance, over a 2- to3-day period of ongoing physiological monitoring of patients, IoMT-based devices are used to recommend physiological exercises and food habits. IoMT devices will continuously observe and store the health data of the patient in a cloud database during this period (Lorincz et al., 2004). This allows doctors to diagnose the health condition of the patient and not only use laboratory tests, but also patient health data obtained from IoMT-based sensors to achieve better results. Sensor data is also most commonly used to take effective action for the recommendation of patient well-being and care, lifestyle decisions, and early diagnosis, which are important for improving the quality of patient health. Conventional computer storage approaches and mechanisms are not enough in the areas where the volume, speed, and variety of data is increasing above described emerging IoMT-based applications. The development of an efficient storage system for storing and processing voluminous data needs to solve this problem.

1.3.4 Personalized treatment

Personalized care is a popular field in recent smart healthcare systems, and this includes the use of AI, data science, processing of big data, and development of statistical analysis. Especially, the use of AI in processing and analyzing big data has greatly helped in reaching meaningful decision-making using captured data from the IoMT-based devices, thus their integration has been of increase in the smart healthcare system. the AI-enabled IoMT-based system can be used in Personalized medicine for clinical practice that will act as a medical expert to prescribe and treat any patient without direct contact with a medical doctor in a clinic or hospital. This brings about more complex models for diagnosis, forecasts, more accurate predictive models, and validated clinical trials (Fröhlich et al., 2018; Schork, 2019). Various research has proved that personalized medicine has come to stay and is very reliable to sustain the medical field. New approaches to advance this medical area are created by the use of AI techniques in customized medications. Together, AI IoMT-based techniques enable doctors to access, capture, store, and enhance the statistical analysis of the condition of patients. Different experiences have been raised by the application of modern innovation for patient care, especially for patients who require greater precision during diagnosis and treatment. IoMT-based monitoring systems have therefore been developed, providing high real-time efficiency. There is a healthcare monitoring model based on the Intelligent IoT system, like BioSenHealth 1.0 (Nayyar, Puri, & Nguyen, 2019).

The truth is that Alzheimer's disease, which requires regular evaluation, is a complicated diagnosis. This method was made simpler and more relaxed by IoT-based. To track Alzheimer's disease conditions, Khan and his companions (Khan et al., 2019) developed an algorithm. To produce a statistical analysis of IoT-based data, this algorithm utilizes a hybrid feature vector combining methodology. Its three-dimensional viewpoint provides a full image of the state of the patients. By using this new algorithm, the experimental results show an average of 99.2% and 99.02% for binary and multi-class classifications. The 5GSmart Diabetes customized care system (Krishnamurthi, Kumar, Gopinathan, Nayyar, & Qureshi, 2020) was developed based incorporates the methods of big data capture using the IoMT-based devices and AI was used for the processing and analysis of the captured data. The 5G-Smart Diabetes device operates for patients with diabetes and offers data from analytical sensors to predict the progress of diabetes.

1.4 Challenges of artificial intelligence-enabled Internet of medical things

Due to the complexities of AI with broad coverage of healthcare system applications, there are various challenges in applying these algorithms ahead for this emerging field. This section discussed a range of potential directions related to AI efficacy, evolving frameworks, convergence of information, and protection of privacy hoped that these would take the relevant research one step further to develop fully data analytics for the smart healthcare system. In general, there are two major challenges present by active users or machine devices: (1) access collision probability increases; and (2) complicated resource management. When users/devices encounter access collisions, they must relaunch radio access processes, resulting in increased issues that limit and a lower quality of service (QoS). Latency is a significant factor in the outcome of underlying transmissions for delay-sensitive data. Dense active users or computer devices can present additional challenges in wireless resource management like scheduling and power control, thus, resulting in increased disruption created by the fundamental wireless channel and inter-user interference. Furthermore, to efficiently manage resources, channel measurements must be performed when there are a lot of active users and devices, getting the underlying channel state information can be expensive. Secondly, IoMT networks must accommodate a diverse set of wireless users and mechanical devices. In IoMT networks, different wireless devices with different QoS specifications and priorities may coexist. IoMT networks, for example, should prioritize ensuring reliable transmissions of emergency information through proper resource management.

The main problem for the growth of the smart healthcare system is the effective and successful incorporation of heterogeneous information about infectious diseases in various sources and forms. The development of useful applications is hampered by all of this contradictory data. Scientific data (including EHRs) gather from various medical data using text mining, and from high-throughput data must all be combined and interact without any hindrances from within or outside the system (Awotunde, Jimoh, Oladipo, & Abdulraheem, 2021; Fagherazzi, Goetzinger, Rashid, Aguayo, & Huiart, 2020). Furthermore, the availability of data depends on the decision-making from the capture data using sensors, big data can be very relevant in this situation to maximize the resource use during this process. AI and the IoMT-based system could all play a role in this (Adly, 2019; Komenda et al., 2020).

AI-assisted big data analysis can be viewed as a method of teaching computers to imitate human thought processes and even replicate human behaviors. Since it is beneficial to conduct AI using real-time data, it is expected that the precision of the achieved results will improve as computing power and data become more readily accessible (Allam & Dhunny, 2019; Triantafyllidis & Tsanas, 2019). When the problems are very complex, AI algorithms can also help in decision-making. Expert systems integration into intelligent systems has been a great success. Expert systems, on the other hand, may find it difficult to obtain and process any disease data. Integrating data mining with intelligent computing systems to analyze the data gathered and the patterns involved, such as clustering algorithms, neural network algorithms, regression algorithms, and Bayesian algorithms, is critical for recognizing the involved patterns and expertise from different fields (Joseph & Thanakumar, 2019).

Privacy violations, legal questions, and a lack of information protection are only a few of the other issues that may arise. The mining of very large disease datasets can pose computational and storage challenges. Combining different types of information in heterogeneous disease datasets with global information systems, for example, can be difficult. The data mining method would also need a large number of disease experts. Also, the level of diversity in any disease dataset influences the accuracy of data mining performance. Data mining can also have a lot of advantages. AI is quickly evolving the way communities run, fund, and handle all foreseeable utilities such as transit, electricity, healthcare, networking, and many others, as shown by the aforementioned discussions. However, choosing the right technologies to integrate with smart healthcare services successfully and efficiently remains a significant challenge.

The administrative bodies' ability to implement these innovations and integrate them into their daily services and divisions is also a major challenge. There is a preconceived idea regarding the initial expenditure on such developments, which will result in a budget rise. However, the fact that once introduced, it will result in a higher economic performance is also a fact that is underappreciated. In smart healthcare, time-sensitive applications necessitate both real-time and non-real-time data streaming. It is also difficult to build applications that incorporate big data and fast data analytics. Machine learning systems are developed into smart devices in many ways, however, most of them are hard (Patel et al., 2020). Another issue to focus on is the need for flexible ML algorithms for resource-constrained applications that ensure protection and privacy (Muhammad et al., 2020). The databases used for deep learning implementations are often not easily accessible and of sufficient scale to validate outcomes by simulations.

Another big problem with smart healthcare is the ever-changing technical progress; as technology develops, so does the principle of smart healthcare (Abiodun et al., 2021; Habibzadeh, Nussbaum, Anjomshoa, Kantarci, & Soyata, 2019). AI models will face difficulties as smart healthcare begins to develop as it may present difficulties in managing the current paradigm of data analytics from future generations of smart healthcare. This is particularly true if AI technology isn't updated to keep up with the evolving smart healthcare concept (Josefsson & Steinthorsson, 2021; Kuru & Khan, 2020; Sejnowski, 2018). It is, therefore, recommend that researchers continue to change AI models as the smart healthcare paradigm evolves to respond to the evolving existence of smart healthcare in the future. Deep learning will thus remain important in the age of technological development by doing so.

Despite smart healthcare's deep learning penetration and impressive accomplishments, other facets of AI in smart healthcare have largely gone untapped. The feedforward neural model, neural abstract machine, memory enhanced neural network, communication guided deep network, sensitive network, and deep intense network are all examples of deep network learning machines are some examples of deep learning concepts not used in smart healthcare. Exploiting these deep learning features in smart healthcare would be important to see how successful they are at solving problems in smart healthcare. The efficiency of The scope of learning varies according to the object model. Researchers need to understand the right AI algorithm for a specific area to prevent expensive and time-consuming trial and error. The preeminent AI design for each smart healthcare area is still a work in progress (Belhadi et al., 2021; Singh et al., 2020). Researchers should conduct a thorough performance review of various AI architectures in various domains of application to determine the paramount Ai method for each smart healthcare field.

Although there is a comprehensive method for obtaining the optimum objective functions, the adjustment of several inputs demanded by deep learning is still an open research problem. Moreover, the AI algorithm's output is contingent on the optimum parameter configurations. Another unresolved issue is the computing time required by AI techniques; AI training typically takes a while before reaching convergence. Conversely, there are certain concerns where the passage of time is crucial in making a decision. For example, in smart healthcare, crime notification to protection authorities, accident notification to the appropriate authority if an accident occurs, fire outbreaks, and health-related concerns, time is important, as any delay will lead to death. Scientists should focus on optimizing the convergence pace of AI techniques in the future used to solve smart healthcare problems.

1.5 Case study for the application of Internet of medical of things-based enabled artificial intelligence for the diagnosis of diabetes mellitus

The Smart Health Management device results are greatly in use by patients and physicians. From the comfort of their home, the patient can check their health condition in real-time and attend hospitals only when they need to. This can be achieved by the proposed method, the product of which is brought big data analytics IoMT-based cloud medical care monitoring system and can be used by any patient from anywhere in the world. The system displays the almost real-time values of different health parameters as it is a prototype model and emulates how the same can be applied in the real world. Doctors may also use the patient's body condition record to analyze and assess the impact of medication or other such products.

In health-related fields, the use of IoMT devices is a relatively recent and rapidly-growing development. IoMT has opened up a wide range of new frontiers for generating, creating, analyzing, and managing large amounts of data frequency, especially online and on other platforms, according to a wide range of research disciplines related to the implementation of IoMT technologies and different devices and sensors Network applications in several medical fields. In this regard, the IoMT-based cloud with big data analytics has created many solutions in the areas of the medical care system, and from the perspective of personalized e-Medical care, it has given convenience as a popular method of handling a large volume of medical care data.

1.5.1 Fuzzy logic

Fuzzy logic is a superset of traditional Probability theory and can handle inconsistent and imperfect data, which is one of the attributes of medical records. It seeks a precise solution to a dilemma, and its capacity to comprehend based on guesswork is similar to human judgment. The phases involved in the care and prognosis of diabetes using fuzzy logic are as follows:

1. Fuzzification of the patient's inputted characteristics
2. The fuzzy rule base structure is developed.
3. Reasoning engine: the fuzzy logic component's decision-making engine.

4. Defuzzification of the inference engine's performance results in crisp values.

Algorithm for fuzzy logic
Step 1: As input crisp values, glucose, INS, BMI, DPF, and age were used.
Step 2: For the fuzzy number, set the triangular membership function.
Step 3: Create fuzzy numbers for the input and output sets' five (5) attributes.
Step 4: Mamdani's fuzzy inference method was used to complete the task.
Step 5: Input the rules and calculate the matching degree of rule with "OR" fuzzy disjunction for fuzzy input set (Glucoselow, Glucosemedium, Glucosehigh, INSlow, INSmedium, INShigh, BMIlow, BMImedium, BMIhigh, DPFlow, DPFmedium, DPFhigh, Ageyoung, Agemedium, Age-old).
Step 6: Compute the aggregation of the fired rules for fuzzy output set DM (DMverylow, DMlow, DMmedium, DMhigh, DMveryhigh).
Step 7: Defuzzify into the crisp values by:

$$z^{*} = \frac{\int \mu A(z).z dz}{\int \mu A(z) dz} \tag{1.1}$$

where, $\int$ is the algebraic integration, $\mu A(z)$ is the number of fuzzy numbers of the output fuzzy variable DM and z represents the weight for $\mu A(z)$.
Step 8: Represent the knowledge in human language form.
End.

1.5.2 Fuzzification

Fuzzification is the first step in the fuzzy inference method. Crisp inputs are converted into fuzzy inputs using this domain conversion. In fuzzification, the fuzzy sets for the indicators and outputs of diabetes diagnosis and management, as well as the membership feature, were specified.

The fuzzy sets for diabetes mellitus indicators and performance are as follows:
Number of pregnancy: {Low, Mild, Severe, Very Severe}.
Diastolic blood pressure: {Low, Mild, Severe, Very Severe}.
Triceps skin thickness: {Low, Mild, Severe, Very Severe}.
Glucose: {Low, Mild, Severe, Very Severe}.
Insulin: {Low, Mild, Severe, Very Severe}.
Body mass index (BMI): {Low, Mild, Severe, Very Severe}.
Diabetes pedigree function (DPF): {Low, Mild, Severe, Very Severe}.
Age: {Young, Medium, Old}.
Output: {Low, Mild, Severe, Very Severe}.

For the output fuzzy set, the system used $0 =$ Low, $\leq 0.4 =$ Mild, $0.4 > x \geq 0.7 =$ Severe and $1 =$ Very Severe.

As displayed in Fig. 1.2, the body temperature, blood glucose, and blood pressure captured data using sensors are initialed, processed, and controlled from the AI processing unit. The captured data from the temperature and blood pressure sensors are stored in the IoMT-based cloud storage system. AI-enabled IoMT can provide a relatively detailed blood pressure reading, and some blood pressure measuring instruments can be connected to a phone or a computer, with all data transferred to an IoMT-based cloud storage system. Corresponding applications will assist people in analyzing data and explaining why their blood pressure has altered. If the specialist determines that something is amiss and the user requires an advanced physical examination, the user will be notified and given the option to schedule an appointment with a specific doctor through the App. Users have a better understanding of why sickness occurs with the usage of an AI model. The device can detect the signs of any disease using an AI model. Researchers can, for example, use association rules to determine the potential relationship between illness genes. Scientists can see a potential causative association between a sequence of genes after studying a patient with Alzheimer's disease.

The range used for the body temperature is in Table 1.1. The temperature range membership feature is possible to clarify as:

$$\text{Low} = \begin{cases} 1, & x \quad < 36^{\circ}\text{C} \\ 0, & x \quad > 36^{\circ}\text{C} \end{cases}$$

$$\text{Mild} = \begin{cases} 1, & 36^{\circ}\text{C} \leq x \leq 37.5^{\circ}\text{C} \\ 0, & x > 37.5^{\circ}\text{C and} < 36^{\circ}\text{C} \end{cases}$$

$$\text{Severe} = \begin{cases} 1, & x > 37.5^{\circ}\text{C} \\ 0, & x < 37.5^{\circ}\text{C} \end{cases}$$

Likewise, various ranges of blood pressure readings are often considered to assess the patient's health status, as in Table 1.2. The blood pressure membership feature is a list as follows:

$$\text{Low} = \begin{cases} 1, & x \quad < 36 \text{ BPM} \\ 0, & x \quad > 60 \text{ BPM} \end{cases}$$

$$\text{Mild} = \begin{cases} 1, & 36 \text{ BPM} \leq x \leq 100 \text{ BPM} \\ 0, & x \quad > 100 \text{ BPM and } x < 60 \text{ BPM} \end{cases}$$

$$\text{Very severe} = \begin{cases} 1, & x > 100 \text{ BPM} \\ 0, & x < 100 \text{ BPM} \end{cases}$$

The rules for diagnosing the patient's health status are carried out based on these various ranges of values. The following membership feature is diagnosed with the output health state: Strong, Sick, Hypothermia, Fever, and diabetes mellitus as required. The overall functions of the state are as follows (Table 1.3):

$$\text{Checkup} = \begin{cases} 1, & x < 20 \\ 0, & x > 20 \end{cases}$$

Table 1.1 The body temperature measurement range.

Range	State
36.0°C–37.5°C	Low
>37.5°C	Very severe
<36.0°C	Low

Table 1.2 The blood pressure measurement range.

Range	State
60 BPM–100 BPM	Mild
>100 BPM	Very severe
<60 BPM	Low

Table 1.3 Classification of blood pressure, diabetes, and obesity.

Risk factors		Normal	Pre-stage	Stage I	Stage II	Stage III
Blood pressure	SBP	<110	110–138	139–160	≥160	
	DBP	<70	70–89	90–99	≥100	
Glucose	PGC	≤71–100	100–121	≥122		
	INS	≤97–138	138–190	≥191		
BMI		≤30	30–39.9	40–44.9	45.0–49.9	≥50
Waist circumference		≤40, ≤30	>40, >30			

SBP, *Systolic Blood Pressure;* DBP, *Diastolic Blood Pressure;* PGC, *Plasma Glucose Concentration;* INS 2, *hour Serum Insulin;* BMI, *Body Mass Index.*

$$\text{Unwell} = \begin{cases} 1, & 20 \leq x \leq 40 \\ 0, & x > 40 \text{ and } x < 20 \end{cases}$$

$$\text{Hypothermia} = \begin{cases} 1, & 40 \leq x \leq 60 \\ 0, & x > 60 \text{ and } x < 40 \end{cases}$$

$$\text{Healthy} = \begin{cases} 1, & x > 80 \\ 0, & x < 80 \end{cases}$$

As shown in Table 1.4, the rules for the diagnosis of performance health status are:

The duration of diabetes is a significant determinant of diabetes risk factors. The inability to follow the physician's prescribed treatment, dietary restrictions on disease control, lack of blood glucose management, blood sugar monitoring, and the impact on their vision problems have all made diabetes patients over 15 years of age feel that their quality of life has been affected significantly during the last period. Nonetheless, patients who have had diabetes for a short to medium period of time (less than 5 years or between 5 and 10 years after diagnosis) tend to be significantly more comfortable with the time required to manage the disease. These diagnosis rules can be summarized by taking into account all the combinations

The body temperature, blood glucose, blood pressure, room temperature, and humidity sensors' values respectively are regulated using the microcontroller. These sensor values are then sent to the IoMT-based cloud server in the database. The data can be accessed from the cloud by registered users using the platform of IoMT application. The patient's illness is diagnosed based on these

Table 1.4 Rules for diagnosing disease.

Blood pressure	Body temperature		
	Low	**Normal**	**High**
Low	Health review	Sick	Health review
Normal	Hypothermia	Strong	Fever
High	Health review	Sick	Health review

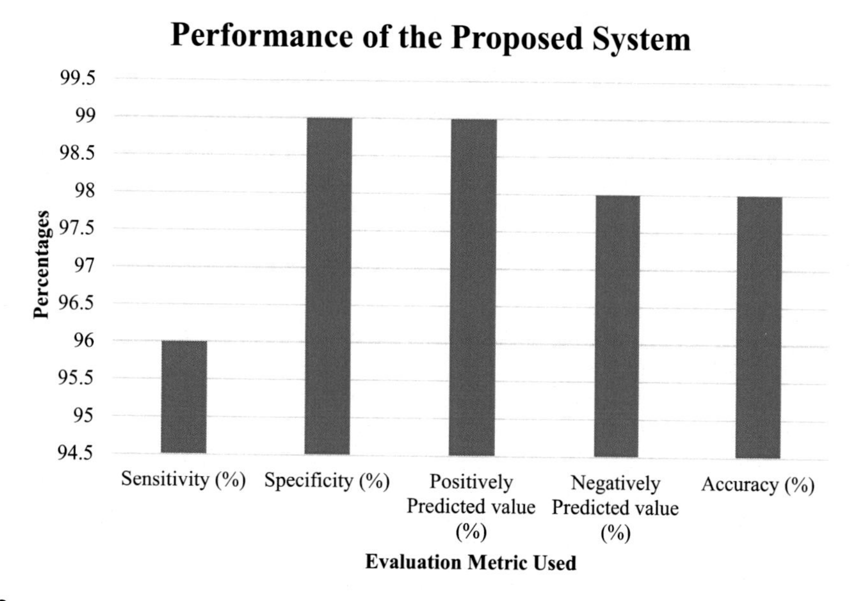

FIGURE 1.3

Performance evaluation of the proposed system.

values obtained by succeeding the rules set out. The health disorder diagnosis is performed by the medical practitioner and expert as shown in the system framework. The medicines can be administered and necessary action can be recommended even from a distance by the physicians and medical experts. The entire proposed work is autonomous, so the medical staff ratio requirement can be minimized and if can use this monitoring device at home, it would also eliminate the need for physical accompaniment to monitor patients. Hospital staffing costs are also exponentially decreased.

Fig. 1.3 shows the performance evaluation of the proposed framework, the results display 96% sensitivity, 99% precision, 99% positively predicted value, 98% negatively predicted value, and 98% accuracy. This demonstrates that the proposed system outperformed the findings shown. Fig. 1.3 shows the overall performance obtained from the proposed system for diabetes mellitus diagnosis. The hybridized method outperformed the single strategy, according to the findings.

1.6 Conclusions

The Introduction of IoMT-based technologies to any disease situation can be used to build a social forum to help individuals access appropriate treatment at home and to develop a robust repository on disease control for government and healthcare organizations. AI has demonstrated great performance and adaptability in various fields for big data processing. These are accomplished with increased accuracy rates and demonstrated usefulness and value in smart healthcare technologies. The algorithm's advancement and the trend for future progress and research have significantly differentiated it from other algorithms. Furthermore, the layer hierarchy and learning supervision are the important key factors for making successful AI applications. The hierarchy is very useful for proper classification and oversight means of maintaining the database in an application. The AI is also used for the optimization of the existing framework, and the processing of the hierarchical layer. In applications like digital image processing and speech recognition, AI can deliver successful results. Therefore the chapter reviews the applicability of AI-enabled IoMT-based systems in the smart healthcare system. The applications of AI have greatly increased the accuracy aspect of decision-making in smart healthcare systems, especially in the analysis of the captured data using IoMT-based devices and sensors. The process of digital image processing has also been enhanced by the use of AI models, hence, AI has become a hot research topic in recent years, and proving to be a real optimization technique in the smart healthcare system. However, the technologies are very new and ephebic, but it is very unique, and it is expected that there is going to be rapid growth in the nearest future. The algorithms have a great boom and prosperity in various fields like natural language processing, healthcare, remote sensing, and speech recognition will inevitably achieve goals and levels of triumph and satisfaction. There are a variety of promising potential directions in the use of AI technologies in smart healthcare. When the feature sets and delivery structures of the training and testing data are identical, it is assumed that the training process delivers accurate outcomes. Transfer learning is a research field in which the delivery of training and testing is adjusted or moved from one framework to another. Scientists can also concentrate on incorporating semantic technology into smart healthcare applications to enhance the interaction between smart devices and their users. The use of virtual objects in combination with AI algorithms will facilitate the development of virtual representations of real objects that could be run automatically. Smart healthcare applications and devices are frequently mobile and wearable, requiring users to touch screens in small spaces, which can be difficult for less technically-savvy users and senior citizens. Integration of speech recognition technologies into smart devices to enable natural language comprehension is also a promising research field. It is critical to recognize that in the procedure of developing such smart devices, we must avoid creating islands in which applications are developed solely to lag non-integration with one another. Finally, the usability of smart devices is very critical. The limitation of this chapter can be enhanced by using hybrid AI techniques for the diagnosis and monitoring of the patient in real-time.

References

Abikoye, O. C., Ojo, U. A., Awotunde, J. B., & Ogundokun, R. O. (2020). A safe and secured iris template using steganography and cryptography. *Multimedia Tools and Applications*, *79*(31–32), 23483–23506, 2020.

Abiodun, M. K., Awotunde, J. B., Ogundokun, R. O., Misra, S., Adeniyi, E. A., Arowolo, M. O., & Jaglan, V. (2021, February). Cloud and big data: A mutual benefit for organization development. In *Journal of physics: Conference series* (Vol. 1767, No. 1, pp. 012020). United Kingdom: IOP Publishing.

Adeniyi, E. A., Ogundokun, R. O., & Awotunde, J. B. (2021). *IoMT-based wearable body sensors network healthcare monitoring system. IoT in healthcare and ambient assisted living* (pp. 103–121). Singapore: Springer.

Adly, A. S. (2019). *Technology trade-offs for IIoT systems and applications from a developing country perspective: Case of Egypt. The internet of things in the industrial sector* (pp. 299–319). Cham: Springer.

Agbehadji, I. E., Awuzie, B. O., Ngowi, A. B., & Millham, R. C. (2020). Review of big data analytics, artificial intelligence and nature-inspired computing models towards accurate detection of COVID-19 pandemic cases and contact tracing. *International Journal of Environmental Research and Public Health, 17*(15), 5330.

Aggour, K. S., Gupta, V. K., Ruscitto, D., Ajdelsztajn, L., Bian, X., Brosnan, K. H., ... Vinciquerra, J. (2019). Artificial intelligence/machine learning in manufacturing and inspection: A GE perspective. *MRS Bulletin, 44*(7), 545–558.

Ahmed, M. B., Boudhir, A. A., Santos, D., El Aroussi, M., & Karas, İ. R. (Eds.). (2020). Innovations in smart cities applications edition 3: In *The proceedings of the 4th international conference on smart city applications*. Germany: Springer Nature.

Alehegn, M., Joshi, R., & Mulay, P. (2018). Analysis and prediction of diabetes mellitus using machine learning algorithm. *International Journal of Pure and Applied Mathematics, 118*(9), 871–878.

Allam, Z., & Dhunny, Z. A. (2019). On big data, artificial intelligence and smart cities. *Cities (London, England), 89*, 80–91.

Alwashmi, M. F. (2020). The use of digital health in the detection and management of COVID-19. *International Journal of Environmental Research and Public Health, 17*(8), 2906.

Amit, K., Chinmay, C., & Wilson, J. (2021). Intelligent healthcare data segregation using fog computing with internet of things and machine learning. *International Journal of Engineering Systems Modelling and Simulation*. Available from https://doi.org/10.1504/IJESMS.2021.10036745.

Arrieta, A. B., Díaz-Rodríguez, N., Del Ser, J., Bennetot, A., Tabik, S., Barbado, A., ... Herrera, F. (2020). Explainable artificial intelligence (XAI): Concepts, taxonomies, opportunities and challenges toward responsible AI. *Information Fusion, 58*, 82–115.

Arunkumar, C., & Ramakrishnan, S. (2018). Prediction of cancer using customised fuzzy rough machine learning approaches. *Healthcare Technology Letters, 6*(1), 13–18.

Attia, Z. I., Kapa, S., Lopez-Jimenez, F., McKie, P. M., Ladewig, D. J., Satam, G., ... Asirvatham, S. J. (2019). Screening for cardiac contractile dysfunction using an artificial intelligence–enabled electrocardiogram. *Nature Medicine, 25*(1), 70–74.

Awotunde, J. B., Adeniyi, A. E., Ogundokun, R. O., Ajamu, G. J., & Adebayo, P. O. (2021). MIoT-based big data analytics architecture, opportunities and challenges for enhanced telemedicine systems. *Studies in Fuzziness and Soft Computing, 2021*(410), 199–220.

Awotunde, J. B., Ayo, F. E., Jimoh, R. G., Ogundokun, R. O., Matiluko, O. E., Oladipo, I. D., & Abdulraheem, M. (2020). Prediction and classification of diabetes mellitus using genomic data. *Intelligent IoT Systems in Personalized Health Care*, 235–292.

Awotunde, J. B., Folorunso, S. O., Bhoi, A. K., Adebayo, P. O., & Ijaz, M. F. (2021). Disease diagnosis system for IoT-based wearable body sensors with machine learning algorithm. *Intelligent Systems Reference Library, 2021*(209), 201–222.

Awotunde, J. B., Folorunso, S. O., Jimoh, R. G., Adeniyi, E. A., Abiodun, K. M., & Ajamu, G. J. (2021). Application of artificial intelligence for COVID-19 epidemic: An exploratory study, opportunities, challenges, and future prospects. *Studies in Systems, Decision and Control, 2021*(358), 47–61.

Awotunde, J. B., Jimoh, R. G., Oladipo, I. D., & Abdulraheem, M. (2021). Prediction of malaria fever using long-short-term memory and big data. *Communications in Computer and Information Science*, *2021*(1350), 41–53.

Awotunde, J. B., Matiluko, O. E., & Fatai, O. W. (2014). Medical diagnosis system using fuzzy logic. *African Journal of Computing & ICT*, *7*(2), 99–106.

Ayo, F. E., Awotunde, J. B., Ogundokun, R. O., Folorunso, S. O., & Adekunle, A. O. (2020). A decision support system for multi-target disease diagnosis: A bioinformatics approach. *Heliyon*, *6*(3), e03657.

Ayo, F. E., Ogundokun, R. O., Awotunde, J. B., Adebiyi, M. O., & Adeniyi, A. E. (2020, July). Severe acne skin disease: A fuzzy-based method for diagnosis. *Lecture notes in computer science (including subseries Lecture notes in artificial intelligence and lecture notes in bioinformatics)*, ICCSA 2020, 12254 LNCS (pp. 320–334).

Barstugan, M., Ozkaya, U., & Ozturk, S. (2020). Coronavirus (covid-19) classification using CT images by machine learning methods. *arXiv preprint arXiv:2003.09424*.

Belhadi, A., Djenouri, Y., Srivastava, G., Djenouri, D., Lin, J. C. W., & Fortino, G. (2021). Deep learning for pedestrian collective behavior analysis in smart cities: A model of group trajectory outlier detection. *Information Fusion*, *65*, 13–20.

Berglund, D. D., Kurowicki, J., Giveans, M. R., Horn, B., & Levy, J. C. (2018). Comorbidity effect on speed of recovery after arthroscopic rotator cuff repair. *JSES Open Access*, *2*(1), 60–68.

Bhadoria, R. K., Saha, J., Biswas, S., & Chowdhury, C. (2020). IoT-based location-aware smart healthcare framework with user mobility support in normal and emergency scenario: A comprehensive survey. *Healthcare Paradigms in the Internet of Things Ecosystem*, 137–161.

Bravo, C. E., Saputelli, L. A., Rivas, F. I., Perez, A. G., Nikolaou, M., Zangl, G., ... Nunez, G. (2012, January). State-of-the-art application of artificial intelligence and trends in the E&P industry: A technology survey. In *SPE intelligent energy international*. Society of Petroleum Engineers.

Brown, D. E., Abbasi, A., & Lau, R. Y. (2015). Predictive analytics: Predictive modeling at the micro level. *IEEE Intelligent Systems*, *30*(3), 6–8.

Chakraborty, C., & Abougreen, A. N. (2021). Intelligent internet of things and advanced machine learning techniques for covid-19. *EAI Endorsed Transactions on Pervasive Health and Technology*, *7*(26), e1.

Das, A., Rad, P., Choo, K. K. R., Nouhi, B., Lish, J., & Martel, J. (2019). Distributed machine learning cloud teleophthalmology IoT for predicting AMD disease progression. *Future Generation Computer Systems*, *93*, 486–498.

Dey, R., Bajpai, V., Gandhi, G., & Dey, B. (2008, December). Application of artificial neural network (ANN) technique for diagnosing diabetes mellitus. In: *2008 IEEE region 10 and the third international conference on industrial and information systems* (pp. 1–4). IEEE.

Dimitrov, D. V. (2016). Medical internet of things and big data in healthcare. *Healthcare Informatics Research*, *22*(3), 156–163.

Fagherazzi, G., Goetzinger, C., Rashid, M. A., Aguayo, G. A., & Huiart, L. (2020). Digital health strategies to fight COVID-19 worldwide: Challenges, recommendations, and a call for papers. *Journal of Medical Internet Research*, *22*(6), e19284.

Folorunso, S. O., Awotunde, J. B., Ayo, F. E., & Abdullah, K. K. A. (2021). RADIoT: The unifying framework for IoT, radiomics and deep learning modeling. *Intelligent Systems Reference Library*, *2021*(209), 109–128.

Fröhlich, H., Balling, R., Beerenwinkel, N., Kohlbacher, O., Kumar, S., Lengauer, T., ... Rebhan, M. (2018). From hype to reality: Data science enabling personalized medicine. *BMC Medicine*, *16*(1), 150.

Ganasegeran, K., & Abdulrahman, S. A. (2020). *Artificial intelligence applications in tracking health behaviors during disease epidemics. Human behaviour analysis using intelligent systems* (pp. 141–155). Cham: Springer.

Gia, T. N., Dhaou, I. B., Ali, M., Rahmani, A. M., Westerlund, T., Liljeberg, P., & Tenhunen, H. (2019). Energy efficient fog-assisted IoT system for monitoring diabetic patients with cardiovascular disease. *Future Generation Computer Systems*, *93*, 198–211.

Goodman, K. E., Lessler, J., Cosgrove, S. E., Harris, A. D., Lautenbach, E., Han, J. H., ... Tamma, P. D. (2016). A clinical decision tree to predict whether a bacteremic patient is infected with an extended-spectrum β-lactamase–producing organism. *Clinical Infectious Diseases*, *63*(7), 896–903.

Guan, Z., Lv, Z., Du, X., Wu, L., & Guizani, M. (2019). Achieving data utility-privacy tradeoff in Internet of medical things: A machine learning approach. *Future Generation Computer Systems*, *98*, 60–68.

Habibzadeh, H., Nussbaum, B. H., Anjomshoa, F., Kantarci, B., & Soyata, T. (2019). A survey on cybersecurity, data privacy, and policy issues in cyber-physical system deployments in smart cities. *Sustainable Cities and Society*, *50*, 101660.

Han, B. A., Schmidt, J. P., Bowden, S. E., & Drake, J. M. (2015). Rodent reservoirs of future zoonotic diseases. *Proceedings of the National Academy of Sciences*, *112*(22), 7039–7044.

Hoofnagle, C. J., van der Sloot, B., & Borgesius, F. Z. (2019). The European Union general data protection regulation: What it is and what it means. *Information & Communications Technology Law*, *28*(1), 65–98.

Jayanthi, N., & Valluvan, K. R. (2017). A review of performance metrics in designing of protocols for wireless sensor networks. *Asian Journal of Research in Social Sciences and Humanities*, *7*(1), 716–730.

Josefsson, M. Y., & Steinthorsson, R. S. (2021). Reflections on a SMART urban ecosystem in a small island state: The case of SMART Reykjavik. *International Journal of Entrepreneurship and Small Business*, *42*(1–2), 93–114.

Joseph, S. I. T., & Thanakumar, I. (2019). Survey of data mining algorithm's for intelligent computing system. *Journal of trends in Computer Science and Smart technology (TCSST)*, *1*(01), 14–24.

Kaur, P., Kumar, R., & Kumar, M. (2019). A healthcare monitoring system using random forest and internet of things (IoT). *Multimedia Tools and Applications*, *78*(14), 19905–19916.

Khan, F. N., Fan, Q., Lu, C., & Lau, A. P. T. (2020). *Machine learning methods for optical communication systems and networks. Optical fiber telecommunications VII* (pp. 921–978). Academic Press.

Khan, U., Ali, A., Khan, S., Aadil, F., Durrani, M. Y., Muhammad, K., ... Lee, J. W. (2019). Internet of medical things–based decision system for automated classification of Alzheimer's using three-dimensional views of magnetic resonance imaging scans. *International Journal of Distributed Sensor Networks*, *15*(3), 1550147719831186.

Kishor, A., Chakraborty, C. H., & Jeberson, W. (2021). A novel fog computing approach for minimization of latency in healthcare using machine learning. *International Journal of Interact Multimed Artif Intell*, *6*(6), 10–20.

Komenda, M., Bulhart, V., Karolyi, M., Jarkovský, J., Mužík, J., Májek, O., ... Dušek, L. (2020). Complex reporting of the COVID-19 epidemic in the Czech republic: Use of an interactive web-based app in practice. *Journal of Medical Internet Research*, *22*(5), e19367.

Krishnamurthi, R., Kumar, A., Gopinathan, D., Nayyar, A., & Qureshi, B. (2020). An overview of IoT sensor data processing, fusion, and analysis techniques. *Sensors*, *20*(21), 6076.

Kumar, A., Sangwan, S. R., & Nayyar, A. (2020). *Multimedia social big data: Mining. Multimedia big data computing for IoT applications* (pp. 289–321). Singapore: Springer.

Kumar, P. M., & Gandhi, U. D. (2018). A novel three-tier internet of things architecture with machine learning algorithm for early detection of heart diseases. *Computers & Electrical Engineering*, *65*, 222–235.

Kurd, Z., Kelly, T., & Austin, J. (2007). Developing artificial neural networks for safety-critical systems. *Neural Computing and Applications*, *16*(1), 11–19.

Kuru, K., & Khan, W. (2020). A framework for the synergistic integration of fully autonomous ground vehicles with smart city. *IEEE Access*, *9*, 923–948.

Li, C., Hu, X., & Zhang, L. (2017). The IoT-based heart disease monitoring system for pervasive healthcare service. *Procedia computer science*, *112*, 2328–2334.

Liberti, L., Lavor, C., Maculan, N., & Mucherino, A. (2014). Euclidean distance geometry and applications. *SIAM review*, *56*(1), 3–69.

Lingaraj, H., Devadass, R., Gopi, V., & Palanisamy, K. (2015). Prediction of diabetes mellitus using data mining techniques: A review. *Journal of Bioinformatics & Cheminformatics*, *1*(1), 1–3.

Londhe, V. Y., & Bhasin, B. (2019). Artificial intelligence and its potential in oncology. *Drug Discovery Today*, *24*(1), 228–232.

Lorincz, K., Malan, D. J., Fulford-Jones, T. R., Nawoj, A., Clavel, A., Shnayder, V., ... Moulton, S. (2004). Sensor networks for emergency response: Challenges and opportunities. *IEEE pervasive Computing*, *3*(4), 16–23.

Ma, H., & Pang, X. (2019). Research and analysis of sport medical data processing algorithms based on deep learning and internet of things. *IEEE Access*, *7*, 118839–118849.

Mahdavinejad, M. S., Rezvan, M., Barekatain, M., Adibi, P., Barnaghi, P., & Sheth, A. P. (2018). Machine learning for internet of things data analysis: A survey. *Digital Communications and Networks*, *4*(3), 161–175.

Mahroum, N., Adawi, M., Sharif, K., Waknin, R., Mahagna, H., Bisharat, B., ... Watad, A. (2018). Public reaction to Chikungunya outbreaks in Italy—Insights from an extensive novel data streams-based structural equation modeling analysis. *PLoS One*, *13*(5), e0197337.

Manogaran, G., Varatharajan, R., Lopez, D., Kumar, P. M., Sundarasekar, R., & Thota, C. (2018). A new architecture of internet of things and big data ecosystem for secured smart healthcare monitoring and alerting system. *Future Generation Computer Systems*, *82*, 375–387.

Marques, G., & Pitarma, R. (2016a). An indoor monitoring system for ambient assisted living based on internet of things architecture. *International Journal of Environmental Research and Public Health*, *13* (11), 1152.

Marques, G., & Pitarma, R. (2018, November). Smartwatch-based application for an enhanced healthy lifestyle in indoor environments. In: *International conference on computational intelligence in information system* (pp. 168–177). Springer, Cham.

Marques, G., Ferreira, C. R., & Pitarma, R. (2019). Indoor air quality assessment using a CO_2 monitoring system based on internet of things. *Journal of Medical Systems*, *43*(3), 1–10.

Marques, G., Roque Ferreira, C., & Pitarma, R. (2018). A system based on the internet of things for real-time particle monitoring in buildings. *International Journal of Environmental Research and Public Health*, *15* (4), 821.

Marques, M. S. G., & Pitarma, R. (2016b). Smartphone application for enhanced indoor health environments. *Journal of Information Systems Engineering & Management*, *1*, 4.

Mehta, V. (2018). *A novel approach to realize internet of intelligent things. Big data analytics* (pp. 413–419). Singapore: Springer.

Miller, D. D., & Brown, E. W. (2018). Artificial intelligence in medical practice: The question to the answer? *The American Journal of Medicine*, *131*(2), 129–133.

Mohammadi, M., Al-Fuqaha, A., Guizani, M., & Oh, J. S. (2017). Semisupervised deep reinforcement learning in support of IoT and smart city services. IEEE Internet of Things. *The Journal*, *5*(2), 624–635.

Muhammad, A. N., Aseere, A. M., Chiroma, H., Shah, H., Gital, A. Y., & Hashem, I. A. T. (2020). Deep learning application in smart cities: Recent development, taxonomy, challenges and research prospects. *Neural Computing and Applications*, 1–37.

Nathani, B., & Vijayvergia, R. (2017, December). The Internet of intelligent things: An overview. In: *2017 International conference on intelligent communication and computational techniques (ICCT)* (pp. 119–122). IEEE.

Nayyar, A., Puri, V., & Nguyen, N. G. (2019). Biosenhealth 1.0: A novel internet of medical things (iomt)-based patient health monitoring system. In: *International conference on innovative computing and communications* (pp. 155–164). Springer, Singapore.

Neri, E., Miele, V., Coppola, F., & Grassi, R. (2020). Use of CT and artificial intelligence in suspected or COVID-19 positive patients: Statement of the Italian Society of Medical and Interventional Radiology. *La Radiologia Medica, 1.*

Neyja, M., Mumtaz, S., Huq, K. M. S., Busari, S. A., Rodriguez, J., & Zhou, Z. (2017, December). An IoT-based e-health monitoring system using ECG signal. In *GLOBECOM 2017-2017 IEEE global communications conference* (pp. 1–6). IEEE.

Nguyen, G., Dlugolinsky, S., Bobák, M., Tran, V., García, Á. L., Heredia, I., ... Hluchý, L. (2019). Machine learning and deep learning frameworks and libraries for large-scale data mining: A survey. *Artificial Intelligence Review, 52*(1), 77–124.

Oladele, T. O., Ogundokun, R. O., Awotunde, J. B., Adebiyi, M. O., & Adeniyi, J. K. (2020, July). Diagmal: A malaria coactive neuro-fuzzy expert system. *Lecture notes in computer science (including subseries lecture notes in artificial intelligence and lecture notes in bioinformatics)*, ICCSA 2020, 12254 LNCS, (pp. 428–441).

Oladipo, I. D., Babatunde, A. O., Aro, T. O., & Awotunde, J. B. (2020). Enhanced neuro-fuzzy inferential system for diagnosis of diabetes mellitus (DM). *International Journal of Information Processing and Communication (IJIPC), 8*(1), 17–25.

Oladipo, I. D., Babatunde, A. O., Awotunde, J. B., & Abdulraheem, M. (2021). An improved hybridization in the diagnosis of diabetes mellitus using selected computational intelligence. *Communications in Computer and Information Science, 2021*(1350), 272–285.

Özdemir, V., & Hekim, N. (2018). Birth of industry 5.0: Making sense of big data with artificial intelligence,"the internet of things" and next-generation technology policy. *Omics: A journal of integrative biology, 22*(1), 65–76.

Panzirsch, M., Weber, B., Rubio, L., Coloma, S., Ferre, M., & Artigas, J. (2017). Tele-healthcare with humanoid robots: A user study on the evaluation of force feedback effects. In: *2017 IEEE world haptics conference (WHC)* (pp. 245–250). IEEE.

Patel, H., Singh Rajput, D., Thippa Reddy, G., Iwendi, C., Kashif Bashir, A., & Jo, O. (2020). A review on classification of imbalanced data for wireless sensor networks. *International Journal of Distributed Sensor Networks, 16*(4), 1550147720916404.

Pati, J. (2018). Gene expression analysis for early lung cancer prediction using machine learning techniques: An eco-genomics approach. *IEEE Access, 7*, 4232–4238.

Pouyanfar, S., Sadiq, S., Yan, Y., Tian, H., Tao, Y., Reyes, M. P., ... Iyengar, S. S. (2018). A survey on deep learning: Algorithms, techniques, and applications. *ACM Computing Surveys (CSUR), 51*(5), 1–36.

Rodgers, W. (2020). *Artificial intelligence in a throughput model: Some major algorithms*. CRC Press.

Rodrigues, J. C. L., Hare, S. S., Edey, A., Devaraj, A., Jacob, J., Johnstone, A., ... Robinson, G. (2020). An update on COVID-19 for the radiologist-A British society of thoracic imaging statement. *Clinical Radiology, 75*(5), 323–325.

Schork, N. J. (2019). *Artificial intelligence and personalized medicine. Precision medicine in cancer therapy* (pp. 265–283). Cham: Springer.

Sejnowski, T. J. (2018). *The deep learning revolution*. MIT Press.

Shaban-Nejad, A., Michalowski, M., & Buckeridge, D. L. (2018). *Health intelligence: How artificial intelligence transforms population and personalized health*. Nature Publishing Group.

Shen, C., Chen, A., Luo, C., Zhang, J., Feng, B., & Liao, W. (2020). Using reports of symptoms and diagnoses on social media to predict COVID-19 case counts in mainland China: Observational infoveillance study. *Journal of Medical Internet Research, 22*(5), e19421.

Singh, S., Sharma, P. K., Yoon, B., Shojafar, M., Cho, G. H., & Ra, I. H. (2020). Convergence of blockchain and artificial intelligence in IoT network for the sustainable smart city. *Sustainable Cities and Society, 63*, 102364.

Solanki, A., & Nayyar, A. (2019). *Green internet of things (G-IoT): ICT technologies, principles, applications, projects, and challenges. Handbook of research on big data and the IoT* (pp. 379–405). IGI Global.

Tokmurzina, D. (2020). Road marking condition monitoring and classification using deep learning for city of Helsinki. *Master's Programme in ICT Innovation*. Aalto University.

Triantafyllidis, A. K., & Tsanas, A. (2019). Applications of machine learning in real-life digital health interventions: Review of the literature. *Journal of Medical Internet Research, 21*(4), e12286.

Wong, Z. S., Zhou, J., & Zhang, Q. (2019). Artificial intelligence for infectious disease big data analytics. *Infection, Disease & Health, 24*(1), 44–48.

Woodhead, R., Stephenson, P., & Morrey, D. (2018). Digital construction: From point solutions to IoT ecosystem. *Automation in Construction, 93*, 35–46.

Xiuqin, P., Zhang, Q., Zhang, H., & Li, S. (2019). A fundus retinal vessels segmentation scheme based on the improved deep learning U-Net model. *IEEE Access, 7*, 122634–122643.

Zhang, B., Qi, S., Monkam, P., Li, C., Yang, F., Yao, Y. D., & Qian, W. (2019). Ensemble learners of multiple deep CNNs for pulmonary nodules classification using CT images. *IEEE Access, 7*, 110358–110371.

Zheng, N., Du, S., Wang, J., Zhang, H., Cui, W., Kang, Z., ... Ma, M. (2020). Predicting COVID-19 in china using hybrid AI model. *IEEE Transactions on Cybernetics, 50*(7), 2891–2904.

Zheng, X., Sun, S., Mukkamala, R. R., Vatrapu, R., & Ordieres-Meré, J. (2019). Accelerating health data sharing: A solution based on the internet of things and distributed ledger technologies. *Journal of Medical Internet Research, 21*(6), e13583.

Further reading

Berglund, E., & Sitte, J. (2006). The parameterless self-organizing map algorithm. *IEEE Transactions on neural networks, 17*(2), 305–316.

Bragazzi, N. L., Alicino, C., Trucchi, C., Paganino, C., Barberis, I., Martini, M., ... Icardi, G. (2017). Global reaction to the recent outbreaks of Zika virus: Insights from a big data analysis. *PLoS One, 12*(9), e0185263.

Folorunso, S. O., Awotunde, J. B., Adeboye, N. O., & Matiluko, O. E. (2022). Data classification model for COVID-19 pandemic. In Advances Studies in Systems. *Decision and Control, 2022*(378), 93–118.

Li, Z., Moran, P., Dong, Q., Shaw, R. J., & Hauser, K. (2017, May). Development of a tele-nursing mobile manipulator for remote care-giving in quarantine areas. In *2017 IEEE international conference on robotics and automation (ICRA)* (pp. 3581–3586). IEEE.

Lukoševičius, M. (2012). *A practical guide to applying echo state networks. Neural networks: Tricks of the trade* (pp. 659–686). Berlin, Heidelberg: Springer.

Pramanik, P. K. D., Solanki, A., Debnath, A., Nayyar, A., El-Sappagh, S., & Kwak, K. S. (2020). Advancing modern healthcare with nanotechnology, nanobiosensors, and internet of nano things: Taxonomies, applications, architecture, and challenges. *IEEE Access, 8*, 65230–65266.

CHAPTER

Sensor and actuators for smart healthcare in post-COVID-19 world

2

Aumnat Tongkaw
Faculty of Science and Technology, Songkhla Rajabhat University, Songkhla, Thailand

2.1 Introduction

Information technology applications are commonly used to serve many human functions; technology has developed into a mechanism for delivering information and assisting users across all aspects, such as ATMs and deposit-withdrawal systems. Syncing personal records will help to reduce the likelihood of human error. Furthermore, data transmission through Information Management Systems is a simple process that retains consistent quality. Essential components are used in health information technology. Various tools are being used to provide health facilities, such as emergency data from health insurance, diagnosis documentation, medication records, X-ray devices, and sensor devices. The system's essential components are hardware, information, and personnel. Hardware covers computing appliances, servers, clients, and networking; software includes operating systems like Windows, system utilities like antivirus, applications like Microsoft Word, and HIS. However, the technology that we have today can replace and solve the problem of a shortage of medical personnel in many parts thanks to the procedure of patient data acquisition (input), data processing (process), and data output (output).

There are various types of health information technologies: HIS, computerized provider order entry (CPOE), electronic health records (EHRs), picture archiving and communication system (PACS), (Garavand et al., 2016), and much other Health IT forms such as m-Health (Istepanian & Lacal, 2003; Istepanian, Laxminarayan, & Pattichis, 2007; Lupton, 2012) health information exchange (HIE) (Kuperman, 2011), biosurveillance information retrieval (Hills, Lober, & Painter, 2008), Telemedicine, and Telehealth.

The minimization of data errors remains a big concern with the health information system, including the fact that the United States' health services have "systematic issues" that contribute to preventable medical errors. In certain countries, the standard of medical care cannot be overlooked. A severe pandemic approach, for example, relies exclusively on systematic fixes, which is inefficient. As a result, information management platforms in hospitals play a vital role in addressing this problem.

COVID-19 is indubitably a threat to the health care system. Because the demand for medical care has increased at a rapid rate while the nursing home's capacity and the medical personnel did not add up due to the lack of equilibrium in the system, the result is a lack of response to the needs of health care.

Implementation of Smart Healthcare Systems using AI, IoT, and Blockchain. DOI: https://doi.org/10.1016/B978-0-323-91916-6.00003-5

There was a large number of serious illnesses and deaths. The cumulative number of infection cases as of March 18, 2021, is around 121 million people, and almost 2.6 million people have died worldwide. For Thailand, the cumulative number of infection cases resulted in 27,494 and 89 deaths.

COVID-19 is undeniably a challenge to the health care sector because the need for medical care has risen at an accelerated pace. In contrast, nursing homes and medical staff's capability has not added up due to a lack of system equilibrium, resulting in a lack of response to health care needs. There were many significant illnesses and fatalities; the total number of infection cases as of March 18, 2021, is estimated to be about 121 million people, with almost 2.6 million deaths worldwide. Thailand's total number of illness cases resulted in 27,494 infections and 89 deaths.

COVID-19 has a far-reaching effect on healthcare professionals, especially in developing countries like Thailand. Its main influence is the infection of many healthcare professionals, resulting in the resignation and a steadily declining number of medical practitioners. Its effects continue to place significant pressure on Thailand's healthcare services (Triukose et al., 2021). However, concerns that may occur, including an uptick in psychiatric admissions as a consequence of the economic crisis, treatment for residual patients that may have progressed illness development, resignation and loss of healthcare staff, and financial loss to the hospital and the health insurance industry as a result of grappling with COVID-19 (Katewongsa, Widyastari, Saonuam, Haemathulin, & Wongsingha, 2021).

This chapter first addresses sensors, actuators, and associated technology used in healthcare systems, such as RFID, Wireless Sensor Network, Near Field Communication, Zigbee, Z-Wave, and Bluetooth LE. The chapter then selected a case study in a general hospital in Thailand with standards and processes. The hospital has sensors and actuators technology for use in the HIS. Many clinics in the hospital use sensors and actuators, including master patient index (MPI), Insurance Eligibility System, Appointment Scheduling, Nursing Application, Pharmacy Applications, Laboratory Information System (LIS), Imaging Applications or PACS, Imaging Application, Billing System, Enterprise Resource Planning (ERP), Electronic Medical Records, Computerized Physician Order Entry (HIE). This chapter then depicts the method of integrating sensors and actuators in a Thai hospital. However, when implementing sensors and actuators, it is often essential to include risk in services. This chapter further discusses the dangers of potential attacks and threats, risk mitigation and monitoring, and future network security checks of those networks. Finally, the chapter depicts the outcomes of user satisfaction surveys and discussions after implementation in the general hospital. To enhance convenience, the system will be enhanced with ANN technology. This method will offer good quality in the screening process, reduce the risk of COVID-19 virus infection, reduce the workload of medical staff, and provide reliable results.

2.2 Sensors for smart healthcare

2.2.1 Radio-frequency identification

Radio-frequency identification tags may be placed on each patient's wrist to identify their medical diagnosis. Suppose a patient has been identified and monitored by medical personnel. In that case, the doctor or nurse may use the mobile reader to check data from tags, informing all healthcare providers that they have been taken care of. Both records, such as medication orders, treatment specifics, and adverse effects, will be shown to the healthcare provider, and this information will be transmitted in a brief amount of time.

All patient records, including medication and infusion records, can be easily maintained since RFID technology is dependable and efficient. RFID is marketed as a cost-effective way of preserving data added to the hospital computer system. The use of RFIDs in the examination would allow for quicker processing of details, which would help emergency patients. (Yao, Chu, & Li, 2010) Not just that, but it will reduce the error in transmitting incomplete information and avoid incidents that may result in medical errors.

Furthermore, in the event of a hospital transfer, the patient's records, including medical history, type of injury, and treatment, may be forwarded to the admitted medical center. The treatment status can be re-labeled with new details and sent to the nearest attending hospital. Since these inputs can be achieved simultaneously by reading the RFID tag, redundant manual inputs are reduced. Avoiding the traps that may emerge from main data errors is also critical (Finkenzeller, 2010).

RFID is often used in neonatal management to ensure the data is compatible with parental data. Because of the resemblance of the newborn's features, infant transitions are a common occurrence. Furthermore, if the child is transferred to another room, they should be properly forwarded. The information gathered about the baby's physical state will also be used to identify patients discharged in a hospital robbery case. During the COVID-19 period, the workload of healthcare workers increased, which makes doctors and nurses tired. Devices linked with RFID to process information quickly help reduce medical staff's work time. The desire of medical professionals to embrace and employ RFID applications in their daily job is influenced by their perceptions of RFID benefits. In other words, doctors can use RFID to give a variety of operational and functional solutions based on their task needs and work requirements (Abugabah, Sanzogni, Houghton, AlZubi, & Abuqabbeh, 2021).

Since RFID tags are inexpensive, they can be applied to a wide range of devices and can find different device locations or medical products to scan and verify the objects. The persistence of such objects, or even materials or devices that are prevented from being moved may track unwanted movements. It may also be connected to alert systems like blinking LED lights or messages to signal movement RFID technology used in pharmaceutical applications. It will alert you in advance and reliably say the storage place for detecting the persistence of medications and medical equipment, assisting in detecting and reducing the spread of illegally prescribed medicines or even expired drugs.

Nowadays, RFID will operate in tandem with the Electronic Product Code (EPC). Each specific product code is assigned an EPC, which is an electronic serial number scheme. Within the RFID, each piece has a unique number that allows it to read and display various product details and show different manufacturing and expiration dates. To make inventory management more effective, reduce errors in keying products to standard, and reduce unexpected supply shortages.

2.2.2 Wireless sensor network

A wireless sensor network is an electronic center that monitors the patient's temperature, blood pressure, and current condition. The data is then transmitted in real-time to the central server via the hospital network. They are most widely used in the evaluation of a patient's well-being.

Wireless sensor networks are classified into two types: wearable and implanted. External usage includes attaching the gadget to the arm like a watch, wearing it on the ankle, or positioning it next to the heart. The implanted type will assist in collecting information from body activity. However, having sensors placed on the body raises concerns over personal rights. (Raghavendra, Sivalingam, & Znati, 2006). During the COVID-19, a study used internet of things (IoT) remote healthcare to monitor a system that provides patient information via a Web browser. Contiki OS with a 6LoWPAN protocol stack and Cooja, the Contiki simulator, were employed in the system (Baig Mohammad & Shitharth, 2021). It not only reduces the contact between healthcare professionals and patients. Instead, it sends information from patients directly into the hospital's database system, reducing measurement errors and recording medical personnel. It also reduces the burden of travel expenses to meet patients at home. Moreover, it can track patient information in real-time as well.

2.2.3 Near field communication

Near field communication (NFC) is a non-contact communication system that incorporates RFID and wireless link technologies to work in the 13.56 MHz range and has a transmission distance of around 10 CM. The transfer speed is 106, 212, or 424 Kbit/s, with the possibility of increasing to 1 Mb in the future. ISO18092, ISO21481, ECMA (340, 352, and 356) standards, and ETSI TS 102 190 are all compliant. As contrasted to other short-range wireless networking systems, the non-contact smart card structure, commonly used in ISO14443A and B protocols, such as Philips' MIFARE technology and Sony's Felica technology, NFC is secure and on time. This shortened response time is suitable for use as an electrolysis technology in a wireless communication environment, allowing quick and safe transactions.

NFC is increasingly accepted by retailers and has become an official standard and payment feature since it is compliant with new contactless smart card technologies. NFC technology allows connectivity between devices. Secure, fast, and automatic. Following COVID-19, NFC will become more relevant and more compliant with RFID, NFC, and RFID integration technologies from the past to eliminate healthcare professionals' mistakes. Also, to allow data forwarding from actuators or other medical instruments capable of receiving information directly into the database and high reliability and faster data saving (Lahtela, Hassinen, & Jylha, 2008).

During COVID-19, it is necessary to reduce direct exposure. Recording the patient's personal information with the hospital and making payments or clearing various rights from the hospital. Therefore, it is necessary to bring the NFC system into use even more. In addition to making it unnecessary to contact the system, the information is more accurate at a hospital in Thailand. There are still some that use the card. NFC for hospital spending and hospital canteens. The advantages of NFC devices are saving battery and cheap, although there are some disadvantages that the data transmission must be very close to the reader. However, it is not a problem for hospitals to send information at close distances.

Furthermore, the trend after COVID-19 may have various applications to support this work. Such as the wearable healthcare devices can monitor heart rate and temperature using a wireless/battery-free sensor and a customized smartphone application—the smartphone application for real-time data acquisition and processing (Kang, Lee, Yun, & Song, 2021).

2.2.4 Zigbee

Zigbee is a wireless networking protocol with low data speeds, low power consumption, and low cost. It aims to develop a device known as the Wireless Sensor Network, which would operate indoors and outdoors, in all weather conditions, and for months or years on small batteries such as 2 AA batteries. It is appropriate for use in a variety of tracking groups. ZigBee has three common working frequency bands: 2.4 GHz, 915, and 868 MHz. Each band has 16 channels, ten channels, and one channel. The wireless data transfer rates are at 250, 40, and 20 Kbps, respectively. ZigBee implements the Physical Layer and MAC Layer of IEEE 802.15.4, the Wireless Personal Area Network (WPAN) standard of wireless communication, working in the lower layers (bottom two layers) such as signal power level, link quality, access control, security, etc., but in the next layer will be a format of Zigbee.

As previously said, ZigBee can be integrated into a network since it is based on the IEEE 802.15.4 protocol and is handled as Zigbee in the following layer. IEEE 802.15.4 divides network devices into two types: Full Function Device (FFD), a device capable of completing all network functions and Reduce Function Device. This network device has been degraded. ZigBee is divided into three types of behavior:

1. The coordinator is responsible for building up communication. Networking between End Device and Router or Coordinator and Coordinator together or Coordinator and Router assigns the same address to the device in the network and manages routing, which is comparable to FFD.
2. The End Device is used to receive the signal from the sensor at the end with low power to work, comparable to Reduced Function Device (RFD) or Full Function Device (FFD) in some cases, depending on the sensor used.
3. Routers are responsible for transmitting and receiving data in various network routes, comparable to FFD.

The expenses of medicines and non-preventive health treatments have an impact on healthcare provider organizations' financial budgets, limiting the financial resources available to deliver improved service and healthcare quality. The problem of connectivity is growing as the number of devices grows. The ZigBee or 6LoWPAN technology has a gateway for connecting medical devices with ZigBee/6LoWPAN modules for detecting blood pressure, SpO_2, blood glucose, and weight. Because the ZigBee protocol allows adding new types of sensors to the platform without modifying the software at the gateway, it can make integrating the existing hospital software in Thailand easier (Gómez-García, Askar-Rodriguez, & Velasco-Medina, 2021).

Using the MG2455 platform, research has adopted ZigBee technology for facilitating patient monitoring and identifying the patient's location. In an emergency, patients should call a nurse. The machine would transmit the patient's location number to the server in real-time, allowing nurses to have more accurate access to the patient (Jihong, 2011).

ZigBee can also work with fuzzy systems, including a real-time patient activity tracking device that can show a patient's physical condition, pinpoint their position, and track their body's movement. The information is stored in a database, and fuzzy logic is used to view the patient's status in graphs on the screen in real-time. Which will make it easier for doctors to diagnose (Wang, 2008).

Fig. 2.1 shows Zigbee architecture in each particular layer.

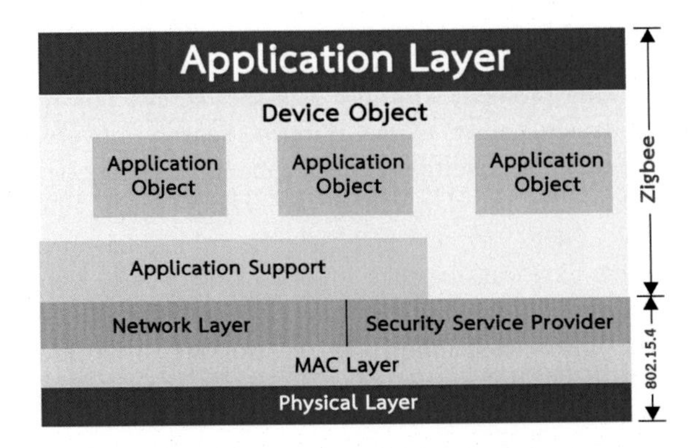

FIGURE 2.1

Zigbee architecture.

2.2.5 Z-wave

Z-wave is a commonly used low-power mesh wireless network standard. Home automation services and monitoring are made possible by machine-to-machine communication (M2M) and smart devices in the IoT network.

Z-wave is the only low-power wireless connectivity system that can guarantee interoperability at the product level. It delivers a total communication solution that ranges from the physical or hardware layer to the application layer. Z-wave is a low-power mesh wireless network protocol that is commonly used. Home control technologies and monitoring can be enhanced with machine-to-machine networking (M2M) and smart devices on the IoT network. One noteworthy feature of the Z-wave is its low power consumption at the hardware level. The ability to install a self-repairing mesh network (self-healing) with long battery life is ideal for devices such as door lock systems that require low latency levels, simple product creation, IP network connectivity, and application-level compatibility with other devices. Z-Wave systems integrate several devices with the production of scenes, which can be done indefinitely. This enables the development of an intelligent system that responds to needs individually while remaining simple in one system. The data transmission in the Z-Wave system is controlled at many speeds. There are also many layers of encryption for information, such as the radio layer, which makes the device incredibly secure (Fouladi & Ghanoun, 2013).

During the COVID-19 period, a concrete Wireless Sensor Network-based health care system can be created using a Z-Wave-based wireless biomedical image analysis system. The Z-Wave protocol can serve as a foundation for a real-world remote healthcare system (Chakraborty, Mali, & Chatterjee, 2021).

2.2.6 Bluetooth low energy

Bluetooth 4.0 is the most recent version of Bluetooth with low energy protocols. Bluetooth SIG update has enhanced Bluetooth low energy, which is a very distinguishing characteristic of this version. It was briefly known as Wibree and Bluetooth Ultra Low Power and its distinguishing features. This Bluetooth variant saw the launch of ultra-low-power Bluetooth technologies, with applications including a phone caller ID display and a system worn by athletes for heart rate tracking while exercising. Bluetooth low energy technology is intended for devices with a battery life of up to a year, has a data transmission rate of 1 MB per second, and connects and transmits data between devices using Advance Encryption Stand-128 (AES-128) encryption. Bluetooth 4.0-enabled devices can operate in both power-saving and standard modes (Pace et al., 2018).

Bluetooth 4.0 has a unique advantage in that it can work with low-power devices such as watches and medical sensors. Bluetooth 4.0 can also relay data at high speeds to devices attached to laptops and handheld devices. It transmits data at a rate of 3 megabits per second (Mbps). This high-speed Bluetooth communication protocol uses the 802.11 g Wi-Fi transmission system to relay data 200 feet from the transmitter. This high-speed Bluetooth transmits general Wi-Fi information and images, videos, audio, and pictures between the phone, computer box, and television.

Health care providers can alter how they detect illness and propose novel treatment options. During a medical emergency, COVID-19 period, real-time monitoring with intelligent sensors can save many lives.

Wireless Body Sensor Networks are commonly used to monitor psychological characteristics like temperature, heart rate, electrocardiogram (ECG), brain activity, and other critical symptoms. Intelligent sensors are embedded in a smartwatch that continuously monitors the patient's vital signs and sends real-time data to the patient's smartphone. Furthermore, Bluetooth LE (Low Energy) is used to detect nearby mobile phones and support contact tracing mApps (Gupta, Tanwar, Rana, & Walia, 2021).

2.3 Actuators for smart healthcare

An actuator is a system that controls and limits the movement or positioning of energy, whether electrical, mechanical, or in the form of a liquid such as air, hydraulic, or other. A motorbike or other moving machinery is usually used to power the actuators connected by ball screws, racks, and other parts. The type of actuator used in the construction of rotational motion, such as linear motion, differs depending on the power configuration. There are several different types of actuators available today. The most common is the Electrical Actuator because electrically driven actuators do not need oil or compressed air pressure, and it is the cleanest system. It can also be powered by DC or AC and can also be converted to mechanical energy. It is popular in many industries, such as multi-turn valves. Electric actuators have a low price. It can handle complex movements but starting up is more complicated than other options.

In smart health, electrical actuators are popular because they are mostly small jobs in hospitals; driving a medical device is entirely electric. It mostly used DC electric current as a safe power source for the patient.

There is research that offers a smart healthcare system. It uses five sensors: a heart rate sensor, room-temperature sensor, carbon monoxide sensor, and a carbon dioxide sensor from the room condition where the patient is staying in real-time. The findings could also analyze the patient's specific health signs. The device will be connected directly to the physician doing the treatment and will be able to assess the actual diagnosis (Islam, Rahaman, & Islam, 2020).

2.4 Sensors and actuators implementation for smart healthcare

The high complexity of information defines healthcare, and there should be little if any inaccuracies. Keeping control of health records is critical; the information must be consistent and still up to date. The consistency of care is determined by the consistency of the records and patient health information interpretation. There is far too much to remember, and doctors only have a limited amount of time to care about each patient. Medical care quality is measured by reliability, timeliness, efficacy, positive results, efficiency, equity, and patient-centeredness. In hospitals, information systems are divided into two types based on the operation's nature: patient service systems (front office) and management systems that are unrelated to service work (back office).

The patient service system (front office) houses electronic medical reports or electronic health records (EHRs), a hospital information system, a clinical information system, and other small applications in each facility.

Management system or Non-service related (back office) manages management information system, ERP research and education information systems, hospital website, intranet, and e-documents. Smart healthcare systems, which are implemented among healthcare institutions across Thailand, consist of various modules as stated below. Some of the applications below interlinked actuators and sensors directly apply to healthcare systems.

2.4.1 Master patient index

The MPI has the following purposes: admission, identification by Hospital Number (HN), patient demographics, and other processes that use information from this system to identify or request information about patients.

Admission-discharge transfer is a mechanism for admitting, discharging, and referring patients to other hospitals. It supports the following purposes: Admission, discharge, and transition, also known as patient management, offers details on the patient's condition if admission is present, including wards that admit and statistics used to measure bed occupancy rate. This system is connected to the financial system and sends details for payment of medical expenses.

Most of the time, the registry's registration system is dependent on scanning barcodes by reading basic information from the Electronic Data Capture (EDC) computer, acting like an actuator, which reads basic information from the NCC card, which is a smart card, such as name, surname, and date of birth. The card reader can download information from the civil registry computer system and enter it into the HIS system to open access to the system.

2.4.2 Insurance eligibility system

In Thailand, all Thai citizens are eligible for the Insurance Eligibility System. This system will decide whether a patient is entitled to medical costs such as health care (30 baht), social security, civil servant privileges, and so on, or whether they have no privileges at all, which necessitates the payment of cash. This method's fundamental premise is to assess if the patient's entitlement to medical bills includes the care that the patient will get. The system must be linked with different departments, such as the National Health Security Office, the Social Security Office, and the

Comptroller General's Department, to quantify the expense. The authentication is carried out by a kiosk system, with the hospital kiosk fitted with an EDC computer for reading ID cards to validate the rights. The IDs are read using the NCC actuator and are equipped with a thermal printer. The billing system is also related to the same kiosk.

2.4.3 Appointment scheduling

Appointment scheduling functions include logging patient schedules, determining the number of patients that may have an appointment with a doctor, or per assessment unit that supports the appointment, canceling the appointment, or displaying a list of patients that are scheduled on a certain day in each unit that will be examined. This will limit the number of patients scheduled or designate a day when appointments are not possible. Appointment systems are also used with NCC ID actuators and sensors.

2.4.4 Nursing application

Nursing applications include nursing assessments, interventions, outcomes, charting, and vital sign recording. Nursing informatics standardization can be used to aid in the documentation of a care plan. Care planning promotes coordination within the inter-carpool team, such as a carpool control system and tracking events that arise for prompt problem-solving. This can involve a forecast system as well as a timely notification system.

2.4.5 Pharmacy application

Workflow assistance from prescribing drug orders/prescriptions to dispensing and pricing is one of the most significant pharmacy applications. This is done to reduce human error, which leads to prescription mistakes and to encourage medication protection. It also aids in the control of medication inventories. Thailand's drug inspection system has implemented an RFID system in the drug room to make pharmacy stock checks easier.

2.4.6 Laboratory information system

After the procedure, where the tubes and specimens are paired with the patients in the system, the Laboratory Information System has received data and processing lab orders. Order collection, specimen registration and processing, lab findings validation and documentation, and specimen preservation in the specimen store are all procedures conducted within the laboratory. Actuators in the LIS module also include the blood tester actuator.

2.4.7 Imaging applications or picture archiving and communication system

An imaging application is a PACS that serves an important purpose. It collects X-ray images from different X-ray modalities over a LAN line and easily saves them to the machine. If the medical facility is situated a long distance away, it can be distributed by high-speed fiber-optic Internet to another hospital. In the event of a joint diagnosis, the device will display X-ray images that are easy to interpret for healthcare practitioners. Mostly used for X-ray images, but it can also be used for Magnetic Resource Imaging (MRI) devices for cardiology, anatomy, and ophthalmology, among other things. The advantage is saving space for X-ray film storage, saving on film costs. Medical personnel can display several images concurrently, view images from a distance, and perform image stabilization and editing to avoid the loss of X-ray film.

2.4.8 Imaging application

Radiology Information System as workflow control, which facilitates job processes, is included in the image application. Patient admission, appointments and preparation, consultations, imaging report printing, and so on are all handled within the radiology department.

2.4.9 Billing system

The Billing System executes the following functions: Determine the cost of the patient's care. Calculate the charges to be paid per the patient's medical benefit/ privileges. Record the cost spent by the patient and the residual balance for potential payment recording, also known as insurance eligibility and reimbursement of specific programs.

Which also transfers the collected balance details to the payroll or back-office system, which also reports the hospital's revenue and reimburses the outstanding care costs from multiple funds insurance claims to government entities.

In certain hospitals, billing networks use a kiosk for authentication. This authentication process also uses an actuator and a sensor to confirm the medical rights of the patient, such as the 30-baht health scheme, using the NCC ID actuator.

Since it avoids interaction with hospital staff, this device can play a larger role after COVID-19. Furthermore, the device will immediately slash costs, with the patient's permits not having to reserve for the first payment if eligible.

2.4.10 Enterprise resource planning

Finance, accounting, budgeting, expense containment, scheduling, material management, sourcing, inventory management, human capital, recruiting, assessment, promotion, disciplinary actions, human resource development and training, payroll, and employee compensation are Enterprise Resource functions Planning.

2.4.11 Electronic medical records

The medical record is a collection of records that are used to track a patient's illness history. In Thailand, physical examination records and health records have the same meaning. Electronic medical records aim to document vital information for future treatment to ensure the quality of care. This is vital for the treatment of patients with chronic conditions such as diabetes and high blood pressure and follow-up appointments such as after surgery for the patients' well-being. This is attributable to the importance of understanding the patient's history to avoid the risks that the patient may have been diagnosed with, such as drug allergies, a list of current medications, and a symptom list.

Individuals may also use the medical record to connect with other medical staff, such as by sending a patient to a specialist doctor, or other medical personnel, or by communicating with other doctors: Communication with physicians and nurses, pharmacists, physical therapists, and other healthcare providers when referring patients from one hospital to another. The medical professional could provide testimony in court in the case of a lawsuit, keeping note of what has been done or how the patient has been treated. Reasons about what was done, who did it, and why it was done included evidence to address whether the service met professional expectations. What services were given to the patients? What are the patient's rights in terms of medical expenses? This is also used in the audit to assess the appropriateness of seeking compensation for medical expenses. The patient's advantage includes, for example, coverage for medical costs from the patient's insurance company. Self-education and self-care are also helpful. Researchers and physicians use the information to gain new insights from health data. Electronic Medical Records (EMRs), also known as EHRs, can contain any scanned medical records or data viewed on a screen.

2.4.12 Computerized physician order entry

If a doctor needs further lab tests, they can prescribe them, whether it is a medical/lab/diagnostic/imaging order, which can be performed using a computer system. Nurses and pharmacists may determine the orders' suitability before approving them. Orders will be entered into EHRs or HISs by the system. The benefit is that the order is not misplaced, allowing you to set the correct dosage, unit, direction, and frequency. Clinic decision support systems can also be used in tandem, such as searching for drug reactions and expediting the transition from ordering to completion. Fig. 2.2

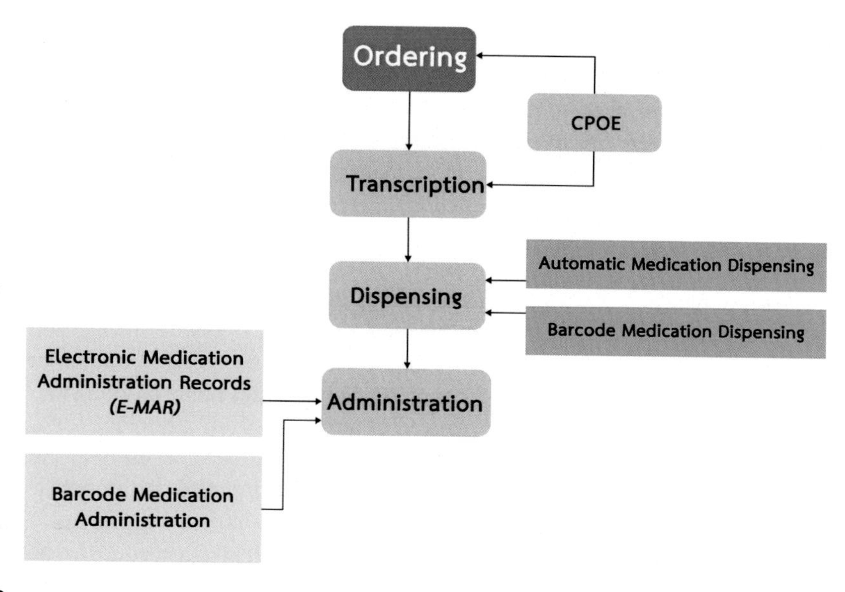

FIGURE 2.2

Computerized provider order entry processes.

illustrates how CPOE collaborates with many stakeholders during the medication process, referred to as stages. CPOE produces ordering or transcription as the first step in the process. Automatic drug dispensing or barcode prescription dispensing are two options for dispensing. The administration process will then go on to the Electronic Drug Administration Records (e-MAR) and barcoded medication administration.

2.4.13 Clinical decision support systems

Clinical decision support systems are systems that assist clinicians in making clinical decisions in many ways, such as expert systems that serve as consultants. This can use artificial intelligence, machine learning, logic, or mathematical approaches to answer questions or provide needed knowledge, such as a device that requires differential diagnosis or recommends care. Physician queuing schemes, notifications, and warnings depending on predetermined circumstances, such as drug allergy scans, drug-drug contact tests, and reminders for preventive measures including smoking reduction, professional procedure guidance incorporation, and knowledge base coordination. A prescription archive is a basic device that facilitates decision-making, such as highlighting irregular test findings, etc., for physicians or pharmacists.

Hospitals in Thailand have limited screening by medical staff as a result of the recent COVID crisis. Instead of allowing the patient to walk into the original screening procedure, a system will be in place that will issue a queue pass to allow the patient to go directly to the facility where they have an appointment. If the patient does not undergo screening, they will need to consult with medical staff first so that this system will eliminate the first encounter with the first arriving patient. It has the potential to decrease the risk of infection. Furthermore, the hospital has protocols to develop technologies, such as intelligence for electronic clinic allocations, which helps the system understand how to queue for a specialist based on initial complaints and instantly route patients to the right clinic. It can also be set up in the patient's examining room if the clinic's consultation room is empty without nurses' need to help in the queue. Furthermore, it integrates a collection of Alerts and Reminders into everyday lives. It's used in the hospital's queuing reminder system, and it's often included to make waiting for service more manageable.

2.4.14 Health information exchange

Health information exchange is the exchange of health information between hospitals where the patient can change the medical facility transfer. Medical treatment rights to other hospitals are subject to various limitations, such as seeing a specialist doctor that the hospital of origin cannot arrange, or HIE may be used to search for information received from other hospitals.

Information stored in the database can be viewed at any time, from any place, by anybody that can be reached, and it would be possible to share information with several users at the same time. This also eliminates the issue of forwarding the profile, as the information in the profile is lost during the transition from one clinic to another. In the past, there have been examples of sending patients to other hospitals and the patient bringing their profile home, causing the hospital to be unable to locate the patient's records or recall the patient's health privileges. Access to medical records will allow health providers to log information in detail, complete and standardize it, and fill in the information quickly due to a system of drug names and frequently used word recommendations. Work process improvement or workflow, also known as process improvement or business process reengineering/redesign, may input different forms of information, either by keying or browsing the request instead. It can also be printed in an easy-to-read format, which helps to minimize errors.

2.5 Internet of things in the healthcare system

Because there is a rising number of elderly individuals, IoT has played an increasingly larger role in healthcare systems. Many patients are frequently left unnoticed or excluded from the healthcare process, which may be owing to a shortage of medical personnel in crowded hospitals, long wait lines when receiving medical checkups, or abrupt sickness. Regular patients, especially during a pandemic, are left out because of an increase in COVID-19 cases. The public service in developing countries' healthcare systems is sometimes weak at the institutional level, making it difficult to offer equitable treatment owing to a shortage of medical staff or facilities to lodge patients. Checking in on patients is more difficult than ever without the necessary equipment, especially if the patient lives in a rural place where the hospital's services are difficult to access. Hospitals in developing countries, such as Thailand, may be congested, and there may not be enough medical personnel to cater to a single patient. According to research in healthcare and IoT, investing in smart solutions in conjunction with current hardware and software will provide individuals with improved access to healthcare, whether at home or in the hospital. With the introduction of the 5G network, internet connectivity is now available at any time and from any location, allowing healthcare facilities to offer frequent patients the treatment they demand daily. This could be accomplished by utilizing the hospital's technology solutions or databases, or by communicating with the patients' attending medical professionals on many topics such as medical appointment scheduling, check-ups, and even emergency dialing.

IoT can help to speed up medical procedures and reduce health risks associated with not being able to access healthcare services in a fast and effective manner. As a result, the IoT plays a role in data gathering and transmission to the hospital. IoT would be required in healthcare facilities, particularly in rural areas, to help patients. Thailand is currently using IoT, both hardware and software, to assist the elderly. The following is how IoT is being used to assist the elderly.

1. IoT gadgets that assist home health, such as electric lighting and bedside doors, that can be linked to a smart home system and controlled by the resident, as well as an automatic chair lift or wheelchair to help the resident's mobility.
2. A wearable (wearable IoT) gadget is one that a patient may wear and that can give healthcare staff the patient's present state, such as blood pressure, heart rate, breathing rate, and oxygen levels in the blood, and that can be directly connected to the hospital system.
3. A mobile application that has been created in a variety of methods. To monitor each person's general health and link to other devices such as a smartwatch for exercise.

IoT is an innovation that will play an even larger role in the future, and the information from the patient will be forwarded to the hospital, where the healthcare facility will be able to access the patients' information and daily needs, such as health checkups or a scheduled appointment, or notify the emergency services, which will reduce the fatality or severity of an accident in an e-health setting.

2.6 Smart healthcare system design and implementation

2.6.1 Process design

Because of the COVID-19 situation, screening patients includes an extra stage of fever assessment using sensors and monitors to sense body heat, which the device can check everyone's temperature. Visitors to the hospital must be led to the required measuring point; if their body temperature exceeds 37.3°C, the device will alert the nurse, and the nurse must screen the patient. The use of sensors and actuators in systems is a 21st-century breakthrough, with medicine being the third largest industry for wireless sensors. The advantages of using these technologies paved the way for telemedicine to become a viable alternative for reducing medical costs and burdens, especially during a global pandemic like COVID-19 (Chakraborty, Gupta, & Ghosh, 2013).

The hospital separates the patient into three groups: new patients, existing patients, and admit patients. The patient may schedule an appointment using the smartphone application or by going directly to the hospital. However, once the patient books an appointment via the smartphone application, they must first be screened by a nurse, and the nurse, like the kiosk, will issue a queue card. If a patient arrives at the hospital without booking via the smartphone application, they must go to the kiosk and insert the National ID card to check the 13-digit national ID. The machine will provide the queue number by generating a temporary table in a database not linked to HIS/HOSxp, the main application in Thailand's hospital. After passing the preliminary test, the patient proceeds to the screening phase, including multiple actuators such as height measurements, weighing scales, and blood pressure monitors. Besides, the measured data is also saved in a temporary database. When this procedure is passed, it will be broken into two cases: if the patient exhibits abnormal signs, such as high blood pressure, the system will alert medical personnel, who will determine the symptoms. Using an artificial neural network, the medical team will anticipate and decide how the patient should be screened to continue with the medical process based on the symptoms. Suppose the patient completes the screening system's check and the general symptom is not feverish. In that case, the patient may be allowed to access the kiosk and collect a queue card, allowing them to begin regular treatment. The kiosk that opens the patient's visit will print the queue card slip and add the patient's details to the actual HIS/HOSxp database so that the hospital staff can see all of the information when the visit is opened. The details connected to HIS/HOSxp are printed on a queue card and forwarded to the assigned clinic via the navigation system. A queue card will be printed by the screening system, whether it is a kiosk or medical staff.

The queue card is a slip that holds data such as screening results and the patient's personal information. Patients will keep the slip and head to the clinic where they were referred; the slip will detail the location, time, and queue number. The parallel scheme is a mobile phone booking system that the patient will use to be transferred to a list first. Still, the nurse will not see the patient's details until it has been through a screening procedure with the hospital actuators initially. If the patient fails to obtain the screening results, they will be subjected to a nurse's preliminary screening. The numbers on the cue card correspond to those used in the screening process.

A monitor will be mounted at two different locations. The first is the screening point in front of the examining room, where the nurse will check the clinic's line, see all of the clinic's queues, and process patients for examinations. The nurse can direct the patient to wait in front of the examination room, where 2–3 more patients will be waiting. Point 2 is the examining room's monitor screen location, which displays the sequence of waiting in front of the examination room. The nurse will now see where the patient will be examined, which will be in a room with a specific number. If the patient is not in the waiting queue, the nurse will call for the next patient in line. Given the patient, there will be a monitor that will display the call sequence and the approximate wait period to minimize the amount of time spent waiting in front of the examination room. The patient is either discharged or referred to another medical process after the treatment.

When a patient has been prescribed drugs from a doctor, a prescription admission order will be put in a queue and sent to the pharmacy room following treatment. There will be a display for the drug queue in the drug room, and when the prescription is packed and ready for pickup, a call will be made according to the drug queue, depending on the dispensing points. Another monitor displays the patient's dispensing queue to see, and pharmacists will double-check the dispensing before the medication is handed over.

The X-ray system and the laboratory inspection system are both connected to the queue system mentioned herein. This system's reminders are as follows: (1) Once the patients entered their ID cards into the machine, they were informed of their rights and privileges. (2) A reminder for the remaining medical history interview queue, sent by the LINE application, informing the patient of the number of patients in line ahead of them. (3) A warning for the examination room's queue line, which functions in the same manner as number 2. (4) A prescription queue notification, which follows the same format as the previous two sections, and (5) A final call for the patient.

Before initiating payments for patients or their families, the hospital e-payment system ensures that the patients' privileges are valid and can be routed to the patient or relative kiosk. The system reads an ID card or the HN, a barcode on the queue slips that patients get as they come to the hospital. The machine would be able to read various cards and link them to the government's authorization database. The machine will inform you how much it will cost, taking into account all benefits and discounts. It will even notify that the patient owes nothing. If you must pay, you can do so using various means, including credit cards and cash, which the system can connect to The Comptroller General's Department, particularly if it is a government license or 30-baht scheme. The kiosk also has a camera, on which an image processing program can be added to recognize a face if identification is an issue. Fig. 2.3 depicts the overall process design.

2.6.2 Database design

This screening system employs MySQL as a database management system (DBMS) connected to the hospital's main application, HIS/HOSxp. With limitations of the HIS/HOSxp system, which sticks to a native MySQL, a NoSQL database could be further introduced which ensures high scalability and performance. It has an additional type of interaction with the hospital's other databases. The systems contain two databases: the configuration database and the kiosk database. The systems are linked to two servers: the kiosk server and the HIS server. The configuration system, queue order service system, patient registry system, and screening system are the four major systems.

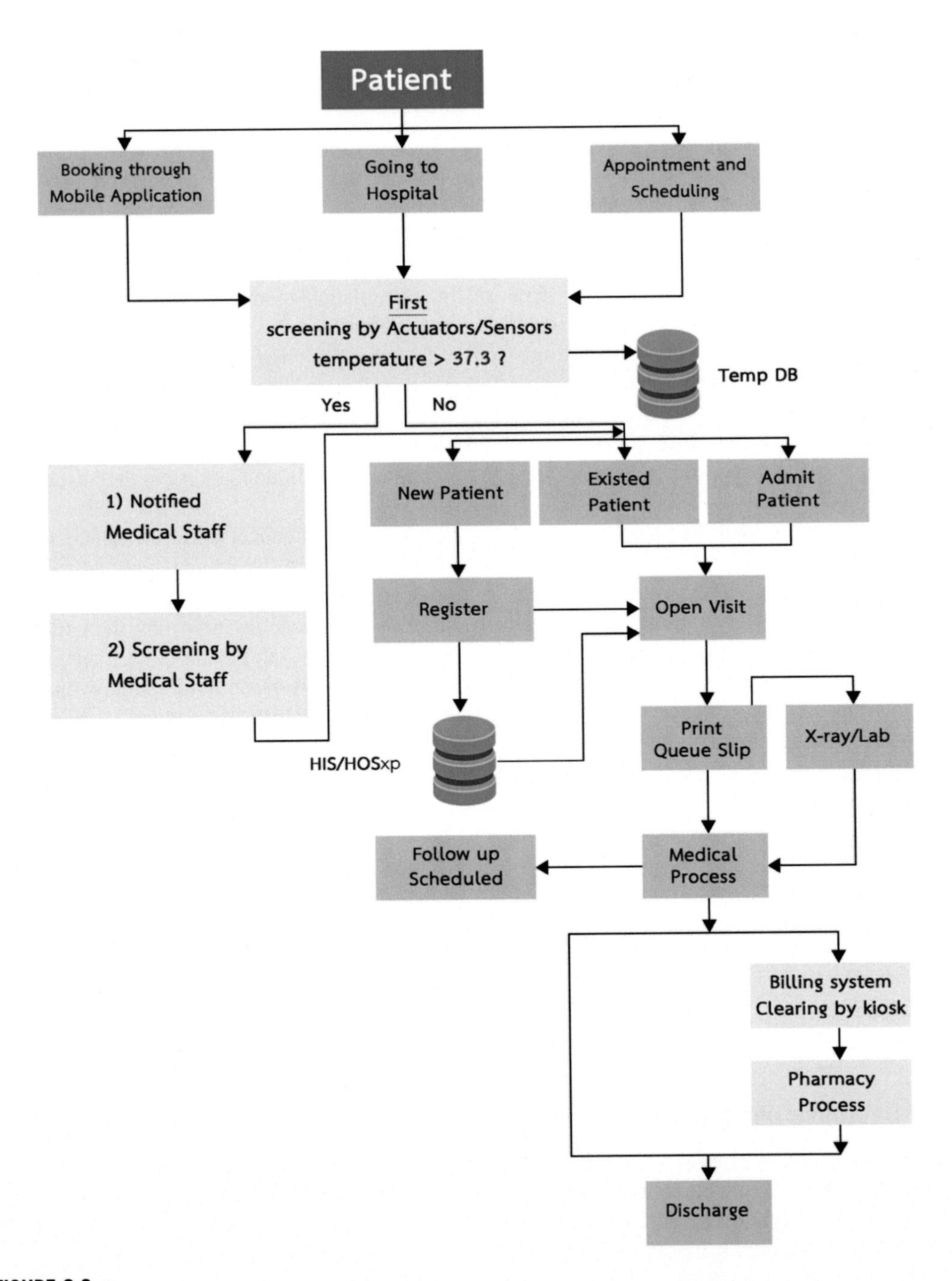

FIGURE 2.3

Process design.

2.6.3 Artificial neural network design for screening process

Nowadays, the operation of screening systems in hospitals employs a manual clinic selection scheme. Patients choose a clinic to book themselves through the screen after screening via the actuators. In the future, an integrated method will be built using Artificial Neural Network (ANN) methods for classification, and the clinic assigned to patients will be automatically printed to queue cards. The method of inserting a National ID card into a kiosk, going through a height screening device, weighing, and measuring pressure is depicted in Fig. 2.4. The results were compared with the model learned from the ANN framework from the initial data structured in the clinic, and the result was a clinic queue card based on the questions that the patient answered. Machine Learning implementation could minimize the tremendous burden on health practitioners, especially during the COVID-19 crisis. It could be used as a secondary mode of diagnosis or prognosis. Patients' signs should be analyzed at the very least to more accurately refer the patient to a specific treatment (Muhammad et al., 2021). The accuracy of the healthcare system for high-risk monitoring patients is crucial for the healthcare industry; with this in mind, the complexity of determining diagnosis puts a lot of burden on health professionals. A framework such as ANN that analyzes large numbers of datasets and characteristics would significantly reduce the workload on health practitioners, avoiding the possibility of mistakes in high-risk situations, thereby reducing the pressure on health professionals (Kishor, Chakraborty, & Jeberson, 2020). The design includes six questions, five of which will be used to feed the ANN network. Some of the questions include: Do you have a current medical issue, or are you on any medications? Have you recently had coughing symptoms? Do you have sinusitis presently? Do you have a lack of taste or smell? The neural network, also known as an artificial neural network, is a predictive construct inspired by how the brain works. These networks are made up of artificial neurons that conduct calculations dependent on the input. The connecting lines represent the output, which is the weighted number of the inputs, and each node represents an input feature. ANN computes the input sum several times, first calculating

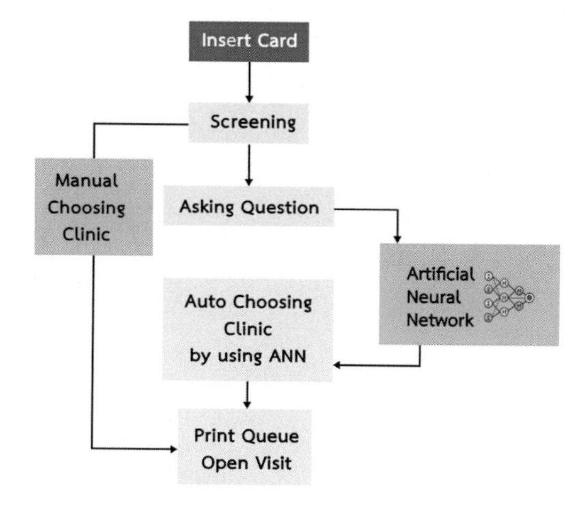

FIGURE 2.4

Artificial neural network design for screening process.

the hidden units inside the hidden layers, then combining the results to produce the final output (Grus, 2015; Müller & Guido, 2016). The output of this design will be a total of 44 clinic choices, some of which include X-rays, Dentistry, and Surgical.

Fig. 2.5 shows an example of the ANN model, the backpropagation technique, and the computation of the ANN technique, with 6 input layers, 4 hidden layers, and 4 output layers. This paper implements the use of MLP or multi-layer perceptron classifier, with an input of patient data.

2.6.4 Read data from RS-232 and RS-423

Many types of hospital actuators are still in use and receive data directly into the HIS system. This screening system uses the principle of reading data from the actuator via ports RS-232 and RS-423 data (get) from actuators received from the scales height meter, temperature meter, and a pressure gage. Fig. 2.6 below shows the ReadScale module used to read data from port RS-232 or RS-423.

2.6.5 Read data from electronic data capture

After the medical procedure, the patient must insert a National ID card into the electronic data capture (EDC) card reader to clear the bill; the HIS device is linked to the EDC unit. The work processes begin with reading data from the National ID card and connecting with the Comptroller General's Department, Social Security, and the EDC system, which links to the computer via USB and serial port. The machine would write data into HIS database using the node.js language's get and post functions. It would differ from vital signs in that it only functions via the get function and does not have a post function. Fig. 2.7 depicts an EDC computer from a bank integrated with a kiosk in a Thai hospital.

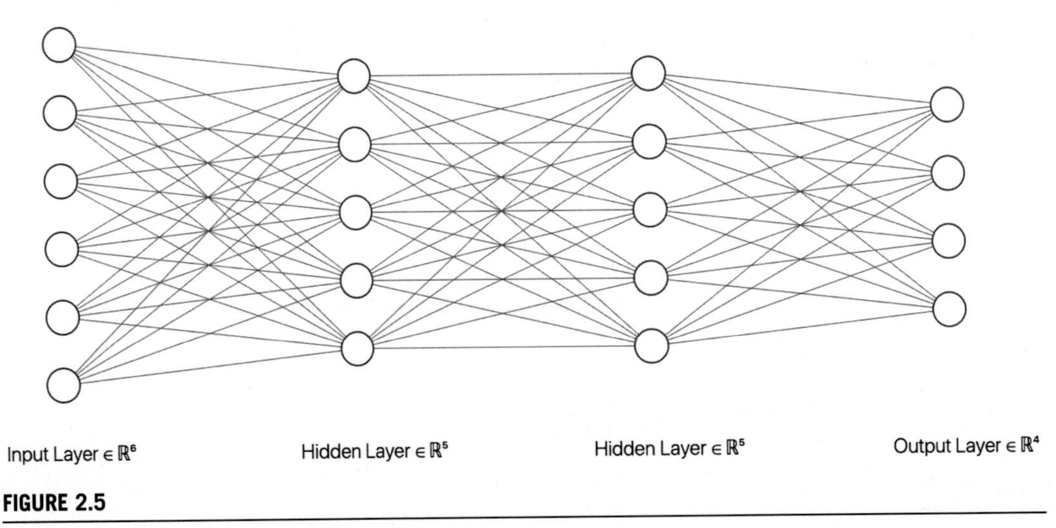

FIGURE 2.5

A 6–5–5–4 backpropagation model.

```
Private Sub ReadScale()
    Try
      'Open Port
      Dim portScale As SerialPort = New SerialPort()
      portScale = New SerialPort("COM3", 9600, Parity.None, 8, StopBits.One)
      portScale.Open()

      'Read Data from Scale
      Dim ScaleData As String = portScale.ReadLine()
      If ScaleData.Length > 0 Then
          Try
            ScaleData = ScaleData.Replace(":", "")
            Dim data() = ScaleData.Split("H")
            'นน. / ส่วนสูง
            Me.lblW.Text = Convert.ToDouble(data(0).Replace("W", "")).ToString()'
            Me.lblH.Text = Convert.ToDouble(data(1)).ToString()
          Catch ex As Exception
          End Try
      End If
    Catch ex As Exception
      MessageBox.Show(ex.Message)
    End Try
  End Sub **
```

FIGURE 2.6

ReadScale function from port RS-232 and port RS-423.

2.7 General potential ICT risks IN healthcare services

Several healthcare departments use Hypertext Transfer Protocol Secure (HTTPS) with 128-bit Secure Sockets Layer (SSL) technology to provide online records, appointment scheduling, and payment services. These billing and scheduling systems include invoices, medicine payments, and medical appointments. When working with online providers, using SSL technologies may offer a reliable communication channel; however, the technology does not provide a way to validate each patient's uniqueness (Monika & Upadhyaya, 2015).

Furthermore, several healthcare agencies have developed internal wireless (Wi-Fi) network infrastructure to deliver online services to their employees and patients. Wi-Fi devices allow real-time web access to department information systems such as patient databases, various medical application applications, internet access, and surveillance cameras. On the other hand, Wi-Fi network technology is seen as less secure than wired network technology (Kosta, Dalal, & Jha, 2010). As a consequence, the number of possible risks to healthcare departments could increase.

Furthermore, IoT network medical devices such as IoMT (internet of medical things) are becoming increasingly popular in healthcare organizations and consumer segments (Amine, Lloret, & Oumnad, 2020). When opposed to conventional medical care, IoMT technologies and software will offer greater options for diagnosing, handling, and maintaining a patient's health and finances. However, as the number of IoT-connected devices grows, so does the potential for IoT-related ICT security threats. There are many risks and threats associated with the use of online services in healthcare systems, including ransomware attacks, patient data theft, phishing, viruses, hackers,

FIGURE 2.7

Electronic data capture machine from a bank.

Denial of Service (DoS), Distributed Denial of Service (DDoS), Domain Name Server (DNS) vulnerability—DNS sinkhole, fake websites, hacked IoMT devices, identity theft or online Identity (ID) theft, and data integrity. Below are some examples of system services and the risks associated with them in healthcare online services:

1. Data integrity: This involves computer corruption and destruction triggered by warm-up contamination or malware viruses, as well as individuals that deceive patients or hospital employees into clicking on links from unknown sources.
2. DoS and DDoS: Because of the hack's simplicity, "Denial of services" is usually a problem when used on many servers. The problem caused the server to stop providing service, forcing the hospital's medical services to be stopped. This compels the doctor to temporarily pause the examination to correct the error, causing considerable damage to the hospital's infrastructure.
3. Data security: In Thailand, there is concern about data protection in hospitals because it contains sensitive information that may be misused, compromised, destroyed, hacked, or intercepted by an intruder via phishing, spyware, or any social media-related links that might not be identified.

4. Misuse of resources: The wireless (Wi-Fi) network of the healthcare company is obsolete, and the system cannot be upgraded. Then it could be jeopardized as a result of simple internet resource misuse by medical professionals, customers, and the general public, such as phishing links, ransomware, harassment, pornography, and illegal websites.
5. System penetration security risks: Healthcare organizations can be the target of attacks, especially ransomware attacks. Such devices cannot be avoided and may cause significant harm if corrective action is not taken. This issue could be triggered by uploading a file from social media without knowing where it came from.
6. IoMT compromise at a network level: The breach happens at the OSI network level, and IoMT and medical device victims' network traffic is diverted to a compromised sensor node, which will result in unauthorized access to the patient records and the disclosure of sensitive patient and personnel details.

Table 2.1 illustrates the attack types that may occur in the hospital, potential threats, assets, impacts, and countermeasures or contingencies. (L represents low, M represents medium, and H represents high).—[Ref: adapted from Sunsern Limwiriyakul (Limwiriyakul, 2012)]

2.8 Results and discussions

By introducing the procedure, patients were able to easily and conveniently access healthcare at healthcare centers, especially during COVID-19. Before the initial medical procedure, patients were required to monitor their temperature at the point of admission to ensure that they did not have the preliminary symptoms of COVID-19. The patients are often sorted based on their associations with the hospital, such as whether they are new patients, current patients, or admitted patients, and their existing identity is reviewed against the HIS/HOSxp database. The patient's records would be held on a temporary table and used during the medical procedure. Following the implementation of a complete screening procedure in a general hospital, a test evaluation for the user's satisfaction form was produced. The research team established goals and reviewed principles, theories, manuals, texts, publications, and related research regarding functionality and customer satisfaction after application deployment and the screening process in the hospital. A physician, two technicians, and twenty-seven patients were among those who tested the systems. An assessment form includes scoring questions with a 5-level Likert Scale checklist. Below are the questionnaire scoring criteria:

1. Level 5 Very Satisfied
2. Level 4 Satisfied
3. Level 3 Neutral
4. Level 2 Unsatisfied
5. Level 1 Very Unsatisfied

The findings obtained from the users are Very Satisfied in all topics in both design and contents, see Table 2.2. The simplicity of the service queue and the reliability of system speed is the most satisfying variables. These findings have important implications for an unrivaled use of IoT in healthcare on a comparable scale. Implementing this technology will reduce the workload on

Table 2.1 A summary of potential risks in healthcare organization online services.

Attack and threats	Assets	Impacts[a]	Countermeasures or contingencies
1. Brute force and dictionary attacks	Email, online web, online medical, database, wireless, and internet of medical things (IoMT) infrastructure systems, and information reputation	M	Checklist trail software to view logins and failed logins; Deploy a 2-factor authentication system; Filter/disable unnecessary important ports and disabled some unused protocols on the Internet gateway and firewall; Operating System passwords change frequency; Security education for the user; Setting a strong password for the system, policy control, and encrypt with complex functions of all communication ports and channels by the organizations' IT teams; Use strong encryption technologies to protect all transmitted and kept information, including patient information and data
2. DoS/DDoS attacks	Email, online web, online medical, staff and patient databases, wireless and IoMT infrastructure systems, and information reputation	M/H	Combine technologies to control the use of encryption of all communication ports and channels by the IT members; Deploy IDS/IPS, alert and monitoring system at Internet gateway and wireless infrastructure systems; Checking filter/disable unused ports and protocols on both the internal firewalls and Internet gateway; Guidelines on regular backups of critical data to be issued audited and verified; Install wireless security filtering system; Download patching to install all the applications that are used in the systems and all OS software of all the online services and check it frequently; Training users about internet security;

Table 2.1 A summary of potential risks in healthcare organization online services. ***Continued***

Attack and threats	Assets	Impacts[a]	Countermeasures or contingencies
			Centralize the control of security only at the computer center with only the IT members
3. Eaves-dropping attacks	Email, online web, online medical, wireless, and IoMT infrastructure systems	M/H	Checking control and encryption of all used communication ports and channels by the IT members; Deploy Email, URL, and social media filtering systems; Filter/disable unnecessary used ports and protocols on both the internal firewalls and Internet gateway; Install wireless security filtering system; Security education for user
4. DNS sinkhole/ Blackhole DNS	online web, online medical systems	M	Incorporate cloud DNS sinkhole services with the Internet gateway firewall system
5. Hacker attacks	Email, online web, online medical, database systems, wireless and IoMT infrastructures, and information reputation	M/H	Deploy a 2-factor authentication system; Checking control and encryption of all used communication ports and channels; Deploy Email, URL, and social media filtering systems; Filter/disable unnecessary used ports and protocol on both the Internal firewalls and Internet gateway; Training guidelines on basic frequency backups of severe data to be issued audited and verified; At the Internet gateway system, install IDS/IPS; Install wireless security filtering system; Install patch regularly for all applications and OS of all systems connect online Educated users about security Centralize the control of security only at the computer center with only the IT members

(*Continued*)

Table 2.1 A summary of potential risks in healthcare organization online services. *Continued*

Attack and threats	Assets	Impacts[a]	Countermeasures or contingencies
6. Interference and jamming and attacks	Wireless and IoMT infrastructures and information reputation	M/H	Implementation of filtering detecting system to follow or monitor for detecting any strong interference issues Educated users about security
7. Internal theft of online systems devices/ equipment at an organization's premises	Email, online web, online medical, database systems, wireless and IoMT infrastructures andinformation reputation	M/H	Access controls within the organization's buildings Checking background and identity checks conducted on staff/ contractors Deploy access tracking system Install CCTV system in sensitive areas Lock up all mobile devices (i.e., laptops, tablets) when not in use Permanently remark and put the sign on all online systems tools in the organization Retrieval message is only important Hardware audit by the IT members in the organization Educated user about security
8. Malicious active code attacks	Email, online web, online medical, and database systems	M	Deploy Email, URL, and social media filtering systems Enable filtering function on firewalls systems in all organizations to limit the use of Java applets and ActiveX control objects Control users' access to use Java applets or ActiveX control objects on online services systems
9. Phishing	Email, social media systems	M	Automatically updated protection software for cleansing malware, virus, and worm. Deploy Email, URL, and social media filtering systems Educated user about security
10. Ransomware	Email, online web, and social media systems, physical hardware, and data	H	Automatically updated protection software for cleansing malware, virus, and worm. Deploy Email, URL, and social media filtering systems Educated user about security

Table 2.1 A summary of potential risks in healthcare organization online services. ***Continued***

Attack and threats	Assets	Impacts[a]	Countermeasures or contingencies
11. Social engineering attacks	Email, online web, online medical, and database systems	L/M	Checking background/identity on staffs/contractors Deploy Email, URL, and social media filtering systems. Restrict/minimize access to serious information related to services that online and DB systems Security education for user Unauthorized personnel information needs to be limited when given out
12. Virus, malware, and worm introduction through the Internet and social media usage	Email, online web, online medical, and database systems	H	Automatically updated protection software for cleansing malware, virus, and worm. Control and encryption of all used communication channels and ports by the IT members Deploy Email, URL, and social media filtering systems Training on basic backups of important data to be issued, audited, and verified Regularly install patch all important applications and OS all online services systems Educated users about security centralize the control of security only at the computer center with only the IT members

[a]*Impact index:* L, *low;* M, *medium;* H, *high.*

healthcare professionals while also speeding up the treatment service in healthcare centers, allowing them to care for more patients. In terms of connecting all healthcare services, the use of actuators and sensors is the first of its kind to be implemented in healthcare. The users observed that the application of this technology has proved to be significant for the development of health services not only in Thailand but also across the world. This paradigm will also minimize potential costs for human capital as well as the preservation of significant health requirements such as hygiene.

Fig. 2.8 depicts the system's design formatting part, which is presented in Table 2.2. As shown in Fig. 2.8, the background color and font size are the most suited for reading within the application, and the overall attractiveness of the design is modern and appealing. The consistency of imagery meaning and context, as well as the design appropriateness of the colors, are the least ranked

Table 2.2 User satisfying results.

Questions	SUM	X̄	S.D	Results
Design and formatting				
1. The beauty of the modern and attractive screen.	30	4.78	0.35	Very satisfied
2. The screen formatting is easy to read and use.	30	4.62	0.49	Very satisfied
3. The colors in the application design are appropriate.	30	4.53	0.59	Very satisfied
4. Easy-to-use menu	30	4.62	0.48	Very satisfied
5. The background color and the font color are appropriate for reading	30	4.73	0.44	Very satisfied
6. Font size and font style easy to read and beautiful	30	4.61	0.32	Very satisfied
7. Images and content are consistent and convey meaning.	30	4.53	0.35	Very satisfied
8. Overall, how satisfied are you with the design of the application?	30	4.58	0.52	Very satisfied
Contents				
1. Ease of booking service queue	30	4.83	0.40	Very satisfied
2. Service queue status tracking	30	4.75	0.41	Very satisfied
3. The suitability of the information within the application	30	4.85	0.34	Very satisfied
4. The convenience and speed of the service	30	4.75	0.40	Very satisfied
5. Accuracy and completeness of the information	30	4.78	0.44	Very satisfied
6. Illustrations can convey meaning.	30	4.79	0.39	Very satisfied
7. Speed of downloading data	30	4.82	0.40	Very satisfied
8. Overall, are you satisfied with the quality of the content?	30	4.92	0.30	Very satisfied

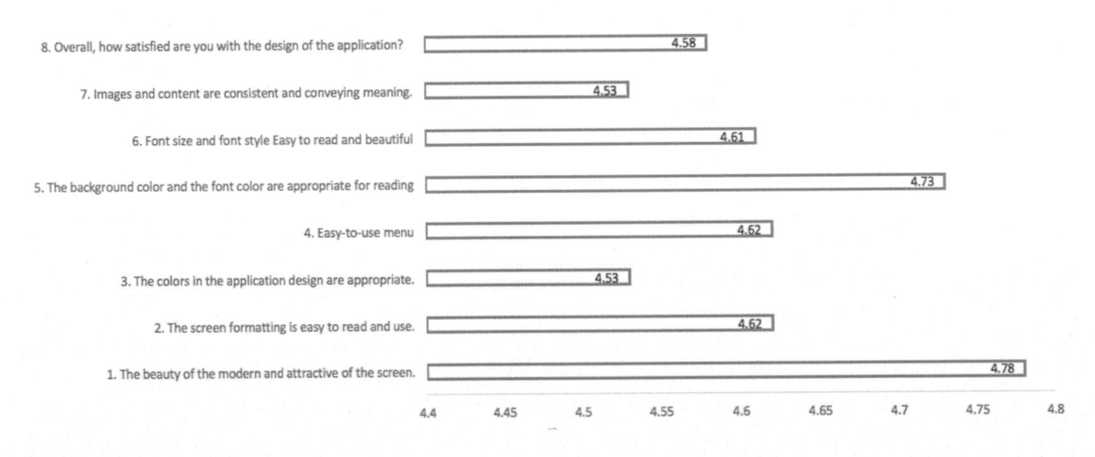

FIGURE 2.8

Mean of user satisfying results—design and formatting.

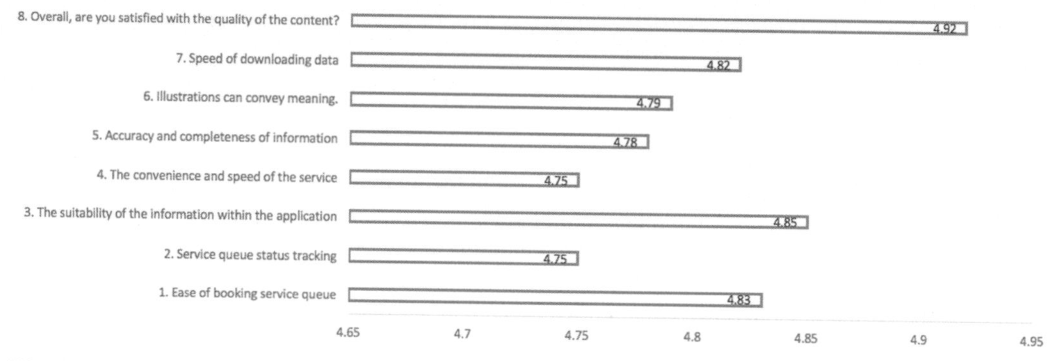

FIGURE 2.9

Mean of user satisfying results—contents.

features in the Design and Formatting section. Fig. 2.9 depicts the data from the Table 2.2 Contents section. The overall content quality is the most pleasing aspect in terms of contents, while the convenience and speed of the device, as well as service queue status tracking, are the least ranked categories.

The scope of this IoT implementation in healthcare centers includes not just the use of standardized routine in hospitals, but also the creation of this scheme for other institutions. The utilities that could require this type of application include government clerical services such as ID or license renewal, as well as businesses that choose to install actuators or sensors for quality control within manufacturing plants on a case-by-case basis. Overall, several industries would benefit significantly from the deployment of these systems, and the study could be extended to address the unique demands of a specific field. The system may also be enhanced with more intelligent capabilities by incorporating machine learning algorithms and other existing infrastructures such as surveillance cameras or internal machinery sensors in the case of industrial plants. Connecting the infrastructure to a larger database is also part of the plan, which will enable patients and data from emergency departments to be transferred more efficiently across all healthcare facilities in the country. In addition, patients have the option of opting in for an all-encompassing program in which healthcare providers can view health records from their smartwatches or smartphones.

2.9 Conclusions

COVID-19 affects people worldwide; thus, many lifestyle choices would be made, with “technology” and “innovation” playing a greater role. Technology has evolved into an invaluable tool for halting and preventing the transmission of infections and facilitating people’s lifestyles. Outposts in health care must be fully prepared in the fight against the epidemic, and they must work around the clock. During the severe COVID-19 health crisis, the study has documented a wide variety of COVID-19 preventive technologies used as a platform for inpatient screening. This approach decreases the risk of infection from treating patients, keeping up with demands, and minimizing infection risk among healthcare workers. The impression is that healthcare workers would work tirelessly to combat the crisis while still risking infection. Smart Healthcare research should consider the well-being of healthcare staff at all levels of the work setting. The treatment process, which begins with screening the patient and screening for infection, is based on principles that must be followed promptly and efficiently. Incorporating a hospital's HIS/HOSxp framework into an IoT system of sensors or actuators enables the system to operate more efficiently during the screening period, eliminating mistakes caused by the nurse's data entry, and patients can do this themselves. Since pathogens transmit quickly, less contact with patients and those who visit the facility is essential. The future work, we would like to introduce, the use of ANN to choose clinics in this scheme would minimize the contact between patients and healthcare staff. This aids in the prevention of COVID-19 infection while further automating the procedure and reducing the pressure on healthcare workers. Smart Healthcare research cutting-edge technology that can help reduce the risk of infection of healthcare workers. Further study seeks to incorporate ANN to be able to deploy in a real-world hospital setting, as well as adjust models based on patient population and community guidelines, as well as healthcare and privacy rules of the local government and the Royal Thai Government. A research route might be explored to create smartwatches or home systems that could be integrated with the healthcare system while adhering to privacy requirements. IoT devices might be used to transmit messages in both directions, from the healthcare facility to the patient and from the patient to the facility, allowing medical personnel to check in on patients and give the best treatment possible. Patients with preexisting conditions, as well as those at high risk, may be regularly monitored by their local hospital and informed of their health benefits and clinic options based on their circumstances directly at home, which may also be performed by deploying an ANN model. The development of software supplements integrating existing sensors and actuators as a basis for innovative research to link hardware, hospital databases, and national databases to make system operations more intelligent and smarter.

References

Abugabah, A., Sanzogni, L., Houghton, L., AlZubi, A. A., & Abuqabbeh, A. (2021). RFID adaption in healthcare organizations: An integrative framework. *Computers, Materials and Continua.*, *70*, 1335–1348. Available from https://doi.org/10.32604/cmc.2022.019097.

Amine, R., Lloret, J., & Oumnad, A. (2020). Big data classification and internet of things in healthcare. *International Journal of E-Health and Medical Communications.*, *11*, 20–37. Available from https://doi.org/10.4018/IJEHMC.2020040102.

Baig Mohammad, G., & Shitharth, S. (2021). Wireless sensor network and IoT based systems for healthcare application. *Materials Today: Proceedings*. Available from https://doi.org/10.1016/j.matpr.2020.11.801.

Chakraborty, C., Gupta, B., & Ghosh, S. K. (2013). A review on telemedicine-based WBAN framework for patient monitoring. *Telemedicine Journal of E-Health Official Journal of American Telemedicine Association*, *19*, 619–626. Available from https://doi.org/10.1089/tmj.2012.0215.

Chakraborty, S., Mali, K., & Chatterjee, S. (2021). Edge computing based conceptual framework for smart health care applications using Z-wave and homebased wireless sensor network. In A. Mukherjee, D. De, S. K. Ghosh, & R. Buyya (Eds.), *Mobile edge computing* (pp. 387–414). Cham: Springer International Publishing. Available from https://doi.org/10.1007/978-3-030-69893-5_16.

Finkenzeller, K. (2010). *RFID handbook: Fundamentals and applications in contactless smart cards, radio frequency identification and near-field communication*. John wiley & sons.

Fouladi, B., & Ghanoun, S. (2013). Security evaluation of the Z-Wave wireless protocol. *Black Hat USA*, *24*, 1–2.

Garavand, A., Mohseni, M., Asadi, H., Etemadi, M., Moradi-Joo, M., & Moosavi, A. (2016). Factors influencing the adoption of health information technologies: A systematic review. *Electronic Physician*, *8*, 2713.

Gómez-García, C. A., Askar-Rodriguez, M., & Velasco-Medina, J. (2021). Platform for healthcare promotion and cardiovascular disease prevention. *IEEE Journal of Biomedicine Health Information*, *25*, 2758–2767. Available from https://doi.org/10.1109/JBHI.2021.3051967.

Grus, J. (2015). *Data science from scratch: First principles with python* (1st ed.). Sebastopol, CA: O'Reilly Media.

Gupta, M., Tanwar, S., Rana, A., Walia, H. (2021). Smart healthcare monitoring system using wireless body area network. In Proceedings of the ninth international conference on reliability, infocom technologies and optimization (trends and future directions) (ICRITO). Presented at the 2021 9th International Conference on Reliability, Infocom Technologies and Optimization (Trends and Future Directions) (ICRITO) (pp. 1–5). <https://doi.org/10.1109/ICRITO51393.2021.9596360>.

Hills, R. A., Lober, W. B., & Painter, I. S. (2008). *Biosurveillance, case reporting, and decision support: Public health interactions with a health information exchange. International workshop on biosurveillance and biosecurity* (pp. 10–21). Springer.

Islam, Md. M., Rahaman, A., & Islam, Md. R. (2020). Development of smart healthcare monitoring system in IoT environment. *SN Computer Science*, *1*, 185. Available from https://doi.org/10.1007/s42979-020-00195-y.

Istepanian, R., Laxminarayan, S., & Pattichis, C. S. (2007). *M-health: Emerging mobile health systems*. Springer Science & Business Media.

Istepanian, R.S., Lacal, J.C. (2003). Emerging mobile communication technologies for health: Some imperative notes on m-health. In *Proceedings of the twenty-fifth annual international conference of the IEEE engineering in medicine and biology society (IEEE Cat. No. 03CH37439)* (pp. 1414–1416). IEEE.

Jihong, C. (2011). Patient positioning system in hospital based on Zigbee. In *2011 international conference on intelligent computation and bio-medical instrumentation. presented at the international conference on intelligent computation and bio-medical instrumentation* (pp. 159–162). <https://doi.org/10.1109/ICBMI.2011.72>.

Kang, M. H., Lee, G. J., Yun, J. H., & Song, Y. M. (2021). NFC-based wearable optoelectronics working with smartphone application for untact healthcare. *Sensors*, *21*, 878. Available from https://doi.org/10.3390/s21030878.

Katewongsa, P., Widyastari, D. A., Saonuam, P., Haemathulin, N., & Wongsingha, N. (2021). The effects of the COVID-19 pandemic on the physical activity of the Thai population: Evidence from Thailand's surveillance on physical activity 2020. *Journal of Sport and Health Science.*, *10*, 341–348. Available from https://doi.org/10.1016/j.jshs.2020.10.001.

Kishor, A., Chakraborty, C., & Jeberson, W. (2020). A novel fog computing approach for minimization of latency in healthcare using machine learning. *International Journal of Interactive Multimedia and Artificial Intelligence*, 1. Available from https://doi.org/10.9781/ijimai.2020.12.004.

Kosta, Y.P., Dalal, U.D., Jha, R.K. (2010). Security comparison of wired and wireless network with firewall and virtual private network (VPN). In *Proceedings of the International conference on recent trends in information, telecommunication and computing. Presented at the international conference on recent trends in information, telecommunication and computing* (pp. 281–283). <https://doi.org/10.1109/ITC.2010.75>.

Kuperman, G. J. (2011). Health-information exchange: Why are we doing it, and what are we doing? *Journal of the American Medical Informatics Association: JAMIA*, *18*, 678–682.

Lahtela, A., Hassinen, M., Jylha, V. (2008). RFID and NFC in healthcare: Safety of hospitals medication care. In Proceedings *second international conference on pervasive computing technologies for healthcare. Presented at the second international conference on pervasive computing technologies for healthcare* (pp. 241–244). <https://doi.org/10.1109/PCTHEALTH.2008.4571079>.

Limwiriyakul, S. (2012). A method for securing online community service: A study of selected Western Australian councils.

Lupton, D. (2012). M-health and health promotion: The digital cyborg and surveillance society. *Social Theory Health, 10*, 229–244.

Monika, Upadhyaya, S. (2015). Secure communication using DNA cryptography with secure socket layer (SSL) protocol in wireless sensor networks. In *Procedia computer science, proceedings of the fourth international conference on eco-friendly computing and communication systems 70*, (pp. 808–813). <https://doi.org/10.1016/j.procs.2015.10.121>.

Muhammad, L. J., Algehyne, E. A., Usman, S. S., Ahmad, A., Chakraborty, C., & Mohammed, I. A. (2021). Supervised machine learning models for prediction of COVID-19 infection using epidemiology dataset. *SN Computer Science*, *2*, 11. Available from https://doi.org/10.1007/s42979-020-00394-7.

Müller, A. C., & Guido, S. (2016). *Introduction to machine learning with python: A guide for data scientists.* (1st ed.). Sebastopol, CA: O'Reilly Media.

Pace, P., Aloi, G., Gravina, R., Caliciuri, G., Fortino, G., & Liotta, A. (2018). An edge-based architecture to support efficient applications for healthcare industry 4.0. *IEEE Transactions on Industrial Informatics*, *15*, 481–489.

Raghavendra, C. S., Sivalingam, K. M., & Znati, T. (2006). *Wireless sensor networks*. Springer.

Triukose, S., Nitinawarat, S., Satian, P., Somboonsavatdee, A., Chotikarn, P., Thammasanya, T., ... Poovorawan, Y. (2021). Effects of public health interventions on the epidemiological spread during the first wave of the COVID-19 outbreak in Thailand. *PLoS One*, *16*, e0246274. Available from https://doi.org/10.1371/journal.pone.0246274.

Wang, P. (2008). The real-time monitoring system for in-patient based on zigbee. In Proceedings of the *second international symposium on intelligent information technology application. Presented at the second international symposium on intelligent information technology application* (pp. 587–590). <https://doi.org/10.1109/IITA.2008.110>.

Yao, W., Chu, C.-H., Li, Z. (2010). The use of RFID in healthcare: Benefits and barriers. In *Proceedings of the IEEE international conference on RFID-technology and applications* (pp. 128–134). IEEE.

CHAPTER

Voice signal-based disease diagnosis using IoT and learning algorithms for healthcare

3

Abhishek Singhal and Devendra Kumar Sharma

Faculty of Engineering and Technology, Department of Electronics and Communication Engineering, SRM Institute of Science & Technology, Delhi NCR Campus, Ghaziabad, Uttar Pradesh, India

3.1 Introduction

The first-word recognition system was developed in 1952 to train the recognize digits at Bell Laboratory. Nowadays, several voice recognition systems are available, such as speaker-independent systems, speaker-dependent systems, spontaneous recognition systems, etc. (Apte, 2012). The general implementations of gender-dependent systems are annotation in multimedia, speaker indexing, and speaker recognition and identification. Voice user interfaces are utilized for voice recognition and synthesizing systems. In the voice recognition systems, there are some common problems such as the continuous character of speech, background disturbance, and speaker variation. The voice signal also varied due to the properties of the speaker, such as sex, age, profession, and diseases (Chaudhari & Kagalkar, 2015a, 2015b).

The voice signal carries the main characteristics of the speaker. With the help of the analysis of the voice signal, the recognition system also identifies the several information about the speaker's effective factor, social factor, and characteristics of the voice signal generator. By using this information, the system can recognize the gender of the speaker when the speaker is invisible or hidden. Several applications are implemented with the help of a voice recognition system. Some important applications are recognition of gender; identifying the age, information of health, sociolect, emotional state, attentional state, language, dialect, and accent. The characteristics of voice signals have many applications in forensic, human-robot interaction, law enforcement, language learning, call routing speech translation, and smart workspaces. The identification of gender, group of age, and declaration of accurate age are difficult tasks using the analysis of voice signals.

Recognition of gender can be classified into two steps, gender identification and gender verification. Identification of gender contains the following phases: Feature Extraction, Training, and Testing. In the gender identification task, also known as 1: N matching, the characteristics of the unidentified speaker are compared to a dataset of N identified speakers and the finest matching of characteristics of the speaker is declared as the decision of recognition. In the gender verification task, also known as 1:1 matching, a decision has been taken according to the voice samples which are claimed by the speaker. If the margin between the dataset sample and the voice sample exceeds a predefined threshold, the claimed speaker is verified otherwise rejected (Ramdinmawii & Mittal, 2016).

Implementation of Smart Healthcare Systems using AI, IoT, and Blockchain. DOI: https://doi.org/10.1016/B978-0-323-91916-6.00005-9

For age identification, there are two approaches. The first approach is Age group classification (young, Adult, and Old), in which one group is allotted to the speaker. The second approach is known as age estimation; the accurate age in years of the speaker is estimated by using the regression technique (Chaudhari & Kagalkar, 2015a, 2015b).

It is a very difficult task to identify the state of the emotion of the speaker from the same set of speakers. Situations play a very important role in emotion recognition. Emotion recognition has several applications such as medical analysis, web interactive services, text-to-speech synthesis, and information retrieval. The recognition of emotion has also been useful in healthcare systems for depressed and anxious patients. For the high accuracy of the recognition of emotions, the database should contain the voice samples of the actor's emotions (Revathi & Jeyalakshmi, 2019).

The recognition of emotions is an important characteristic of human-computer interaction as well as human robotics interaction. Therefore, human robotics interaction has become an issue in the currently ongoing research area of mechatronics and robotics. The recognition of the emotion also depends on the type of emotion and the application (Kumar, Gudmalwar, Rao, & Dutta, 2019). Such as, the call center monitors the response of the customer care representative during a conversation with an angry customer (Yacoub, Simske, Lin, & Burns, 2003).

The generation of voice signals is a complex combination of the various body systems. The lung has the most important role in the generation of voice signals. In the case of Asthma, the volume of the lung reduces, and stress in the muscles of the tongue and neck increases. Asthma can change the sound quality of the voice signals due to the airway swelling caused by allergies. The disorganized change of airflow in the vocal tract creates spectral noise, so unmannered vocal folds vibration causes hoarseness (Sonu & Sharma, 2011).

Similarly, the larynx is the combination of muscles that are surrounded by the vessels of blood. So, the vocal cord parameters are also dynamically correlated with blood vessels. When the blood pressure of the speaker varies, the vibration of the vocal folds is also affected (Mesleh, Skopin, Baglikov, & Quteishat, 2012). So, some features of the voice signals are also affected which are related to paralinguistics (Saloni, Sharma, & Gupta, 2013).

James Parkinson firstly recognized Parkinson's disease in London in 1817. Nearly 70% of Parkinson's patients show shaking that is most noticeable in movement in the arm and fingers. It is very difficult to diagnose Parkinson's disease because no tests are available in the diagnostic lab. With the help of analysis of voice signals, the diagnosis of Parkinson's disease can be recognized at an earlier stage because the voice signal of the person varies in the starting phase of the disease. The diagnosis of Parkinson's disease through analysis of voice signals is a very efficient, reliable, low-cost method. In this method, no medical professionals are essential because this is computerized (Saloni, Sharma, & Gupta, 2015).

The efficiency of the diagnosis in the healthcare field can be improved with the applications of AI. AI is used for all predefined specific goals which are diagnosis, prognosis, and treatment. AI program, which is developed for the field of healthcare, is useful for the diagnosis and management or treatment of diseases. AI is mainly useful for the diagnosis of diseases. Data mining is highly admired in the field of healthcare. Data mining is utilized to identify the diseases and act as a warning alarm system for the early stage of the diseases. In the current pandemic situation, the Internet of things (IoT) has unique and vital applications in the area of the healthcare sector to enhance medical facilities like early diagnosis of diseases, monitoring health, and managing fitness programs. (Islam, Kwak, Kabir, Hossain, & Kwak, 2015; Vasanth & Sbert, 2016). IoT has several applications in many fields such as healthcare, disasters, intelligent transport systems, etc. By using the IoT in the field of healthcare, the movement for the daily routine activity and parameters of the health, including intake of medicine for the senior persons, can be monitored with privacy (Gupta, Chinmay, & Gupta, 2019). In the field of medicare, several diseases can be diagnosed with the help of voice signals (Salhi, Mourad, & Cherif, 2010). The Voice signal of a patient is an important parameter that is changed due to unhealthy habits and the nature of the work (Bone, Lee, Chaspari, Gibson, & Narayanan, 2017; Hariharan, Paulraj, & Yaacob, 2010). So, the diagnosis and recovery time of the diseases can be optimized by the regular follow-up procedure to check the changes in the vocal tract and the quality of the voice signals with the help of IoT. Due to the implementation of IoT, the quality of healthcare services will be enriched and enhanced as well as the cost of medical services will be reduced. Although IoT has many features in health services, it is also having some considerable issues such as risk, security, etc. A concern of a breach is available with any online data. Internet hackers steal or manipulate the sensitive data of patients. There are several cases in which confidential health data is manipulated by unauthorized users. No doubt, this activity does not happen regularly, but this process may be used to disturb the process of treatment of the disease. If important health information is extracted from the online formats by the unwanted person, the consequences may be dangerous. Hackers may create trouble for the patient in the form of finance, reputation, or any other means.

The architecture of an IoT healthcare (IoTM) is the stepwise connection of several components of an IoTM system that are pertinent and connected in a medical environment. An IoTM system is the combination of mainly three components such as subscriber, broker, and publisher. The publisher is a set of networks that is a combination of voice analysis devices and other medical devices that may perform simultaneously or individually to best-ever information of the patient such as gender, age, information, blood pressure, heart rate, and so on. The doctor will continuously observe the patient's information with the help of the smartphone, computer, tablet, etc. After this, the publisher will transfer the feedback which is generated by the doctor, to the patient. The topologies for IoTM are depended on the demand of the healthcare system as well as application. Several changes in the architecture of IoTM have already been implemented in the past few years, so it is difficult to design a common architecture for the IoTM.

This chapter aims to use the analysis of the voice signal for the identification of gender, recognition of emotions, and the diagnosis of diseases by using the IoT and learning algorithms which are very helpful in the field of healthcare. A prologue for the analysis of voice signals is described in Section 3.2. Section 3.3 includes different algorithms for the extraction of features from the voice signals. The classifiers for the analysis are described in Section 3.4. Section 3.5 explains the processing of the voice signals. Section 3.6 describes the IoT-based healthcare sector. Section 3.7 discussed the accuracy of the detection of age and gender from the voice signals. The accuracy of the recognition of the emotions of the speaker is discussed in Section 3.8. Section 3.9 describes the diagnosis of the diseases. Section 3.10 presents the results and discussion followed by the conclusion in Section 3.11.

3.2 A prologue for analysis of voice signals

For the generation of the voice signal, the lungs work as an activator and control the airflow which generates the source of power to produce the vibration in the vocal fold. The plasticity of the expressing and singing voice is dependent on the movement of the vocal folds which is controlled accurately due to the vibration of vocal folds. The characteristics of the voice signal depend upon the resonators (Chest, Sinus, and face) of the body (Saloni et al., 2013).

For finding the stability in the voice signal, the voice signal is divided into a small span of time. By using short-time spectral analysis, the voice signal can be characterized (Benmalek et al., 2017a, 2017b; Kumar & Mallikarjuna, 2011). The starting period of the voice signal is always similar for all speakers and the remaining time of the voice signal is different for each speaker (Ramdinmawii & Mittal, 2016). Generally, the frequency of the recorded sample voice signal is 44.1 kHz. After recording, the frequency of the voice signal is reduced to 16,000 Hz due to the conspicuous information of emotion within the 8 kHz signal. The Nyquist theorem is applied to avoid the complexity issue (Koolagudi & Rao, 2012; Koolagudi, Murthy, & Bhaskar, 2018).

During the conversation, various information about the speaker is available on the voice signal. Voice signal has a unique capability to pass on the information of the message as well as the emotion of the speaker. The listener can create misunderstanding if the emotion is ignored (Kumar et al., 2019). The suprasegmental or prosodic features are those features that are extracted from the longer segment of the voice signals. These features represent the several characteristics of the voice signal such as intonation, duration, stress, rhythm, and loudness. The specific information about the emotion from the voice signal can be captured by using the spectral feature. The specific information for emotion recognition is the movement of articulators, responsible for producing different sound units and the shape and size of the vocal tract (Bahari & Hamme, 2011; Chaudhari & Kagalkar, 2015a, 2015b; Khan & Bhaiya, 2012; Solera-Urena, Garcia-Moral, Pelaez-Moreno, Martinez-Ramon, & Diaz-de-Maria, 2012).

Paralinguistic information of the speaker is indirectly shown by the emotional state of the speaker. The Voice signal of the emotion of the speaker such as fear, anger, and happiness is publicized by the psychological behavior of the voice signals generator such as high heart rate, high blood pressure, etc. These voice signals contain strong, loud, and fast frequency energy. In every emotional state, the dominant frequency in every frame of the voice signal varies due to the modulation features of the speaker and the movement of articulators (Revathi & Jeyalakshmi, 2019).

For the analysis of the voice signal, the voice recognition system is divided into two blocks that are Front-end Block and the Back-End Block. Front-End Block of the voice recognition system contains the preprocessing and feature extraction block. Extraction of the features from a voice signal is a process of pulling out a specific feature from the preprocessed voice signal. There are several features and techniques available for Front-End Blocks such as Cepstrum analysis, Cepstrum method (CEP), data reduction method, spectrogram, linear predictive coding, mel-frequency cepstrum coefficient, etc. A modified feature extraction technique is developed, in which MFCC is extracted with frequency sub-band decomposition to reduce the side effect of the preemphasis signal from voiced sound signals.

Back-End Blocks contain the classification of voice signals. In the classification process, the extracted features from the voice signal are classified and find the best-fitted space from the database. According to the classification, Back—The end Block declares the result. Artificial neural network (ANN), hidden markov model (HMM), vector quantization (VQ), dynamic time warping (DTW), support vector machine (SVM), and some other classification techniques are available to classify the voice signals (Rabiner & Schafer, 2007).

3.3 Extraction of parameters from voice signal

The recognition system is divided into two phases. The first phase is the training phase. In the training phase, the dataset of the feature vector contains unique features after extracting a large number of voice signals. The second phase is the testing phase. In the testing phase, the features of the voice signal will classify according to the database. For the reduction of the dimensions of the feature vector, the following points show the importance: (1) by using the whole feature set, the processes of the identification consume a very large time period (2) Classification processes are also easier if the numbers of the best features reduce (Kumar et al., 2019). The accuracy of the system can be improved by preprocessing before feature extraction. Preprocessing provides the vigor of the frames with enhanced spectral energy.

For the analysis of voice signals, the short-term spectrum is required because the auditory nerve of the human ear works as a quasi–frequency analysis. The analyses of the auditory nerve are based on the mel-frequency scale. The time duration of the voice signal, which is generated by the same speaker with the same text, is different. For the analysis of voice signals, the features of the voice signals are the major parameter for the selection of the classifier (Koolagudi et al., 2018). The frequency-domain features of the voice signal like MFCC, fundamental frequency, spectral density, etc., are more useful features of the voice signal for identification of the characteristics of the speaker because the frequency-domain feature has less noise. By the utilization of a combination of the cepstral coefficient with other acoustic features, the classification can be optimized (Benmalek et al., 2017a, 2017b). The extraction of features of the voice signals is possible if the voice signal is considered stationary for a very short span (Koolagudi et al., 2018).

The selection of features is important for the recognition of emotions for the prevention of misclassification of data and vagueness of fundamentals of emotion. The accuracy of the recognition of emotion can be improved by creating two groups of emotions like soft and arousal emotions (Revathi & Jeyalakshmi, 2019). In the processes of the detection of diseases, only those features are picked up, which shows the changes in the characteristics (Saloni et al., 2015).

3.3.1 Formants

The characteristics of the human vocal tract can be represented by formants. Every formant contains a high degree of energy. The characteristics of the human vocal tract also depend on gender, emotion, age, and diseases (Koolagudi et al., 2009, 2018). The formant frequency also depends on the physiological movement and structure of the vocal tract. Thus, formants are very helpful to segregate vowel sound signals (Kumar et al., 2019).

3.3.2 Mel-frequency cepstral coefficients

For recognition of the voice/ speaker or language, MFCC is a unique feature (Revathi & Jeyalakshmi, 2019). Mel-frequency cepstrum is the equally spaced frequency band on the mel scale and resembles the human auditory system's response (Malode & Sahare, 2012). The voice signals are converted into N samples frame-block, with neighboring frames being separated by M ($M < N$).

N samples are available in the first frame. The neighboring iframe contains the M samples. The second frame overlapped with the first frame by N-M samples (Benmalek et al., 2017a, 2017b).

MFFC is the combination of two types of filters. The first type of filter is a linearly spaced filter for a frequency below 1000 Hz. Another type of filter is logarithmic spaced filters for high frequencies above 1000 Hz. The sampling rate is nearly 16,000 Hz during the recording of the voice samples. By using this sampling signal, the aliasing effect is minimized in the processes of analog to digital conversion (Muda, Begam, & Elamvazuthi, 2010).

For the extraction of MFCC, preemphasis is the first step. In the preemphasis, the signal passes through a filter in which the signal emphasizes higher frequency. The energy of the voice signal at a higher frequency is increased. In the process of framing, the voice samples are segmented into small size frames with the help of an analog to digital converter. The voice samples are distributed into N samples. The size of the frame is 20–40 Ms. The next step of the extraction of the feature is windowing. By using the Hamming window, the discontinuities are removed. So, the edges of the signal become smooth. To obtain the frequency and magnitude response of every frame, the frame is converted from the time domain to the frequency domain by performing a fast Fourier transform (FFT). For extraction of the feature, FFT is calculated after framing and windowing. The output of the FFT is applied to the filter bank which is arranged according to the MEL scale. In the last step, the log mel spectrum converts into the time domain by using the Discrete Cosine Transform. The final conversion is known as MFCC (Bahari & Hamme, 2011; Muda et al., 2010; Solera-Urena et al., 2012; Sonu & Sharma, 2011).

3.3.3 Signal energy

The energy from the voice signal can be extracted by the feature selection process. The energy in the male voice signal is less than the energy in the female voice signal. To recognize the audible segment from the voice signal, the maximum energy frame is extracted from the utterance.

3.3.4 Pitch

The fundamental frequency (F_0) of the vocal fold's vibration is pitch. Several methods are available to calculate the pitch like autocorrelation, zero crossing, *etc*. (Koolagudi et al., 2018). To achieve a noiseless pitch, the autocorrelation method is very useful. Pitch is the major parameter in the voice signal. Pitch adds naturalness to the voice signal. The various information of the speaker can be identified with the help of pitch such as accent, gender, emotional state, way of speaking, etc. (Wenjing, Haifeng, & Chunyu, 2009). If the emotions of the speaker will change, the pitch of the voice signals also changes. So, the emotional state of the speaker may be identified. There is one disadvantage of pitch it depends on the quality of the voice signal.

3.3.4.1 Mean pitch

The average speaking pitch or the fundamental speaking frequency (SF_0) is also known as Mean Pitch. The value of the mean pitch is 128 cycles per second for the adult male. For the adult female, this value is 225 Hz. The maximum value of mean pitch is 260 Hz which is generated by a child whose age is less than 10 years. Standard deviation is a statistical estimate to find the pitch variation from the fundamental speaking frequency (Saloni et al., 2013; Shirvan & Tahami, 2011; Viswanathan, Zhan, & Lim, 2012).

3.3.4.2 Zero-crossing rate

ZCR is one of the frequency-domain features of the voice signal (Giannakopoulos, Pikrakis, & Theodoridis, 2009). The high value of the ZCR shows the high-frequency voice signal and the low value of the ZCR implies a low-frequency voice signal. For example, neutral and sad emotion voice signal has low frequency, and anger and happy voice signal has a high frequency. ZCR can be calculated by using Eq. (3.1) (Atassi & Esposito, 2008).

$$Z_n = \sum_{m=-inf}^{inf} 0.5 * \left|\text{sgn}(x[m]) - \text{sgn}(x[m-1])\right| * w[n-m] \tag{3.1}$$

Here, $\text{sgn}(x)$ is known as the signum function which represents the sign of the signal.

3.4 Classifiers for voice analysis

The selection of the classifier, which is based on the features of the voice signals, is also important. The accuracy and performance of the classifier are depended on the model of the classifier. The dataset of the voice signals is also very important for measuring the performance of the classifier (Koolagudi et al., 2018; Reddy, Singh, Kumar, & Sruthi, 2011).

3.4.1 Gaussian mixture model

A model which is utilized to classify the voice signals and represented by the weighted sum of all Gaussian component densities is known as the Gaussian mixer model (GMM). The mixture weight, covariance matrices and mean vectors from all component densities provide guidelines to create a complete Gaussian mixer model. For multidimensional features, the Gaussian mixer model is the most statistical maturation method.

The efficiency of GMM can be increased if the data distribution is normal. The collection of more than one normal distribution is known as the multivariate normal distribution. In each class, several factors affect the accuracy of the GMM such as the size of the dataset, distribution of the data, the numbers of Gaussian, etc. The best accuracy is achieved by using the normal distribution (Koolagudi et al., 2018).

M is represented by component Gaussian mixture for dimensional input vectors (b) which is obtained by Eqs. (3.2) and (3.3).

$$P(x/\lambda) = \sum\nolimits_{i=1}^{M} W_i b_i(x) \tag{3.2}$$

and

$$b_i(x) = \frac{1}{(2\pi)^{\frac{d}{2}}|\Sigma_i|^{1/2}} e^{\left\{-\frac{1}{2}(x-\mu_i)^T \sum_i^{-1}(x-\mu_i)\right\}} \tag{3.3}$$

For the super vector for each speaker, Gaussian mixture weights are extricated and combined. By using these super vectors of the training dataset, hybrid models can be designed. These hybrid architectures are general regression neural network and weighted supervised nonnegative matrix factorization (Chaudhari & Kagalkar, 2015a, 2015b).

3.4.2 Vector quantization systems

In the VQ approach, code vectors are a set of vectors that are mapped with a finite number of vectors. The efficiency of VQ depends on the generation of an effective codebook (Koolagudi et al., 2018). In the VQ, the sets of L disjoints are created by splitting the entire training vectors. Each set is represented by a single vector which is called the centroid (Djemili, Bourouba, & Korba, 2012).

The number of the codebook of size N in the VQ depends on the number of emotions or states. For computing deviation, the similar feature vectors which are generated by the testing signal are compared with the code vectors. Less deviation represents more matches (Koolagudi et al., 2018).

3.4.3 Support vector machine

For the classification of the voice signals, SVM creates a model with the help of training samples. As shown in Fig. 3.1, support vectors are the closest data points to the hyperplane. It is very difficult to classify the support vector in comparison to other data points that are available near the hyperplane. Margin is the perpendicular line to the hyperplane (Krishnan, Chinmay, Banerjee, Chakraborty, & Ray, 2009; Ramdinmawii & Mittal, 2016). Only one hyperplane is selected from the available hyperplanes which shows the maximum margin (Muhammad et al., 2021; Saloni et al., 2015).

The SVM aims to decide the position of the hyperplane in the multidimensional space. The position of the decision boundary is used to separate different classes which are based on statistical learning techniques. Binary classification can be done by the SVM. Such a type classifier provides the result in the form of true or false. For nonlinear separation, kernel functions like polynomial, Sigmoidal, and RBF can be used (Saloni, Sharma, & Gupta, 2014). The accuracy of the SVM is high. Gaussian radial function SVM shows the best accuracy compared to other kernels of SVM. The SVM classification techniques are also utilized in the real-time analysis with some limitations.

3.4.4 K-means clustering

MacQueen proposed the K-mean algorithm in 1967. This technique follows the unsupervised learning procedure. By using this technique, the selected object has a minimum distance to the centroid (Saloni et al., 2013). In K- mean clustering, a plan of action for the testing and training is similar to VQ. The codebook is replaced by the K centroids. Initially, the K centroid is selected randomly for each emotion. The potion of the centroid may be changed according to the number of executions of K-mean clustering algorithms. The final position of the centroid will be converged either by stopping the execution whenever the minimum movement of the centroid is received or by executing the algorithm times (Koolagudi et al., 2018). For the creation of a cluster, a large number of data are required. The number of K is always fixed before starting the process. If the number of data is less, the real cluster possibility is very less (Li, Yao, & Huang, 2011; Sonkamble & Doye, 2012).

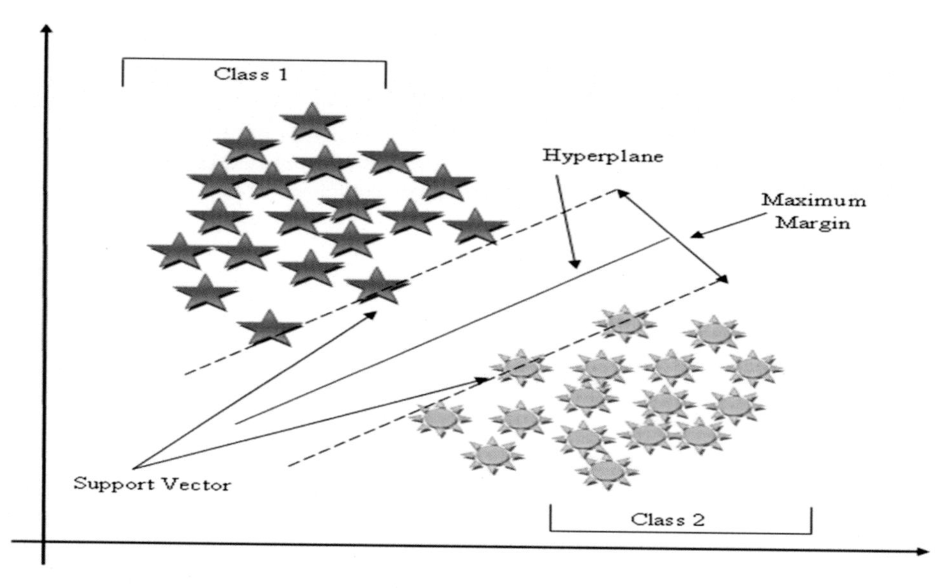

FIGURE 3.1

Support vector machine.

3.4.5 Artificial neural networks

The ANN is a classification technique that is used to classify voice signals according to the applications. ANN is also used in the real-time analysis for the classification of voice signals. ANN is used to process the information with the help of several processing units which are called neurons. ANN is the combination of linear and nonlinear units. At the start of the process, all biases and weights are randomly initialized. Backpropagation with a gradient algorithm is used for the learning. Just like a human brain, the complexity of the nonlinear relationships in data can be captured by the ANN (Kumar et al., 2019).

An ANN is a combination of three layers: input, hidden, and output layer. Every layer contains several neurons. The numbers of neurons depend on the states of the input & output and the related factors of features. The number of nodes at the input layer is equal to the number of feature vectors (Dobry, Hecht, Avigal, & Zigel, 2011). The sigmoid activation functions are available in the hidden layer with nodes. The nodes of the output layer are equal to the states of the desired output (Kumar et al., 2019).

3.4.6 Multilayer perceptron

Multilayer perceptron (MLP) reflects the organization of the human brain. MLP is also equal to the feed-forward ANN. MLP has multiple hidden layers between the input and output. The number of hidden layers is depended on the data mining task. Every neuron in the hidden layer is connected with the neurons of the next layer. The connecting wires between the neurons are known as weights whose values are updated with the help of the learning phase. The learning phase is continuously repeated until the value of the error will be less than the threshold level. The input layer is the combination of the values of the features. The output layer will predict the classification which is based on the information which is passed on by the input layer. The classified output compares with the observed one and calculates the error. According to the error, network weights are updated from the output layer toward the input layer through the intermediate layer. Transmitted information can be calculated by the combination of the connecting weights, node value, and activation function (Tradigo et al., 2015).

3.4.7 Convolutional support vector machine

To learn the lower-level parameters in the convolutional models, the softmax activation function is important. Prediction of the Softmax is the same as predication of SVM. So, these two approaches can be combined. Thus, softmax can be replaced by SVM. The function of the new architecture is the same as a convolutional neural network (CNN). In the hybrid model, the learning of the low-level parameters is performed with the help of the SVM. This model is called a convolutional SVM and its architecture is shown in Fig. 3.2 (Passricha & Aggarwal, 2019).

3.5 Processing of voice signal

The voice processing system can be divided into two phases that are training phase and the testing phase, as shown in Fig. 3.3. In the training phase, a very large number of voice samples are used to train the system. In the first step, preprocessing is performed on the voice samples before extraction of the features. After extraction, the feature vectors are created for learning the system. In the testing phase, preprocessing and feature extraction are similar to that in the training phase. Extracted features from the voice signal and feature vector are applied to the classifier. In the last stage of the system, the classifier performs as a decision-maker. The output of the classifier provides the final decision regarding the identification of the gender, recognition of the emotions, and diagnosis of the diseases.

3.6 Internet of things–based healthcare sector

In the current pandemic situation, the healthcare sector is the most challenging area in the world. In several countries, the healthcare sector is dependent on the medical records of the patients which are maintained on paper. These papers can generate a complex path between the several departments of the hospitals for sharing the information of the patients. With the implantation of IoT in the healthcare sector, a panel of doctors can provide a diagnosis of the diseases, treatment, and monitoring of health very easily after being accessible to the records of the patients with less cost and time (TurcuC & Turcu, 2013). By gathering the data of the patients from several IoT devices, IoT can perform as a lifesaving application and patients can receive a real-time diagnosis (Kulkarni & Sathe, 2014) without physical consultation. IoT techniques have several medicare applications

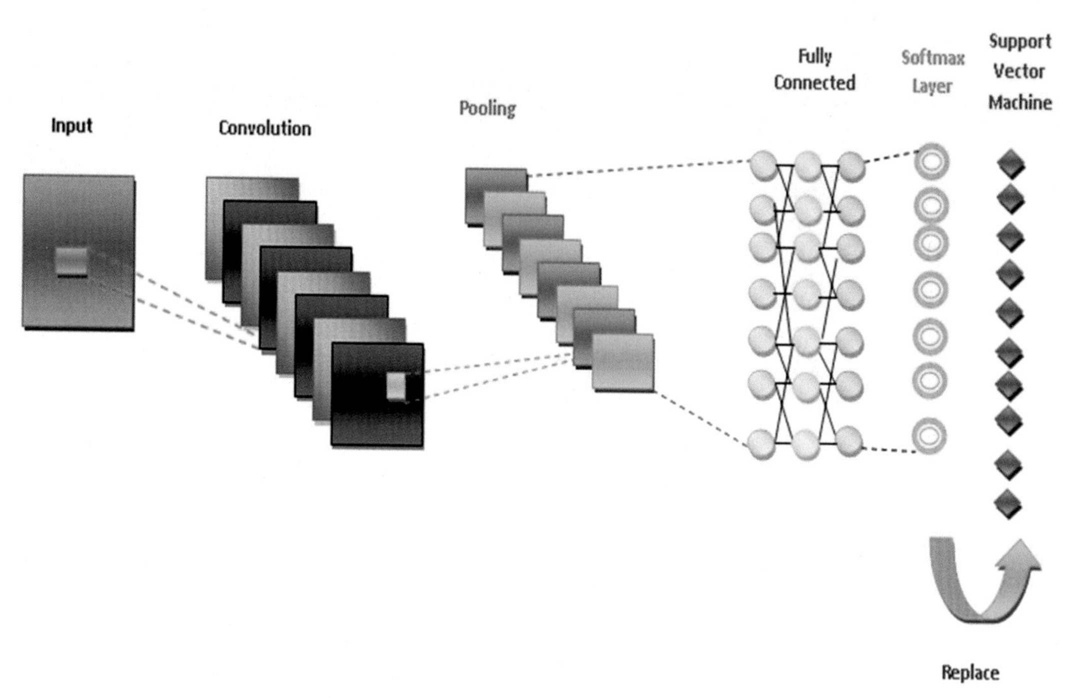

FIGURE 3.2

Architecture of convolutional support vector machine.

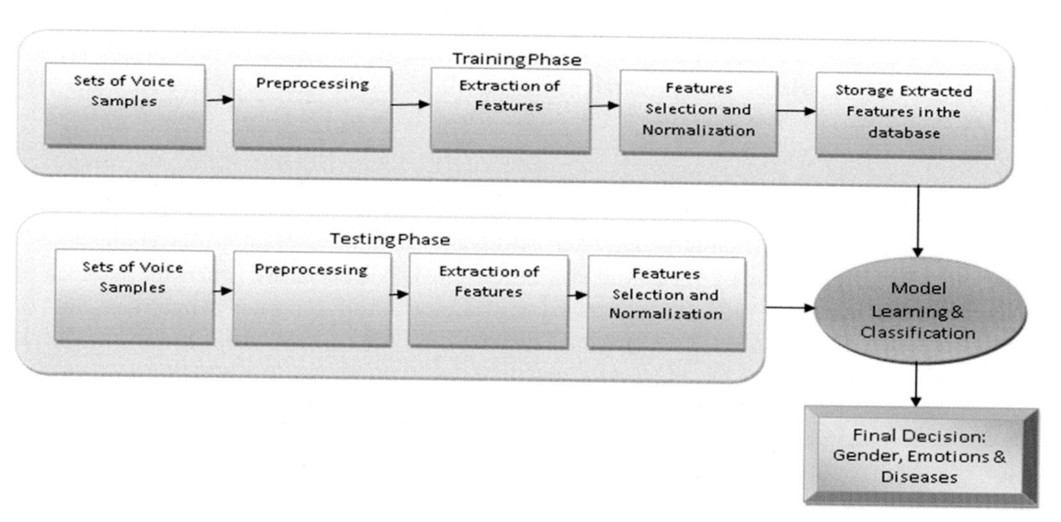

FIGURE 3.3

Model for voice signal processing.

such as monitoring the blood pressure, stages of Parkinson's disease, etc., with the help of voice signals. Several healthcare applications are available for smartphones for medical consultation and other services (Ventola, 2014). Thus, IoT plays an important role in telemedicine.

3.7 Detection of age and identification of gender using voice signal

Gender Recognition systems archive their operation in two phases that are training and identification. In the training phase, the first step is feature extraction from the voice sample of a male or female speaker. The feature extractor transforms the digital voice signal into a train of numerical descriptors, called feature vectors. A classifier is trained a model in a supervised manner for males and females separately. In the identification phase, the first step is similar to the training phase to feature extraction. But in the identification phase, the extraction of the features is applied to the unknown voice signal. After extraction of the feature, a classifier will be matched with two models (male and female). Classifiers find similarities and dissimilarities and provide the final decision based on the calculated measures.

For better accuracy in the identification of gender, the classifier has more number data from the longest utterance length in the training phase. So, the classifier easily discriminates between two groups of speakers. The male group has less accuracy in comparison with the female group for identification systems of gender. The female group has 3.7% more accuracy compared to the male group (Djemili et al., 2012). By using voice signal recognition, the average accuracy is 89.3% for recognition of age and 80.3% for identification of gender (Chaudhari & Kagalkar, 2015a, 2015b). After the implementation of the Gaussian mixture model, the accuracy of identification of gender is achieved at 96.4% (Djemili et al., 2012).

The contour (F0) of the female speaker is higher than the male speaker. The desired pitch frequency of female and male speakers always falls within the range which is 165 to 300 Hz for females and 100 to 180 Hz for males (Ramdinmawii & Mittal, 2016). The energy of the voice signal of the male speaker is less than the energy of the voice signal of the female speaker in the feature extraction.

MLP also provides the best accuracy rate for the largest utterance. The accuracy for the identification of gender depends on the number of epochs and the number of the hidden layer. It means the accuracy for the identification of gender increased when the number of hidden layers increased. On the other hand, the accuracy for the identification of gender decreased when the number of epochs increased. By using MLP, the accuracy for identification of gender is achieved at 96.4%. The same accuracy for identification of gender, 96.4%, is also achieved by using VQ. If the neural network approach is applied to the VQ, a new modified classifier, known as learning vector quantization, is available for classification. LVQ contains several algorithms to execute a competitive scheme. But accuracy for identification is 94.6% (Djemili et al., 2012).

In a comparative study of the Gaussian Mixer Model, MLPs, VQ, and learning vector quantization, LVQ shows the lowest accuracy in comparison to others. GMM, MLP, and VQ have the same accuracy for the identification of gender (Djemili et al., 2012). The accuracy for the identification of gender and age also depends on the features which are extracted from the voice signals of the speakers. The highest accuracy is achieved with the help of MFCC in comparison to the energy and pitch of the signal (Ramdinmawii & Mittal, 2016).

3.8 Recognition of emotions through voice signal

Recognition of emotion from the voice signal of a human provides helps to identify the state of a human. There are several available states of emotions such as happiness, boredom, anger, sadness, neutral, etc. Due to the uncertainty of emotion, the recognition of emotion from voice signals always produces difficulty. The emotions recognition model has several applications such as medical analysis and text to speech synthesis, information retrieval, and web interactive services. For the treatment of patients with anxiety and depression, emotional recognition structures are very useful for the treatment (Revathi & Jeyalakshmi, 2019).

The characteristic of the voice signals varies according to the emotions and individual characteristics of speakers (Kumar et al., 2019). Emotions such as anger and happiness in voice signals depend on psychological behavior, for example, high heart rate and high blood pressure. For the voice signal in different emotions, the principal frequency in every frame is found to be different due to unusual movements or characteristics of the speaker (Revathi & Jeyalakshmi, 2019). Several factors affected the process of classification of the emotions, such as improper emotional database, extraction of proper features, etc. In comparison to images and text, voice signals are more reliable for the recognition of emotions (Koolagudi et al., 2018). The accuracy of the emotion recognition systems is enhanced by using the database which has the recorded voice samples of the actors in the emotion (Revathi & Jeyalakshmi, 2019).

The pitch of the voice signal also depends on the state of the emotions of the speaker. High reaction emotion such as happiness and anger and low reaction emotions such as neutral and sadness has average high and low pitch frequency, respectively. In high reaction emotions, the speaker generates loud, fast, and high pitch frequency speech because the characteristics of the speaker like heart rate, and blood pressure (BP) are increased. In the low reaction emotions, the speaker generates low pitch frequency due to low BP and low heart rate.

The overall accuracy for the recognition of emotion for speaker-dependent systems is high in comparison to the speaker-independent systems. For the speaker-dependent system, the accuracy of recognition of fear emotions is 100% and the accuracy is less than 100% for speaker-independent systems. With the utilization of the combination of features, the accuracy of the speaker-dependent system also improved by 6.2% (Revathi & Jeyalakshmi, 2019). The accuracy for the recognition of emotions is 60% by using a probabilistic neural network with a linguistic data consortium (LDC) database (Cen, Ser, & Yu, 2008). Gaussian mixture model provides better accuracy of 84%, for one type of database, which is extracted by movies, in comparison to an accuracy of 81%, for the other database. The 63% accuracy is achieved by the Gaussian mixture model with Berlin emotional database (Kumar et al., 2019).

The 78.4% accuracy for the recognition of emotions is achieved by using the HMM with the Berlin emotional database. Similarly, 51% accuracy was achieved by MLPs neural network (Kumar et al., 2019). Using Berlin emotional voice database, 55% and 83.17% accuracy was achieved by ANNs for speaker-independent system and speaker-dependent system, respectively. While using a random forest classifier for the same database, the accuracy for the speaker-dependent system was 77.19% and the accuracy for the speaker-independent system was achieved at 48% (Iliou & Anagnostopoulos, 2009; Revathi & Jeyalakshmi, 2019).

To recognize the emotion between two groups: group1 (neutral and sadness) and group 2 (hot and cold anger), the accuracy is 87% for the SVM (Tato, Santos, Kompe, & Pardo, 2002; Yacoub et al., 2003). By utilizing the confusion matrix for the recognition of emotions with the LDC emotional database, the 84.5% average accuracy is achieved. On the other side, boredom has 94.1% highest accuracy and happy emotion has the 72% lowest accuracy (Kumar et al., 2019).

If the dataset is changed, the accuracy for recognition of emotion will change (Batliner, Fischer, Huber, Spilker, & Noth, 2000). Performance of the emotion recognition system, which is based on recognition rate or accuracy, is calculated by the Eqs. (3.4) and (3.5) (Koolagudi et al., 2018; Kumar et al., 2019).

$$\text{Recognition rate} = \frac{\text{No.of successfullyrecognizedsamples in the emotion}}{\text{Total number of samples in the emotion}} \tag{3.4}$$

or

$$\text{Performance accuracy} = \left(\frac{I_e}{T_e} * 100\right) \tag{3.5}$$

where I_e represents the number of correctly identified emotional clips and T_e shows the total number of emotional clips.

3.9 Disease diagnosis using voice signal

The performance parameter of the voice signal can be calculated in the cepstral, spectral, and time domains. Processing of the voice signal based on cepstral analysis is a unique high-performance tool for the detection of unusual variations in the voice. There are several methods available for spectral analysis such as bandwidth, frequency, and amplitude of formants, including sub-band. Pitch variation, shimmer, jitter, etc., are the time domain parameters of the voice signal. To differentiate between the voices of normal and pathological speakers, the methods, which are based on temporal characteristics, are applied. For the detection of the unusual variation in the voice signal, the excellent analysis is the spectral analysis (Sonu & Sharma, 2011).

The sound quality of the voice signal of the speaker is influenced by allergies and Asthma. A chronic inflammatory disorder of the airways, known as Asthma, is concerned with recurrent episodes of wheezing, chest tightness, coughing, and breathlessness in the early morning. Due to the unusual vibration in vocal folds, a noise is generated which is directly related to the turbulences of airflows in the vocal tract. However, for accurate analysis, endoscopic analysis is required (Saloni et al., 2015). The accuracy for the detection of Asthma depends on the section of the features/ analysis. The accuracy for the detection is 62.5% for MFCC/ DTW and 85% for the acoustics analysis (Sonu & Sharma, 2011).

The neck region of the human body is surrounded by blood vessels. The circulatory system of the human provides the ability to control the blood flow in the vessels. High blood pressure causes the main reason for the unusual vibration of the organs (Deshpande, Thakur, Zadgaonkar, & Prof,

2012; Mesleh et al., 2012). So, the voice signal of the speaker is also affected. With the help of analysis of voice, the system can differentiate between high blood pressure v/s normal blood pressure speakers. The characteristics of the features of the high BP speaker spread over a broad range. But for a normal person, the value of features spread over a narrow region (Saloni et al., 2013). For the diagnoses of blood pressure, the accuracy of the pathological result and normal result of the classifier is 80% and 78.51%, respectively. So, the overall accuracy is achieved by 79% (Saloni et al., 2013).

In 1817, James Parkinson recognized a disease that is related to dopamine. This disease is known as Parkinson's disease. The neurotransmitter, known as dopamine, is a very important connecting device between the brain and body. Dopamine realizes a signal in the form of chemicals. This signal contains the planning and programming information of the movements. Shivering in the fingers and hand is a common symptom in the Parkinson-affected person. Rigor in muscles, delay in movements, and less coordination during routine activities are also prominent symptoms of Parkinson's. The probability of occurring Parkinson's disease depends on the age of the person. There is no pathological test available for diagnosing Parkinson's disease. Only brain scans and neurological tests are available for diagnosis. These tests were also identified after a long time of observation. The only voice of the speaker affected in the early stage of Parkinson's disease. So, with the help of an analysis of voice signals, the diagnosis of Parkinson's disease can be done in the early stage. This method is unique and reliable at a low cost. For diagnosing the disease, no medical staff is required. Jitter, Shimmer, and Harmonic to Noise ratios have a high value in comparison to Parkinson's patients (Saloni et al., 2015).

SVM shows the highest accuracy of 95.9% for diagnosing Parkinson's disease. The transit and linear artificial network classifier achieved 87.7% accuracy which is less than in comparison to SVM (Saloni et al., 2015). The local learning-based feature selection algorithm with cepstral coefficients of PLP is achieved as 84.3% accuracy, where 87% accuracy for a healthy person, 89% accuracy for the early stage of Parkinson's, 78% accuracy for the intermediate stage of Parkinson's, and only 71% accuracy is achieved by the advanced stage of Parkinson (Benmalek et al., 2017a, 2017b).

The MFCC is used with a local learning-based feature selection algorithm to diagnose Parkinson's disease. The accuracy for healthy class, early stage class, intermediate stage class, and advanced stage class is achieved as 73%, 95%, 91%, and 38%, respectively. The average accuracy is 86.7% (Benmalek et al., 2017a, 2017b). In an ANN, the accuracy is achieved at 93.2% (Gharehchopogh & Mohammadi, 2013). During the classification consider that 70% of data is used as training data and 30% of data is used as testing data. The 83.3% accuracy was declared for the backpropagation MLP network. 94.5% accuracy was achieved by implementing a genetic algorithm and SVM as a classifier (Khemphila & Boonjing, 2012; Shahbakhi, Far, & Tahami, 2014).

For diagnosing the diseases, the performance of the classifiers was also checked with several measures such as sensitivity (SE), specificity (SP), and overall accuracy (AUC) (Godino-Llorente et al., 2005), as shown in Eqs. (3.6), (3.7), and (3.8).

$$\text{Sensitivity} = \frac{\text{TP}}{\text{TP} + \text{FN}} * 100 \tag{3.6}$$

$$\text{Specificity} = \frac{\text{TN}}{\text{TN} + \text{FP}} * 100 \tag{3.7}$$

$$\text{Overall Accuracy} = \frac{\text{TP} + \text{TN}}{\text{TP} + \text{TN} + \text{FP} + \text{FN}} * 100 \tag{3.8}$$

Where TP represents True Positive, the classifier declared as a patient classified when symptoms of the disease are found. TN is True Negative; the classifier is declared as normal when normal samples are present. FN indicates the value of False Negative; the classifier declares as normal when the symptom of the disease is present and FP is False Positive, the classifier declares as a patient when normal samples are present (Saloni et al., 2013).

3.10 Results and discussion

The voice signals contain a very wide range of information about the speaker such as gender, emotions, and status health. With the help of voice analysis, the healthcare professional can provide the treatment and make a schedule based on age, gender, emotions, and diseases. For accurate treatment, the identification of the gender and age of the patient is the first and most important requirement. After a thorough review of several research articles, it is observed that the identification of the female group is higher than the male group (Djemili et al., 2012). The accuracy of the identification of the gender is also dependent on the selection of features, classifiers, epochs, and the duration of the voice signals. For comparison of the classifier's accuracy is carried out with respective any feature. The accuracy of the system will change as the data changes. The Gaussian Mixer Model, MLPs, and VQ show the highest accuracy in comparison to other classifiers, as shown in Graph 3.1.

Similarly, for the management of the treatment of diseases, recognition of the emotions is also an important consideration because the emotions of the speakers also varied according to their blood pressure and heart rate. The accuracy of the emotions is also dependent on the dataset of the samples. Graph 3.2 shows that the SVM has the highest accuracy for the recognition of the emotions with the LDC database while Hidden Markova Model (HMM) shows the highest accuracy in comparison to other classifiers with Berlin emotional database as shown in Graph 3.3.

For the diagnosis of Parkinson's disease, the accuracy of recognition also depends on the stage of the disease. The intermediate stage has the highest accuracy for the diagnosis of Parkinson's disease which is not easily archived by the pathological tests. Similarly, the diagnosis accuracy is also varied according to the selection of the classifier. SVM has the highest accuracy in comparison to other classifiers as shown in Graph 3.4.

The analysis of voice signals is a significant area in the field of healthcare. Several patients have severe diseases that can be identified by the analysis of voice with IoT (Bhattacharyya, 2014). The voice signals monitoring can be done with high accuracy and in less computation time after receiving the signals from different IoT sensors. Every voice signal contains a watermark to prevent the security and authenticity of the voice signals (Muhammad, Rahman, Alelaiwi, & Alamri, 2017). Voice Pathology Detection (VPD) approach is developed for the smart healthcare sector. VPD is the combination of two signals, which are voice signal and electroglottography (EGG) signal. These signals are analyzed and classified as normal or pathological through the IoT. After the

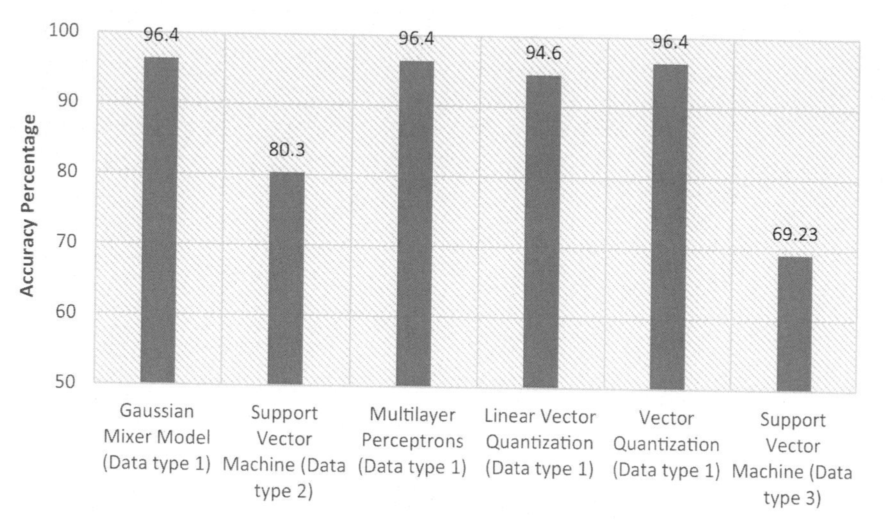

GRAPH 3.1

Accuracy comparison for identification of gender (Djemili et al., 2012, Chaudhari & Kagalkar, 2015a, 2015b; Ramdinmawii & Mittal, 2016).

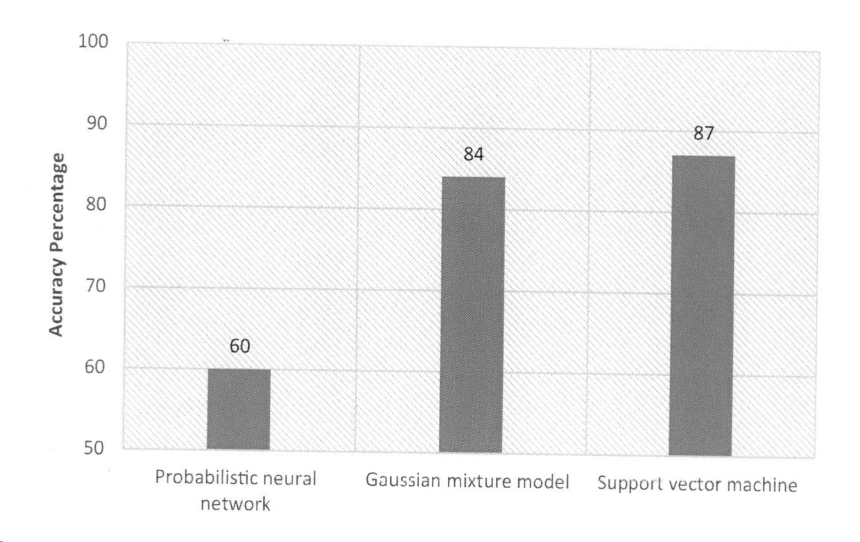

GRAPH 3.2

Accuracy comparison for the recognition of emotions with linguistic data consortium database (Cen et al., 2008; Tato et al., 2002; Yacoub et al., 2003).

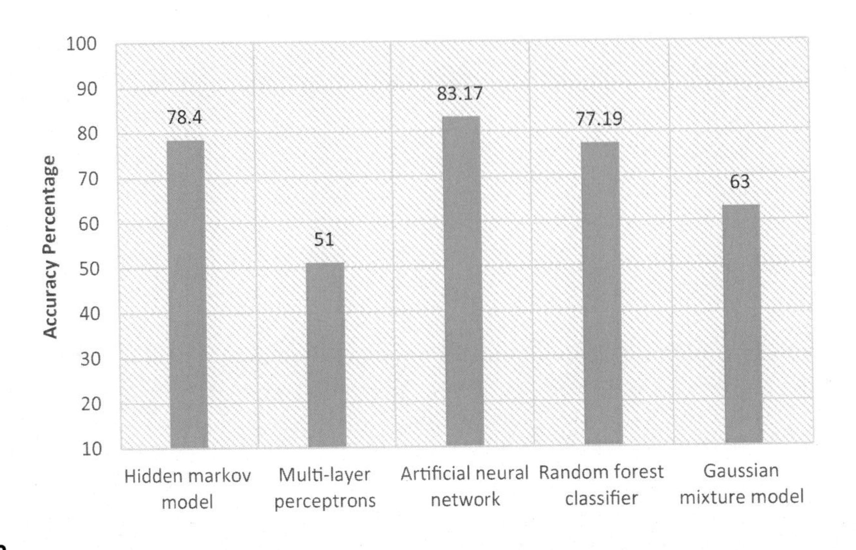

GRAPH 3.3

Accuracy comparison for the recognition of emotions with the Berlin emotional database (Iliou & Anagnostopoulos, 2009; Kumar et al., 2019, Revathi & Jeyalakshmi, 2019).

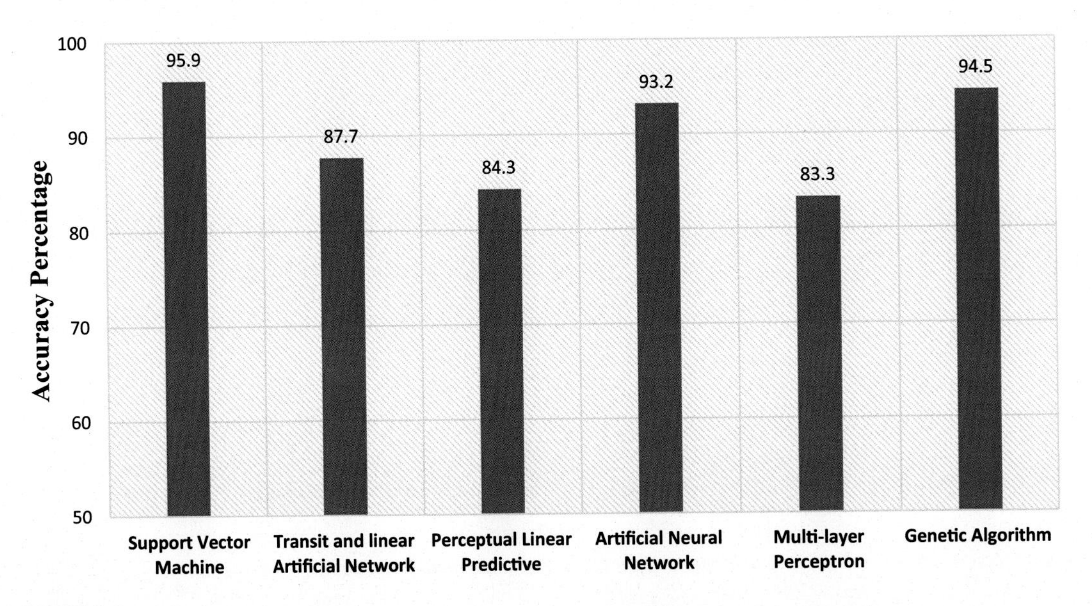

GRAPH 3.4

Accuracy comparison for the diagnosis of Parkinson's disease (Benmalek et al., 2017a, 2017b; Gharehchopogh & Mohammadi, 2013; Khemphila & Boonjing, 2012; Saloni et al., 2015; Shahbakhi et al., 2014).

analysis of signals, the proper treatment is provided to the patient. For the classification of a patient, a Gaussian mixture model is used. VPD is achieved more than 93 percent accuracy.

3.11 Conclusion

The voice signal analysis for the identification of gender, emotions, and diseases is discussed. Various parameters of the voice signals are explored for the purpose with the help of classifiers. The accuracy of gender identity can be increased by using a large number of data of the longest utterance in the training phase. The system efficiency for recognition of age and gender also depends on the extracted features of the voice signals. It is found that MFCC shows the highest accuracy in comparison to other features of the voice signals. The female speakers have a high value of features such as contour F0, pitch, energy, etc., so the accuracy of the female speakers is greater than male speakers. The accuracy of gender identification is varied according to the type of classifier.

The recognition of the emotions of the speakers is a very challenging task because the emotions have highly variable characteristics of the speaker according to the situation. The features of the voice signals also vary according to the emotions. The recognition structure of the emotion is also useful for the treatment of patients. The efficiency of the recognition system also depends on the features and classifiers. The accuracy for recognition of similar kinds of emotions is less as compared to the different emotions.

The vibrations in the voice signal are produced by the excellent and synchronized movement of the vocal folds that provide permission for the elasticity of the voice. Several body systems serve as a resonator such as a chest, face, and sinuses. The analysis of the voice signals can also be helpful to diagnose several diseases. The accuracy of the detection of Asthma depends on the selection of the features/ analysis. For the diagnoses of Blood pressure, the accuracy of the pathological result is greater than the normal result.

The characteristic of the speaker, which is affected in the early stage of Parkinson's disease, is the voice. SVM provides good efficiency for diagnosing Parkinson's disease in comparison to other classifiers. The accuracy of the recognition of Parkinson's disease is also varied according to the stage of the disease that is early stage has the highest accuracy as compared to other stages and can be diagnosed through voice signal analysis.

Many classifiers have been discussed in this chapter. It is found that Gaussian Mixer Model shows the maximum accuracy for the identification of the gender. The highest accuracy for the recognition of the emotions can be achieved with the help of using either VQ or Gaussian mixer model or MLP. SVM shows the highest accuracy for the diagnosis of the diseases.

AI systems will also provide help for diagnosing serious diseases in the early stage such as Parkinson's, Blood Pressure, etc. The diagnosis and treatment of the diseases are less risky with the help of AI/IoT systems. Pathological diagnosis is time-consuming as well as expensive, but these drawbacks may be reduced with the help of AI/IoT.

Due to AI/IoT, it is necessary to take security as a serious issue for data management organizations. Data management companies are trying to apply the securities in their software to eliminate the risk. For the security issues, several measures are considered by the data handling organizations such as updated policies, procedures, and proper guidance regarding the security and privacy of the health data. The data management organization should follow the verification and identification process. By using this process, the organization provides the personal identification number and the password for the authorized user. It is proposed that such organizations should identify the weakness of their architecture to recover the risk and also set privacy filters. To reduce the risk factor, guidelines will be followed for the disposal of unused data. By using cryptography, the security of the data can be increased.

Nowadays, the smartphone can interface with the IoT in the healthcare sector. But healthcare applications for several diseases, which can be identified with the analysis of the voice signals, are limited. IoT provides the capability to consult with doctors anywhere and anytime. In the available application on the smartphone, unauthorized access is the main and important issue.

Several areas require attention to enrich the healthcare sector. Several developmental and research studies are required to develop a method to diagnose rare and new diseases at an early stage with the help of voice signals. For the implementation of the IoT, technical securities and policy measures should be developed for the privacy of the voice signals. By using IoT, a new approach will be developed for regular monitoring, diagnosis, and treatment in remote areas where the availability of doctors is limited.

References

Apte, S. D., Dr. (2012). *Speech and audio processing. Wiley India Edition.*

Atassi, H. & Esposito, A. (2008). A speaker independent approach to the classification of emotional vocal expressions. In: *20th IEEE international conference on tools with artificial intelligence* (pp. 147–152). Dayton, OH, USA.

Bahari, M. H. & Hamme, H. Van (2011). Speaker age estimation and gender detection based on supervised non-negative matrix factorization. In: *IEEE workshop on biometric measurements and systems for security and medical applications* (pp. 1–6). Milan, Italy.

Batliner, A., Fischer, K., Huber, R., Spilker, J., & Noth,E. (2000). Desperately seeking emotions: Actors, wizards, and human beings. In: *The ISCA Workshop on speech and Emotion*, (pp. 195–200). Northern Ireland: the Queen's university of Belfast.

Benmalek, E., Elmhamdi, J., & Jilbab, A. (2017a). Multiclass classification of Parkinson's disease using cepstral analysis. *International Journal of Speech Technology*, *21*(1), 39–49.

Benmalek, E., Elmhamdi, J., & Jilbab, A. (2017b). Multiclass classification of Parkinson's disease using different classifiers and LLBFS feature selection algorithm. *International Journal of Speech Technology*, *20*(1), 179–184.

Bhattacharyya, N. (2014). The prevalence of voice problems among adults in theunited states. *The Laryngoscope*, *124*(10), 2359–2362.

Bone, D., Lee, C.-C., Chaspari, T., Gibson, J., & Narayanan, S. (2017). Signalprocessing and machine learning for mental health research and clinicalapplications [Perspectives]. *IEEE Signal Processing Magazine*, *34* (5), 195–196.

Cen, L. Ser, W. & Yu, Z. L. (2008). Speech emotion recognition using canonical correlation analysis and probabilistic neural network. In: *Seventh international conference on machine learning and applications* (pp. 859–862). San Diego, CA, USA,.

Chaudhari, S. J., & Kagalkar, R. M. (2015a). Automatic speaker age estimation and gender dependent emotion recognition. *International Journal of Computer Applications*, *117*(17), 05–10.

Chaudhari, S. J., & Kagalkar, R. M. (2015b). Methodology for efficient gender dependent speaker age and emotion identification system. *International Journal of Advanced Research in Computer and Communication Engineering*, *4*(7), 58–63.

Deshpande, N. Thakur, Dr. K. & Zadgaonkar, A. S. (2012). Assessment of systolic and diastolic cycle duration from speech analysis in the state of anger and fear. In: *ITCS, SIP, workshop on software engineering & application, CS & IT 04*, (pp. 137–141).

Djemili, R. Bourouba, H. & Korba, M. C. A. (2012). A speech signal based gender identification system using four classifiers. In: *International conference on multimedia computing and systems* (pp. 184–187). Tangiers, Morocco.

Dobry, G., Hecht, R. M., Avigal, M., & Zigel, Y. (2011). Supervector dimension reduction for efficient speaker age estimation based on the acoustic speech signal. IEEE Transactions on. *Audio, Speech, and Language Processing*, *19*(7), 1975–1985.

Gharehchopogh, F. S., & Mohammadi, P. (2013). A case study of parkinson disease using artificial neural network. *International Journal of Computer Application*, *73*(19), 1–6.

Giannakopoulos, T. Pikrakis, A. & Theodoridis, S. (2009). A dimensional approach to emotion recognition of speech from movies. In: *IEEE international conference on acoustics, speech and signal processing* (pp. 65–68). Taipei, Taiwan.

Godino-Llorente, J. I. Gómez-Vilda, P. Sáenz-Lechón, N. Blanco-Velasco, M. Cruz-Roldán, F. & Ferrer-Ballester, M. A. (2005). Discriminative methods for the detection of voice disorders. In: *ITRW on nonlinear speech processing, ISCA tutorial and research workshop* (pp. 158–167). Barcelona, Spain.

Gupta, A., Chinmay, C., & Gupta, B. (2019). Monitoring of Epileptical Patients Using Cloud-Enabled Health-IoT System. *Traitement du Signal, IIETA*, *36*(5), 425–431.

Hariharan, M., Paulraj, M. P., & Yaacob, S. (2010). Time-domain features andprobabilistic neural network for the detection of vocal fold pathology. *Malaysian Journal of Computer Science*, *23*(1), 60–67.

Iliou, T. & Anagnostopoulos, C. (2009). Comparison of different classifiers for emotion recognition. In: *13th Panhellenic conference on informatics* (pp. 102–106). Corfu, Greece.

Islam, S. M. R., Kwak, D., Kabir, M. H., Hossain, M., & Kwak, K.-S. (2015). The Internet of Things for health care: A comprehensivesurvey. *IEEE Access*, *3*, 678–708.

Khan, A., & Bhaiya, L. (2012). Text dependent method for person identification through voice segment. *International Journal of Electronics and Computer Science Engineering*, ISSN- 2277-1956.

Khemphila, A., & Boonjing, V. (2012). Parkinson disease classification using neural network and feature selection. *World Academy of Science, Engineering and Technology: International Journal of Mathematical and Computational Sciences*, *6*(4), 15–18.

Koolagudi, S. G., Maity, S., Kumar, V. A., Chakrabarti, S., & Rao, K. S. (2009). IITKGP-SESC: Speech database for emotion analysis. In S. Ranka, et al. (Eds.), Contemporary computing. IC3 2009, Communications in *computer* and information science (40, pp. 485–492). *Berlin, Heidelberg: Springer*.

Koolagudi, S. G., Murthy, Y. V. S., & Bhaskar, S. P. (2018). Choice of a classifier, based on properties of a dataset: Case study-speech emotion recognition. *International Journal of Speech Technology*, *21*(1), 167–183.

Koolagudi, S. G., & Rao, K. S. (2012). Emotion recognition from speech: A review. *International Journal of Speech Technology*, *15*(2), 99–117.

Krishnan, M., Chinmay, C., Banerjee, S., Chakraborty, C., & Ray, A. K. (2009). Statistical analysis of mammographic features and its classification using support vector machine. *Expert Systems with Applications*, *37*, 470–478.

Kulkarni, A., & Sathe, S. (2014). Healthcare applications of the internetof things: A review. *International Journal Comput. Sci. Inf. Technol.*, *5*, 6229–6232.

Kumar, A., Gudmalwar, P., Rao, Ch. V. R., & Dutta, A. (2019). Improving the performance of the speaker emotion recognition based on low dimension prosody features vector. *International Journal of Speech Technology*, *22*(3), 521–531.

Kumar, C. S., & Mallikarjuna, P. R. (2011). Design of an automatic speaker recognition system using MFCC, vector quantization and LBG algorithm. *International Journal on Computer Science and Engineering, 3* (8), 2942–2954.

Li, X. Yao, M. & Huang, W. (2011). Speech recognition based on k-means clustering and neural network ensembles. In: *2011 Seventh international conference on natural computation* (pp. 614–617). Shanghai, China.

Malode, A. A., & Sahare, S. (2012). Advanced speaker recognition. *International Journal of Advances in Engineering and Technology*, *4*(1), 443–455.

Mesleh, A., Skopin, D., Baglikov, S., & Quteishat, A. (2012). Heart rate feature extraction from vowel speech signal. *Journal of Computer Science and Technology*, *27*(6), 1243–1251.

Muda, L., Begam, M., & Elamvazuthi, I. (2010). Voice recognition algorithms using mel frequency cepstral coefficient (MFCC) & dynamic time warping (DTW) techniques. *Journal of Computing*, *2*(3), 138–143.

Muhammad, G., Rahman, S. M. M., Alelaiwi, A., & Alamri, A. (2017). Smarthealth solution integrating IoT and cloud: A case study of voice pathologymonitoring. *IEEE Commun. Mag.*, *55*(1), 69–73.

Muhammad, L. J., Ebrahem, A. A., Sani, S. U., Abdulkadir, A., Chinmay, C., & Mohammed, I. A. (2021). Supervised machine learning models for prediction of COVID-19 infection using epidemiology dataset. *SN Computer Science*, *2*(11), 1–13.

Passricha, V., & Aggarwal, R. K. (2019). Convolutional support vector machines for speech recognition. *International Journal of Speech Technology*, *22*(3), 601–609.

Ramdinmawii, E. & Mittal, V. K. (2016). Gender identification from speech signal by examining the speech production characteristics. In: *International conference on signal processing and communication (ICSC)* (pp. 244–249). Noida.

Reddy, S. A. Singh, A. Kumar, N. S. & Sruthi, K. S. (2011). The decisive emotion identifier? In: *Third International conference on electronics computer technology* (pp. 28–32). Kanyakumari, India.

Rabiner, L. R., & Schafer, R. W. (2007). Digital processing of speech signals. In: Prentice *hall signal processing series. Pearson.*

Revathi, A., & Jeyalakshmi, C. (2019). Emotions recognition: Different sets of features and models. *International Journal of Speech Technology*, *22*(3), 473–482.

Salhi, L., Mourad, T., & Cherif, A. (2010). Voice disorders identicationusingmultilayer neural network. *International Arab Journal of Information Technology*, *7*(2), 8.

Saloni., Sharma, R. K., & Gupta, A. K. (2014). Disease detection using voice analysis: A review. *International Journal of Medical Engineering and Informatics*, *6*(3), 189–210.

Saloni., Sharma, R. K., & Gupta, A. K. (2015). Voice analysis for telediagnosis of Parkinson disease using artificial neural networks and support vector machines. *International Journal of Intelligent Systems and Applications*, *7*(6), 41–47.

Saloni., Sharma, R. K., & Gupta, A. K. (2013). Classification of high blood pressure persons vs normal blood pressure persons using voice analysis. *International Journal of Image, Graphics and Signal Processing*, *6* (1), 47–52.

Shahbakhi, M., Far, D. T., & Tahami, E. (2014). Speech analysis for diagnosis of Parkinson's disease using genetic algorithm and support vector machine. *Journal of Biomedical Science and Engineering*, *7*(4), 147–156.

Shirvan, R. A. & Tahami, E. (2011). Voice analysis for detecting Parkinson's disease using genetic algorithm and KNN classification method. In: *18th Iranian conference of biomedical engineering* (pp. 278–283). Tehran, Iran.

Solera-Urena, R., Garcia-Moral, A. I., Pelaez-Moreno, C., Martinez-Ramon, M., & Diaz-de-Maria, F. (2012). Real-time robust automatic speech recognition using compact support vector machines. *IEEE Transactions on. Audio, Speech, and Language Processing*, *20*(4), 1347–1361.

Sonkamble, B. A., & Doye, D. D. (2012). Speech recognition using vector quantization through modified K-means LBG algorithm. *Computer Engineering and Intelligent Systems*, *3*(7), 137–145.

Sonu, & Sharma, R.K. (2011). Disease detection using analysis of voice parameters. In: *5th IEEE international conference on advanced computing & communication technologies* (pp. 416–420).

Tato, R. Santos, R. Kompe, R. & Pardo, J.M. (2002). Emotional space improves emotion recognition. In: *7th International conference on spoken language processing*, (pp. 2029–2032). Colorado: INTERSPEECH 2002.

Tradigo, G., Calabrese, B., Macrí, M., Vocaturo, E., Lombardo, N., & Veltri, P. (2015). Voice signal features analysis and classification: Looking for new diseases related parameters. *Conference on Bioinformatics, Computational Biology and Biomedicine*, 589–596.

TurcuC, E., & Turcu, C. O. (2013). Internet of things as key enabler forsustainable healthcare delivery. *Procedia Social Behaviour Science*, *73*, 251–256.

Vasanth, K., & Sbert, J. (2016). Creating solutions for health through technologyinnovation. *Texas Instrum*, *1*, 1–5.

Ventola, C. L. (2014). Mobile devices and apps for health care professionals:Uses and benefits. *Pharmacy Therapeutics*, *39*(5), 356–364.

Viswanathan, M. Zhan, Z.-X. & Lim, J. S. (2012). Emotional-speech recognition using the neuro-fuzzy network. In: *6th International conference on ubiquitous management and communication* (pp. 1–5). Kuala Lumpur, Malaysia.

Wenjing, H., Haifeng, L., & Chunyu, G. (2009). A hybrid speech emotion perception method of VQ-based feature processing and ANN recognition. *WRI Global Congress on Intelligent Systems, Xiamen, China*, 145–149.

Yacoub, S. Simske, S. Lin, X. & Burns, J. (2003). Recognition of emotions in interactive voice response systems. In: *8th European conference on speech communication and technology* (pp. 729–732).

CHAPTER

Intelligent and sustainable approaches for medical big data management

4

Anubha Dubey[1] and Apurva Saxena Verma[2]

[1]Independent Researcher and Analyst, Noida, Uttar Pradesh, India [2]Researcher Computer Science, Bhopal, Madhya Pradesh, India

4.1 Introduction

Big data, it's a trend in every company whether it is the healthcare or the IT industry. A huge amount of data is generated everywhere which needs to be managed properly. These managed data are important for interpretation and analysis in the future. These will enhance the new computational techniques to work on big "V"s that is volume, velocity, veracity, variety, and value. In healthcare too, lots of data are generated which is stored in electronic form. The evolution of data on daily basis requires proper handling and management. As big data to knowledge is the latest demand (Margolis et al., 2014). The term big data is introduced in 1997 (Cox & Ellsworth, 1997) by John R. Mashey. And is a big challenge in biomedical research. All the big databases like international cancer genome consortium, some disease databases like international rare disease consortium, and the international human epigenome consortium, are the ones that are part of data resources. These analytics require capabilities for the representation and modeling of the health data, optimization of different algorithms, and computational power. In the healthcare industry, five distinctive capabilities of data are required: identification of patterns in data that provides care, analysis of unstructured data, providing decision support, better prediction, and traceability (Wang, Kung, & Byrd, 2018). Data generation and collection are faster than data preprocessing and analysis. To cover this gap there is a need for technological progress in various kinds of data acquisition. All this information generated and fed to computers is of utmost importance whether it is molecular information, phenotypic information of an individual patient, or others. These biomedical data need protection, storage capacity, etc. Hence cloud computing techniques come into existence. The cloud computing (CC) standard has engrossed a lot of attraction from both industry and scholars. It proposes diverse services including asset pooling, multi-tenancy, and flexibility (Chkirbene et al., 2020). While the cloud computing standard raises economic efficiency, security is one of the important concerns in adopting thc cloud computing model (Chkirbene, Erbad, & Hamila, 2019).

Cloud computing is a new operating technology that advanced the information technology (IT) industry, it is an expansion of equivalent computing, distributed computing, and grid computing over the same or different networks, (Sniezynski, Nawrocki, & Wilk, 2019). Cloud computing technology is the combination and development of virtualization, utility computing, and all the

Implementation of Smart Healthcare Systems using AI, IoT, and Blockchain. DOI: https://doi.org/10.1016/B978-0-323-91916-6.00010-2

measures. Cloud computing provides Infrastructure-as-a-Service (IaaS), Platform-as-a-Service (PaaS), and Software-as-a-Service (SaaS) as required in different industries (Chkirbene et al., 2020; Hojabri & Rao, 2013). In IaaS clients can get the infrastructure and virtual machines, and networks at the rental cost. PaaS is designed to make a platform for the consumer to expand a web application setup behind it. SaaS is used to develop software for the client. IBM cloud works on the entire services of the cloud. Many cloud providers host and manage the application, and essential infrastructure and handle maintenance too. However, machine learning techniques used for security in different technologies have been established to produce real-time automatic inconsistency detection. But the detection and classification accuracy in the machine learning approach can be achieved only through a large amount of data collected (García, Lucas, Antonai, & David, 2020).

4.1.1 Artificial intelligence and Internet of things/IoTM in healthcare

Artificial intelligence (AI) is one of the intelligent and sustainable technologies for improving medical data management. Machine learning is a part of artificial intelligence that implies the statistical methods to improve the algorithms that are involved in the diagnosis and treatment of diseases. Machine learning makes the system learn from the previous interpretations and artificial intelligence makes it intelligent in the decision-making process. For making AI successful, this technology consists of cloud computing, big data of health records of different types, mobile technologies, sensors, different biotechnology and bioinformatics infrastructure, etc. Not only technology but also a grouping of people like patients, doctors, hospital staff, and different research laboratories are also part of this smart healthcare technology. AI plays a very important role in disease diagnosis where some of the tests seem too inaccurate (Chinmay, 2017). These are only helping medical care providers to identify diseases in real-time. Improving the health issues by previously identifying patterns and symptoms. Cloud computing not only secures data but also manages and stores electronic health data. AI is very important in diagnosis through radiology. These can be virtual assistants for improving difficult health situations in faraway places (Jiang, Jiang, & Zhi, 2017). Digital technology has revolutionized the world of the Internet and so healthcare system. These wearable technologies, sensors, big data, information, and the introduction of the Internet of things (IoT) (Manogaran, Varatharajan, & Lopeza, 2018) or Internet of electronics. These new technologies have changed the path of traditional healthcare way. Now these modern technologies are a part of day-to-day life. Automation has touched the lives of a person in one or another way. These digital technologies are automated has revolutionized the way of scientific thinking and reduced the cost of health needs. The person will achieve new medical services for patients. That means consistent tracking of disease and early detection of any diseases (Dubey, Pant, & Adlakha, 2010; Mandler et al., 2016) and many more.

Hence AI with IoT (Internet of things) or IoE (Internet of electronics) including cloud computing has changed the world. Now, this is IoTM, which brings all the communication technology to one place in healthcare. In this new era of technology mobile health become the key to personal healthcare services. This new perception of m-health in the environment of the Internet of Things (m-IoT) fetches the guarantee of giving a better health management system to patients. IoT-based m-health services will promote the people by saving their time, especially those who live in a rural part of the country, and money from visiting hospitals countless times.

The important parts of m-health and IoT are equipped these medical devices with sensors, communication devices like video calls, Bluetooth technology, and many more. These medical devices work as smart devices that can be used to monitor different parameters such as sugar level, blood pressure, heart rate, sleep monitor, calorie count, distance covered through steps, and physical activity trackers like running, cycling, and swimming, etc.

Therefore in this chapter Kerberos authentication protocol is implemented for data security in the cloud with the auto-machine learning technique (Hojabri & Rao, 2013) for the diagnosis of diseases in time and big data management. When the Kerberos method is implemented on the network, increases the legitimacy and reliability to protect the data of the user by a token allowed through the ticket-granting token (TGT) system. It has time-bound and encrypts it by using the secret key in the ticket-granting service. Data analysis under security combines data from different outlets and looks for any relations and anomalies. Security enhancers are viable tools to make the data as safe as possible. This will make the data threat free and any kind of data traffic is monitored and it is better explained in further Sections (4.2.3).

4.1.2 Contribution

This chapter contributes to developing big data monitoring system:

1. All the recordings of data are done on time due to the implementation of automated machine learning (autoML).
2. The big data architecture with databases is maintained and computing is provided simultaneously.
3. This system provides accurate predictions and an online patient dashboard on demand.
4. A secure cloud environment is developed for protecting highly risky disease data.
5. A sustainable model is described for big data management of different sources and diagnosis of diseases.

The chapter is organized as follows: In Section 4.1.1 provides a brief account of the current research work that is performed in the field of machine learning in healthcare and cloud computing. Section 4.1.2 motivation behind the work. Section 4.2 elucidates the Method for security model development with data, data management, security methods, IoT, and auto-ML. Section 4.4 illustrates SMHC model development with a case study discussed in 4.3, followed by results and discussion in Section 4.5. Section 4.8 describes the conclusion of the chapter with the pros and cons of the model developed in Section 4.6, applications in 4.7, and future scope in 4.9.

4.1.3 Related work

Chkirbene et al. (2019) uses the words "node" and "user." Decisions are shared of the machine learning methods to approximate the final attack. These approaches increase the performance in learning and maintain the system's accuracy. Here results showed that EICD recovers the anomalies detection by 24% as compared to the complex tree. Rovnyagin, Timofeev, Elenkin, and Shipugin (2019) explore a method of deployment architecture for machine learning in the clouds. Zhu and Fan (2019) tried evaluating the Google Cluster dataset and showed that the accuracy has been enhanced compared to the current traditional methods. Moreover, to obtain a better understanding an unsupervised hierarchical clustering algorithm, BIRCH, is used. (García et al., 2020) proposed a distributed architecture to provide machine learning practices with a set of tools and cloud services that wrap the whole machine learning development cycle: ranging from the models' creation, training, validation, and testing to the models serving as a service, sharing and publication.

Chkirbene et al. (2020) presented the DEEP-Hybrid-Data Cloud framework, which showed a transparent approach to existing e-Infrastructures, productively utilizing the distributed resources for the most compute-intensive tasks coming from the machine learning implementation and tried to enhance the accuracy of the developed method. The machine learning algorithm is used to generate a classifier that distinguishes between the attacks. Li, Gibson, Ho, Zhou, and Kim (2013) explained that their research evaluates various machine learning algorithms on the widely adopted cloud framework and Graph Lab framework is approached. His work focused on a problem-based approach to architecture selection.

The concept of auto-machine learning is explained with real world applications. Bio-medical scientists and researchers are developing tools for real-time alerts while monitoring patient's data. Naoual and Benhamila (2018) present a detailed description of big data in healthcare management with Spark and MongoDB tools for real-time alerts while monitoring patients' data. Cirilio and Valancia (2019) explained the use of big data in biomedical research. Automated collection of big data, analysis, and interpretation requires computational approaches for personalized medicine.

4.1.4 Motivation

Health data is extremely very important to store, analyze, and secure is prime importance. As a voluminous amount of health data is generated daily, it requires a proper analysis that would be beneficial for both the patients and doctors. Hence it is tried to develop such a systematic approach to secure all the health records for further analysis or reuse. To achieve the target, Surveillance Machine Learning Health Care (SMHC) Model is proposed to develop. That secures data from breaching and is highly automated to perform all the tasks on its own to save time in computer processing.

4.2 Method

4.2.1 Data

Here HIV dataset is selected, as it is sensitive and requires better care, we have selected it as a study. Many Govt. health plans are programmed and need access and response to every new data. So HIV data needs the care to provide better assessment and address the gaps in utilization and retention in care, retention in care, through data to care activities. These data can be used to study any viral outbreaks. Now data use and data sharing require confidentiality. As it is monitored by the hospital authority how much data is shared and for what purpose. Hence patient data privacy is of utmost importance and needs to be maintained properly. Treatment and research are very important to understand the disease outcomes and health benefits. Data security is important to stop data breaching.

Everyday collection of patient information is raising, the volume and complexity of the data. The data is ranging from petabytes to terabytes. Genomic data alone are exceeding 2.3 trillion gigabytes. The data generated are structured and unstructured (description of symptoms) data, this is planned to provide all the details to the medical scientists or clinicians, a relevant scheme for knowledge and model evolution to development (Genes, 2018). Massive data of genomes if genomic sequencing in case of deadly diseases like cancer, or other experiments like transcriptomics or proteome profiling, liquid biopsy, etc which makes biomedical facts and figures accessible at a fast pace. All the imaging, multi-omics data, health records, and data generated from wearable devices or implantable devices are also increasing day by day. Specific to customized medicine (Collins, 2015), the health and therapy record of a particular patient is also part of electronic health data which needs to be monitored and tracked on time (Collins, 2015). Real-time sensor data like glucose monitoring are also part of stream computing etc.

4.2.2 Data management

The improved data management in healthcare reduces operational costs. The healthcare industry is strictly regulated as data protection is a must. Cloud systems and new technologies are nowadays very cost-effective. The cloud system offers the health care data to offload the majority of their data into a centralized space that is more secure and less priced. The automatic cloud platform allows the cleaning of irrelevant data to make the space free. The second advantage of data management is it reduces patient costs. Systematic data collection and data retrieval help healthcare industries or hospitals with the preventive initiatives taken. If data is provided in databases are in the correct format then areas of improvement can be done on time and maintained properly. Better data needs to be stored for further analysis. Individual patients can be accessed more effectively and accurately. These better data provide good decisions for doctors to provide good treatment on time. A relational database is used. As machine learning is employed, Storm (Evans, 2015) from yahoo if social media data is used. Spark (Zaharia, Xin, & Wendell, 2016) provides a powerful framework to access many data sources. It consists of the Mllib library that has the amalgamation of all the machine learning methods that come up with well-organized algorithms with good speed. The structured data has shown analysis having graph processing unit installed on GraphX and SparkSQL, restores data from different resources. Apache Flink is an open platform for analyzing data in both modes that is batch mode and real-time mode (Ellen & Kostas, 2016). It provides a machine learning library that uses ML approaches for speedy and extensible health data record utilization. Similarly MongoDB (Hows, Membrey, Plugge, & Hawkins, 2015) NoSQL database capable of storing huge amounts of data. It can open, human and machine-readable formats.

4.2.3 Data security

Key features of Kerberos that are considered an important security enhancer are discussed as follows:

- Every client and user has a password.
- Symmetric keys are used both to encrypt and decrypt information or data. A secret key is the same on both sides to encrypt and decrypt it.
- The secret key and information of the key are kept only in the Key Distribution Center (KDC).
- The file server is physically secure (no unauthorized user has access to them).

- The user gives a secret key only once.
- The password is not sent over the network in plain text or encrypted form.
- The user or client requires a ticket-granting ticket for each access.

4.2.3.1 Security by Kerberos

It is a computer network justification protocol to all main operating systems, such as Microsoft Windows, Apple OS X, FreeBSD, and Linux that works on tickets to permit users equivalent over an insecure network to verify their uniqueness to one more in a secure manner. Kerberos protocol is an authentication method in Windows. This protocol name Kerberos was named after the Greek mythology "a three-headed dog" that guards the gates of Hades. These three heads symbolize a client, a server, and a Key Distribution Center (KDC). KDC works as a third-party validation service. By default, Kerberos uses UDP port 88 (Hojabri & Rao, 2013). Kerberos model builds on symmetric key cryptography and needed a trusted third party (KDC) that uses public-key cryptography. The three parts of the Kerberos are as follows:

(i) Key Distribution Center (KDC): It holds all the key information about clients and the secret key for the exact service to authenticate the user. It is said to be a Domain Controller (DC) which is used to generate TGT and SGT. TGT is a ticket-granting ticket that is generated for the user. It is dependable to issue a ticket for the user to get the service from the file server.

(ii) Client: It's a client who is there in the network, to allocate the data. In the Kerberos, only the client must be validated by generating the TGT.

(iii) File Server: Whomsoever the user in the network wants to share the data by taking permission from the server. Functioning of the Kerberos model (shown in Fig. 13.1) in the network, as NTP (Network Time Protocol) is used to coordinate the time within the whole network. (a) Key Distribution Center (KDC) which gets the request generated from the client then (b) the client gets the TGT from the KDC. When any of the authorized clients/users want to share the data in the given network. (c) It sends request + TGT to the KDC. In reply (d) KDC will send TGT with a timestamp to the client. (e) The client sends the TGT with a timestamp to the file server for the resources. (f) NFS (Network file server) verifies this client with the KDC. (g)An authentication gets successful. (h) NFS then allows the resource which is to be shared by the client (Fig. 4.1).

4.2.3.2 Cloud environment

Here in any of the cloud environments using Kerberos and SELinux are installed in all the servers and clients of the given network. Kerberos offers security but using a cloud supplier, strengthens the authentication through confidentiality, reliability, availability, and maintenance at the least expenditure (Hojabri & Rao, 2013; Sutradhar, Sultana, Dey, & Arif, 2018). Any hacker or attacker cannot fully access the system but hacking can be possible. Kerberos provides a solution to a security issue. Kerberos is installed in all the instances of any cloud surrounding and servers with SELinux (Saxena & Dubey, 2019). The SELinux (Security-Enhanced Linux) is a component of the Linux security kernel that acts as a protecting representative on servers. In the Linux kernel, SELinux anticipates mandatory access controls (MAC) that limit users to regulations and strategies set by the system administrator. MAC is a top-level access control than the standard flexible direct

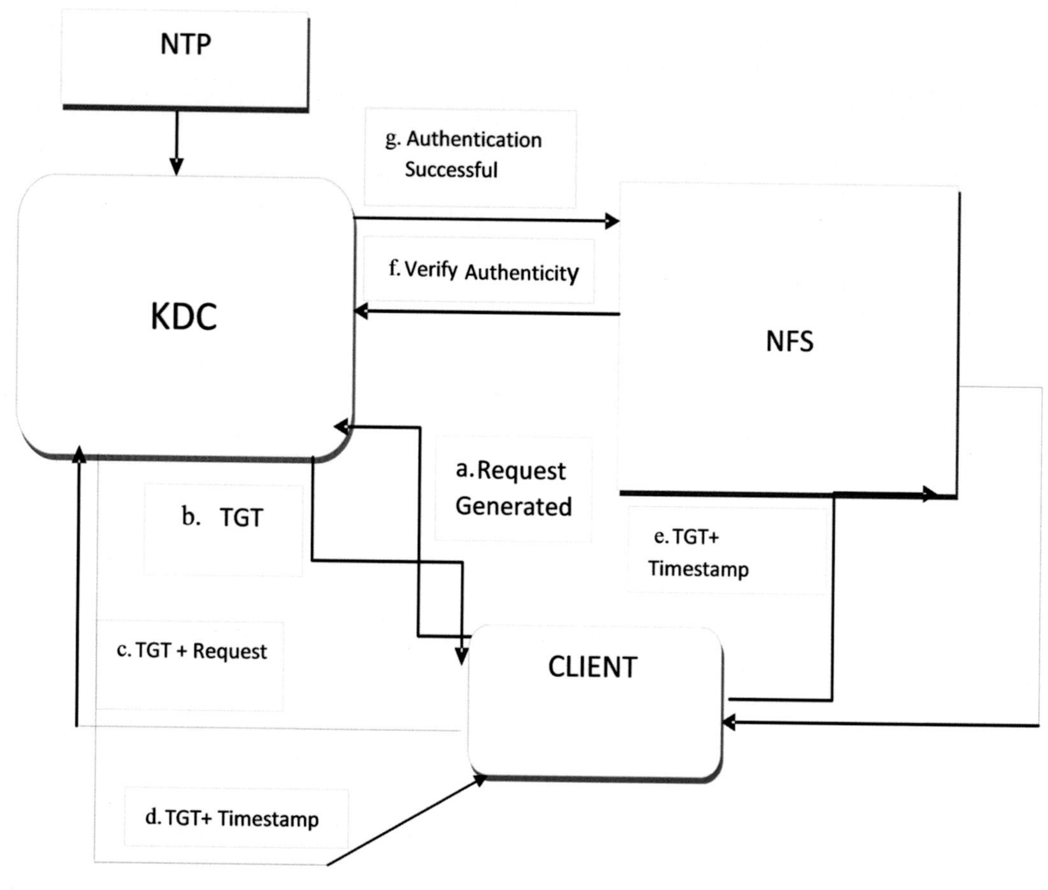

FIGURE 4.1

Kerberos authentication system applied in the cloud.

access control and averts security violates in the system mode by solely taking care of crucial files that the administrator pre-approves.

4.2.4 Data analytics

Steps of data analysis in Machine learning or autoML (Fig. 4.2).

The steps of data analytics are achieved by autoML, model selection, and evaluation are discussed below:

4.2.4.1 Automated machine learning

For automating data analysis autoML comes into existence. It is the next technology upgraded machine learning automation hence the name autoML. It is said that marriage between automation

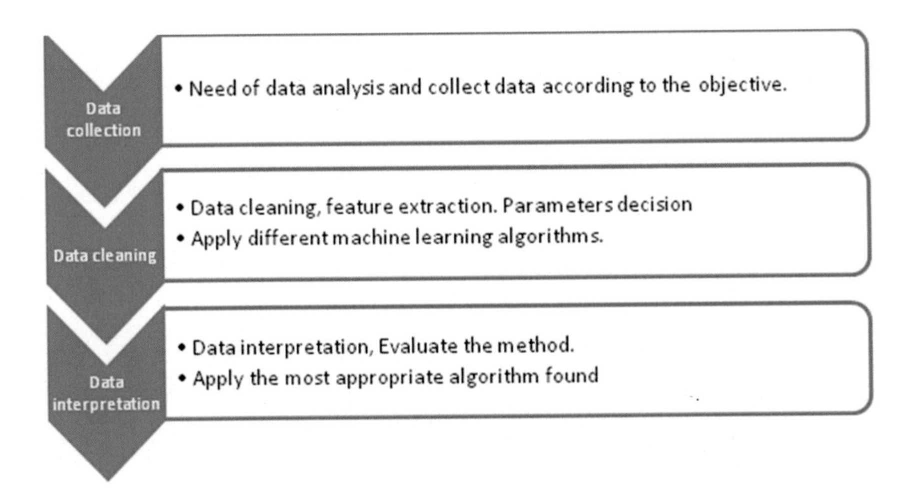

FIGURE 4.2

A simple way of data analysis.

and machine learning. Machine learning is a segment of AI artificial intelligence that allows machines or computers to impulsively incorporate previous data without being exactly programmed. The AI achieves the concept of a SMART computer system to solve complex problems. In other words, these methods can analyze highly complex data. Hence now these AI-enabled machine learning techniques are involved in healthcare.

As in hospitals lots of data are entered within seconds, these need to be analyzed from time to time. For this autoML is implemented in hospitals' data centers. Hence automation is everywhere so, we are trying to develop an autoML model in the cloud for data security and analysis. Here autoML emphasizes using machine learning tools and control is done by the cloud monitoring system. The idea of illustration is given in the figure below: Automation is to use control methods for operation underneath components of data (Phillips & Habor, 1995). Here machine learning tools are adapted to do various tasks as given by input data. The computer is programmed accordingly. AutoML wants good learning performance of mL tools and less human intervention. Because in classical machine learning humans have the job of labor and performance-knowledge intensive, but in autoML, all things are done by computer programs. The biological data prediction requires Shanon entropy prediction to quantify (Barreiro, Munteanu, & Cruzmonteagudo, 2018) and A new autoML approach is proposed for hierarchical planning to achieve competitive results by AutoWEKA (Kotthoff, Thornton, Hoos, Hutter, & Leyton-Brown, 2017), TPOT, Scikit learns (Mohr, Wever, & Hüllermeier, 2018). The difference between classical machine learning and autoML is illustrated as (Table 4.1).

4.2.4.2 Working of automated machine learning

Human intuition and intelligence are involved in data analysis. Automation is inspired by human intervention (Kuo & Golnaraghi, 1995). In this chapter, a framework for autoML is summarized as a figure. In this figure the controller works as a human in finding proper learning tools with the

Table 4.1 Difference between classical machine learning and automated machine learning.

Basis of differentiation	Classical machine learning	Automated machine learning
Speed	It is not too smart or cognitive	Smart and cognitive
Data	Data format needs time and specification, preprocessing is a little tedious	Feature engineering with a new character for fast data retrieval
Analysis	Statistical analysis	Inbuilt module in hardware. Deep learning with neural network combination with reinforcement learning.
Model selection	Simple machine learning algorithms are used like decision trees, random forest, SVM, supervised and unsupervised learning algorithms	Automation is implemented in machine learning algorithms.
Efficiency	Efficiency is achieved by statistical calculations that is accuracy is calculated for each algorithm. WEKA inbuilt software is widely used.	Hardware has the strong computing power to calculate complex biological data problems. that is AUTOWEKA, Autosk-learn etc.

help of the learning process. In the controller, the optimizer and evaluator work simultaneously. The optimizer performs the learning process and the evaluator estimates the performance of the learning tools. The evaluator trains the model based on input data which is cumbersome. Important points to be considered: (1) Feature extraction: as more features are included, the learning performance is enhanced. (2)The raw features are identified and then construct as the most required features. (3) These important features are the ones for model selection and algorithm design. DSM (Kanter & Veeramachaneni, 2015) and explore it (Katz, Shin, & Song, 2016) are available which removes human assistance and new features are automatically constructed. (4) Auto-sklearn is a collection of classifiers implemented in autoML, that chooses the proper classifier for the data, and hyperparameters are decided without the assistance of humans. Since autoML works on supervised learning having noise-free scenarios (Wolpert & Macready, 1997; Wolpert, 1996). All learning algorithms working as the same make empirical method into reality. Once the model is set up, machine learning is deployed as problem definition (need of data analysis), data collection, and then deployment by machine learning methods Fig. 13. These features make ML easy to apply and accessible for healthcare (Fig. 4.3).

4.2.4.3 Basic framework

As shown in Fig. 4.2, in autoML two techniques work: the first is a basic technique and the second is an experienced technique. In basic technique: the optimizer searches and optimize the configurations of all the possible methods for data analysis. It may be grid search or random search (Bergstra & Bengio, 2012) if data is complex reinforcement learning and automatic differentiation is also applied (Baydin, Pearlmutter, Radul, & Siskind, 2017). Now evaluator works on the performance measurement of the learning tools by performing the identification of correct parameters, Hyperparameter optimization can be done successfully (Feurer & Hutter, 2018). Second, the experience techniques work on learning and accumulating knowledge from the past searches of the particular data. Hence both these techniques work according to the need and objective of data analysis.

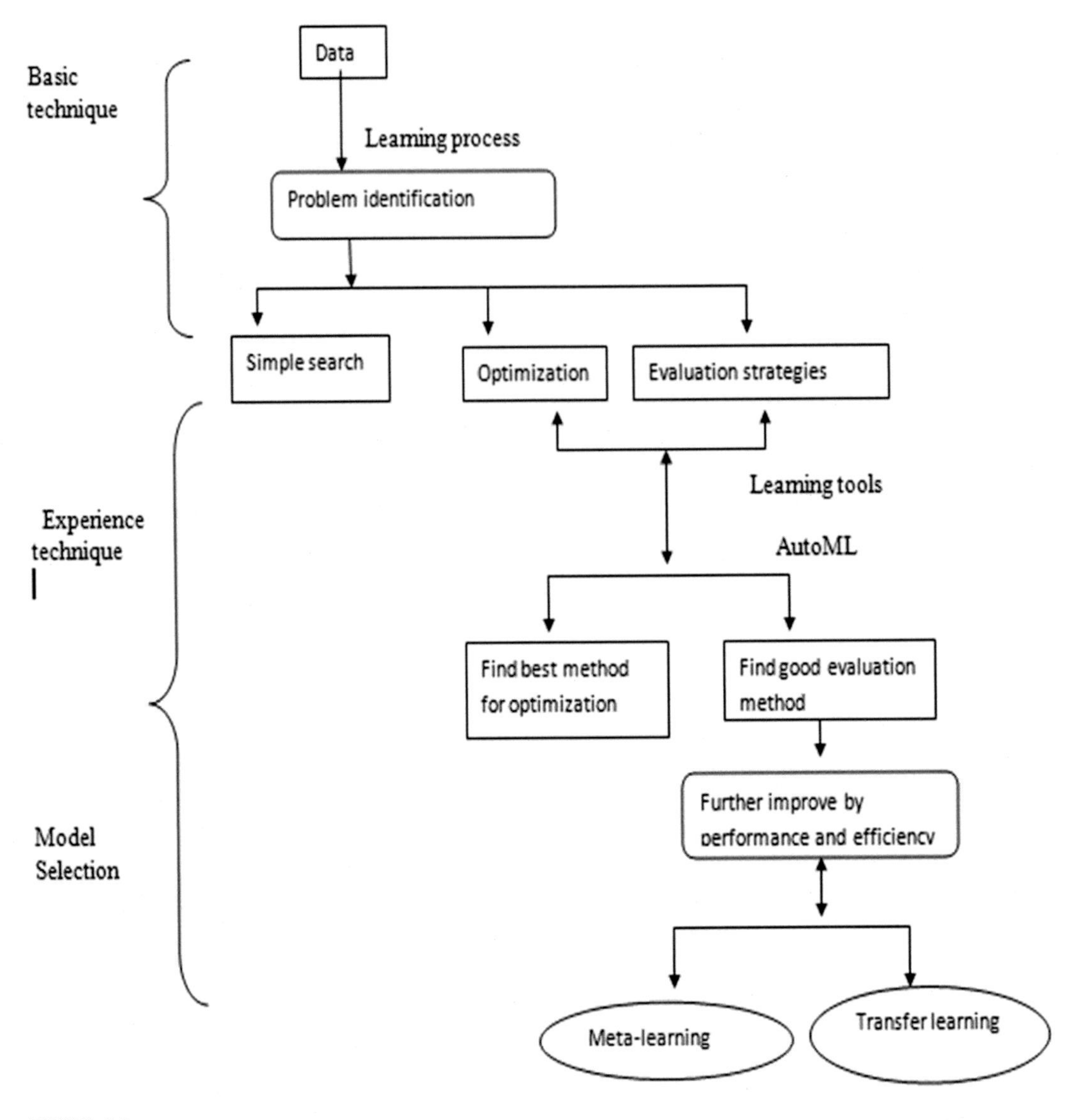

FIGURE 4.3

Schematic representation of working of automated machine learning model.

Basically in autoML, metalearning (Lemke, Budka, & Gabrys, 2015; Vilalta & Drissi, 2002) and transfer learning (Pan & Yang, 2009) work.

- Metalearning: In this, the learning algorithms learn from the output of other machine learning algorithms. This means the basic technique and experience techniques will work simultaneously (Vanschoren, 2019).

- Transfer learning: It is a fact-finding problem in machine learning that informs knowledge gained in one problem and we can apply it to a related problem or maybe a different problem if possible. For example, any of the disease diagnoses by the parameters identified during the study is used to identify other diseases too.

4.2.4.4 Model selection

Classification is the most used well-known technique in data mining for the analysis of healthcare data. As per the literature, there are many classification tools, like tree classifiers, linear classifiers, deep networks, neural networks, and support vector machines. Scikit learn is the choice as it has all the classification tools implemented in autoML. Previously human interpretation is done for obtaining a good classifier by trial and error method based on experience. Hence in model selection, a good classifier is searched and optimized with great learning performance and efficiency is also balanced. *Optimizing Algorithms*: For any learning tool, lots of algorithms are used and optimized at a particular level (Salvador, Budka, & Gabrys, 2018). In classification, a decision tree is mostly used. It is a predictive modeling approach with supervised learning. In healthcare data analysis, a decision tree is used. A decision tree is a step-by-step learning schema where it is identified whether the symptoms of any patient showing is a disease or some infection? for example, a patient can be categorized as high risk or low risk of any disease depending on the disease pattern identified. Many classification methods are in use like K-nearest neighbor, support vector machine, etc. Hence it is said that classification is a supervised learning process where class categories are known. Any supervised learning employed requires evaluation of the process. ***Evaluation of Decision tree generation***: In classification, decision trees are mostly used. Because it is a supervised algorithm as always gives better results. For successful decision tree formation entropy and information, the gain is used which is inbuilt in software. After the development of the decision tree, accuracy, precision, and recall are calculated to have a good classifier for classification. Hence if a decision tree is developed for any disease then accuracy is calculated that how much the decision is correct.

Entropy (class): It is also called Shanon Entropy and is denoted by H(s) for a finite set S. It shows the evaluation of the amount of unpredictability in data. Here P is the possibility of yes, and N shows the possibility of no in the dataset.

$$\mathrm{H(s)} = -\frac{P}{P+N}\log_2\left(\frac{P}{P+N}\right) - \left(\frac{N}{N+P}\right)\log_2\left(\frac{N}{N+P}\right) \tag{4.1}$$

Information Gain: Information gain is also known as Kullback–Leibler divergence indicated by IG (S, A) for set S. It's the successful change in entropy after a distinct attribute-A. It calculates the relative change in entropy with respect to the independent variables.

$$\mathrm{I}_{(\mathrm{Pi},\ \mathrm{Ni})} = -\frac{P}{P+N}\log_2\left(\frac{P}{P+N}\right) - \left(\frac{N}{N+P}\right)\log_2\left(\frac{N}{N+P}\right) \tag{4.2}$$

$$\textbf{Entropy (attribute)} = E\left\{\frac{Pi+Ni}{P+N}\right\} * \left(\mathrm{I}_{(\mathrm{Pi},\ \mathrm{Ni})}\right) \tag{4.3}$$

$$\textbf{Gain} = \mathrm{Entropy}_{(\mathrm{class})} - \mathrm{Entropy}_{(\mathrm{attribute})} \tag{4.4}$$

Through these calculations able to find the attributes for making the final decision tree.

$$\textbf{Accuracy} = \frac{\text{No. of Samples predicted Correctly}}{\text{Total No. of Samples}} \tag{4.5}$$

$$\textbf{Precision} = \frac{\text{TP}}{\text{TP} + \text{FP}} \tag{4.6}$$

$$\text{Recall} = \frac{\text{TP}}{\text{TP} + \text{FN}} \tag{4.7}$$

where *TP*, true positive; *TN*, true negative; *FP*, false positive; *FN*, false negative.

A true positive is a result of the model that correctly estimates the positive class. As same as true negative is the result of the model or classifier correctly providing the negative class. A false positive shows classifier predicts incorrectly. Hence precision is the fraction of the same instances among all the retrieved instances while recall is the fraction of all the same occurrences that are recovered.

4.3 Case study

Suppose data of patients are entered into the system and need to diagnose. The server collects all the data of patients. If symptoms are asymptotic and a doctor visit and his note recommend a p24 test (Quinn, Kline, Moss, Livingston, & Hutton, 1993), cd4 + count, viral load detection (Rich, Merriman, & Mylonakis, 1999). All these data are arranged according to these parameters. Data is processed as discussed above and ready for identification of disease. Then first it is recommended to classify. All the good classifiers are used and results are saved till the best result is reached as discussed in the data analytics section. If it is found that the good classifier for given data is a decision tree. Then the tree will develop like: Patient undergoes diagnosis like p24 antibody test if the test is negative then no need for further identification. If the antibody test is positive, then it is recommended for further tests just to ensure the particular disease. CD4 + and viral load tests are performed through ELISA or RT-PCR. If these test results are positive means the patient is suffering from HIV-AIDS. Hence all these data are recorded and doctors can see this updated information for their treatment decision. Fig. 15.5 clearly explains how the decision tree works. As the system is autoML based then with the help of software the time is saved and the report generates in less time. All the data and reports, tests are case sensitive, so protected by cloud-based system security. The parameters P24, CD4 +, and viral load have come into the criteria for decision tree generation. As CD4 + cells count falls towards 200 then HIV is detected in patients' blood. Hence it indicates the rise of viral load. This means the person or patient is suffering from HIV-AIDS. Each node of the decision tree indicates yes or no. if yes then what to do? And if no means, need not proceed. Yes said to proceed till the objective of classification is achieved. Hence if CD4 + is less than 200 is reached in the decision tree means the person has HIV infection and needs treatment urgently. If in the asymptotic phase. If it is diagnosed that person has an HIV infection then life longevity is increased due to antiretroviral therapies. And in the symptomatic phase, CD4 + cells level down rapidly; it is also tried by medical practitioners to provide proper treatment on time so that somehow the life expectancy of that person could be increased. Hence machine learning

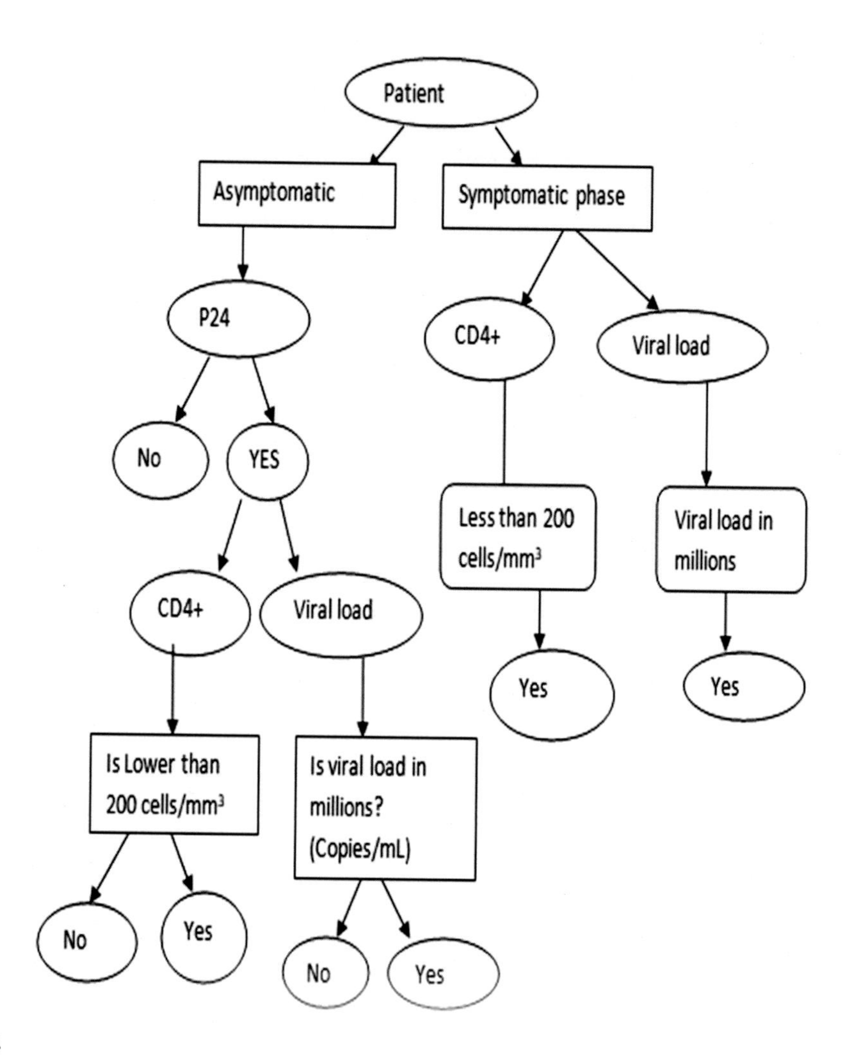

FIGURE 4.4

Decision tree for HIV disease detection in surveillance machine learning health care MODEL.

models are of utmost importance for any disease diagnosis and treatment. In the symptomatic phase of HIV infection graphs like Figs. 13.6 and 13.7 are observed that shows a decline in CD4 + and continuous enhancement of viral load (Fig. 4.4).

4.4 Surveillance machine learning healthcare model development

In this chapter the SMHC model is developed, a cloud-based security system that can be advantageous to secure and store the data properly. This system is specific and can deal with the circumstances like a diagnosis of a particular disease and emergency detection of disease and any kind of breach at any point of the safety line. This simple approach is developed for handling big medical data in different formats.

As new data is entered into the system both the layers data and streaming layer get activated. Here the batch mode is working, which means all the new data is stored in data nodes and then distributed to a semantic component that tries to find meaning in it, and filtering operations are executed. The next step deals with data processing. After it, the data is prepared for analysis through feature selection and feature extraction. Now the facts are ready to develop the model. Hence our health data is ready to predict patients' future health conditions. This is done on an offline basis. There are a variety of data available through many sources like medical sensors or wearables implanted in a patient body that calculates many health issues like blood pressure, weight, blood test report, etc., The assembled data is coexistent based on time and any values that are missed is controlled.

All this work is under cloud security. As shown in Fig. 4.5. Memory sources are challenging, real applications produce noisy data and redundant features are maintained simultaneously which requires computational time.

Big data mining requires classification, sometimes mining patterns frequently and the clustering or grouping method mitigates all the computing efforts through fast extraction of the most relevant knowledge.

Fig. 4.5 is the model which implements the security of electronic health data. This work focuses on the complete study of patient data and is helpful for doctors. Time-saving and environment protected as everything is inside system administrator. This has been called "*khoofia* camera," inside the system. The SMHC model consists of three layers. This has been described as the First—Cloud terminal: Continuous monitoring through the cloud network consists of Kerberos authentication with SE Linux installed in the system and each step of data entry to the data processing to data analysis is under control. If any breaching happens the system generates an alarm for being shut down for time being. Second- Cloud computing and IoT: Cloud is connected through IoT and all the services like literature search to disease diagnosis to the patient are under control. IoT data is also collected through the cloud and data analysis is done by machine learning algorithms if needed assistance. The bonding of cloud computing and IoT big data enabled the system to store, process, and manage data. This not only helps in disease prediction and diagnosis but also in collecting data on frequent changes in patterns of health. It is well maintained and regularly updates the severity of the medical issues over time (Kumar et al., 2018). And it is tried to infer all the information's to reach medical professionals without any delay.

IOTM: Mobile health or m-health is used as a mobile device in accumulating data related to health in real-time from patients, and store it on secure servers linked to the Internet. The data can be accessed by various clusters of consumers like hospitals, pathology labs, health insurance companies, etc. This m-health data is mainly used and accessed by doctors to monitor, diagnosed, and treat patients through administrator authority. It provides a good opportunity for remote health monitoring systems. The important parts of m-health and IOT are equipped these medical devices with sensors, communication devices like video calls by whats-app, Bluetooth technology, etc. These smart devices help recognize certain patterns and raise different alert levels such as normal, cautious, emergency, etc., for patients.

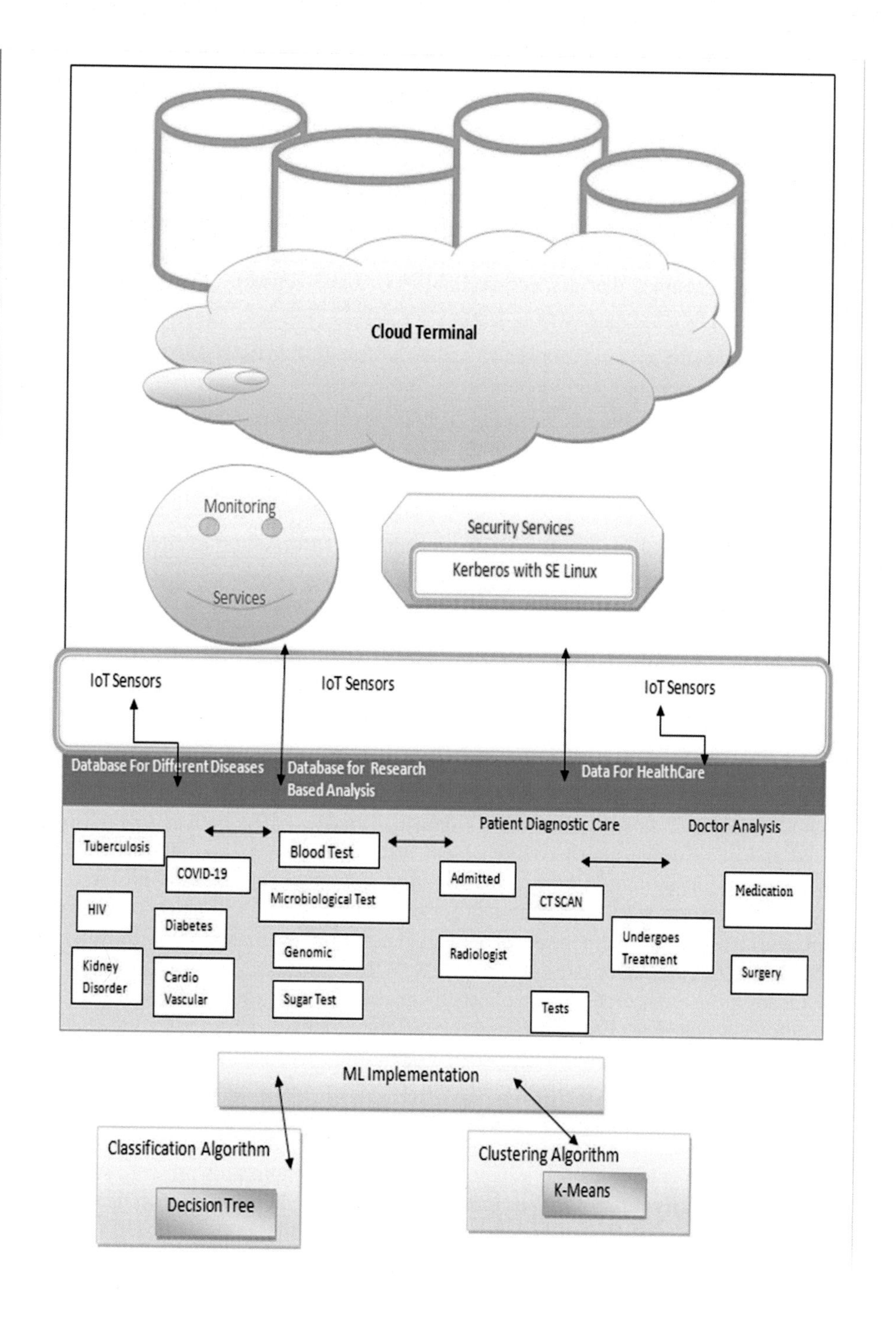

FIGURE 4.5

Surveillance machine learning healthcare model.

The capacity of data of IoT servers is very large, it is connected to the Internet and it is said to be based on a cloud-based application that captures, analyzes, and displays the operating conditions of physical resources like memory, and network bandwidth. Data through sensors is transmitted to the cloud for storage, analysis, and visualization (Amit et al., 2020). Through the Internet, IoTM provides computing resources that are not available to on-site sensors. The fundamental of IoTM systems include:

(a) **Sensors:** One or more sensors are used to gather information from types of equipment like laptops, mobile Phones, physical resources, and processes.
(b) **Data Storage:** IOT gives the opportunities to implement security, maintenance, redundancy, reliability, and user access policies (Amit et al., 2020).
(c) **Web Visualization**: To visualize the sensor information, a user interface (UI) is built to display the sensor data in a report.

This information retrieval system is reliable and filtering information available in this SMART-Cloud system. The emergency system can also run on this within time.

Third-Machine learning Implementation: As the data is prepared for analysis batch computing is executed on drawn-out data from different databases. Here in the healthcare system, disease identification is done by the previous data or databases or the patterns generated. It consists of the following steps:

1. Health data collection:
   a. The continuous monitoring of patient health conditions generates lots of data. These data may be structured (electronic health records) or semi-structured data (medical devices generated data) and unstructured data (biomedical images). The patient's health records are stored in digital format which contains medications, diagnoses, laboratory tests, doctor's visits, doctors' health notes, clinical information, and payment notes for all these are recorded. These are very important data for the analysis. This will help doctors for further treatment and proper treatment on time. This data can be exchanged in biomedical centers if needed (Amit et al., 2020; Amit, Chinmay, & Wilson, 2021).
   b. Data from different sources: Biomedical images are the best tools for disease detection and providing care on time. But processing the different kinds of images is a big challenge, hence a machine learning-based diagnosis system improves every noise and gives doctors better results so that their decision could not affect.
   c. Some of the data are collected through sensors like wrist-watch which monitor patients' health as they measure body temperature, cardiovascular status, etc. For some patients with high-risk disease conditions surveillance cameras, microphones, or pressure sensors are activated in their living room. These are the voluminous data that need to be processed nicely so doctors provide specific medication on time.
   d. Cell phone constitutes one of the best sources to collect early data. These can include monitoring the pregnancy time period (Bachiri et al., 2016), nutrition of a child (Guyon, Buback, & Knittel, 2016), and tracking heart rate (Pelegris et al., 2010). These four phases are the data acquisition phase which collects all kinds of data in a proper format.
2. Extraction of knowledge from the data: The data generated provides fully automated knowledge and interoperability is possible. All this information is stored in databases. The proposed model

defines specific ontology-based databases which allow sharing and reiterating of data in between different cloud environments if needed. The web ontology language shows the worth interchange configuration that requires XML syntax (ref).

3. Processing raw data: This consists of data filtering and data cleaning. It needs data normalization, noise reduction, manages missing data. Expectation maximization (EM), multiple imputation algorithm, and noise treatment is one the most affected algorithm.
4. Data preparation for analysis: after all the removal of noise, the feature or parameters are selected for data analysis.
5. Designing model for prediction: Now the data is ready for prediction. Different machine learning algorithms are employed for identifying proper diseases based on data input. If the quality of data is good it will give better results. This means if all the required parameters are identified by classification or clustering the disease prediction is done properly and gives doctors a way to an effective decision (Muhammad et al., 2021). The classification and clustering algorithm is discussed in the section.

4.5 Result and discussion

As the model of SMHC is implemented in the case of study, to identify the disease if data is demanding. So first the data is collected in electronic form and the patient's previous history is identified as asked to the patient or by old reports. Doctors visit and test for disease diagnosis that is ELISA or other blood tests. P24 antibody test if any thought of HIV assumption. And follow Fig. 4.4 to proceed. As the automation (autoML) is updated simultaneously, the concerned person is diagnosed whether the person is affected with HIV or not. ML algorithms are performed and all the data is collected and stored in databases that are connected through the cloud. With an IoT M system involved in the SMHC model, the data is stored and continuous monitoring is possible through the cloud. Hence this will helps doctors or healthcare providers to take the proper decisions of treatment on time. It is said that point-of-care diagnosis is possible through these approaches.

As the decision is about to take which disease a person is suffering from? Classification is done based on the data provided by the patient. For different classifiers available in system software like Bayes net, Naïve Bayes, logistic, bagging, rotation forest, and J48 a comparative analysis is made.

As autoML is discussed above, the accuracy of the decision tree (J48 Classifier) is calculated by AUTOWEKA (Kotthoff et al., 2017) as (Table 4.2).

Of the various classifiers available like Bayes net, Naïve Bayes, etc. a comparative chart is prepared for accuracies in less time. This table shows that J48 classifiers predict better results as accuracy is 98.9362%. Of the 94 instances, 93 were classified correctly. Detailed accuracy obtained by the J48 classifier is shown in Table 4.3 with the confusion matrix explained in Table 13.4. This will clearly explain the classification of HIV and non-HIV. By this, it is concluded that J48 is a good classifier for the classification or prediction of diseases.

This will help to diagnose a proper disease. In Fig. 13.5 the last node shows yes for CD4 + and viral load means the person is HIV infected or suffering from HIV-AIDS. This model is further evaluated by Entropy, and information gain and the model is judged by accuracy, precision, and

Table 4.2 Comparative classifiers for classification of data.

S. no.	Algorithms	Time (in seconds)	Accuracy	Average
1.	Bayes net	0	98.9200%	0.942
2.	Naive Bayes	0	95.7447%	0.926
3.	Logistic	0.06	97.8723%	0.937
4.	Bagging	0.02	98.9362%	0.937
5.	Rotation forest	0.08	95.7447%	0.925
6.	J48	0	98.9362%	0.937

Table 4.3 Detailed accuracy obtained by J48 classifier through AUTOWEKA.

TP	FP	Precision	Recall	F-measure	Class
0.933	0	1	0.933	0.966	HIV positive

recall. All these calculations are done by inbuilt software in the system as discussed in the method section. Accuracy shows that the method is considered for further evaluation and decisions should be taken by the medical practice. This model helps answer any further issues. Each and every step is carefully monitored by the cloud terminal and all the precautions is maintained while handling the data like HIV. It is always observed that if a person is suffering from HIV then his or her CD4 + cells are going below 200 mm^3.

Hence if CD4 + falls below 200 mm^3 per mL of blood the person is HIV positive. And No is observed means the person is not suffering from HIV positive. Clinicians decide whether to go for further checkups after some time in case of any other symptoms are found. The study is done in all the months of the year 2020.

The Below figures show the representation of CD4 decline and viral load enhancement as HIV disease progresses (Fig. 4.6).

In Fig. 4.7 viral loads are increasing rapidly from January to December 2020. This means the person suffering from chronic HIV needs urgent treatment and proper monitoring from time to time. The IoT sensors are well suited to observe patients' health in real-time. This P24, CD4 +, and viral load are the major biomarkers used to diagnose HIV disease. Same as HIV, other diseases are also observed and detected on time. Data handling and data analysis of various methods are already given in the method section with an evaluation of methods. HIV case is discussed in the case study and the further result shows the effective working of the model. This SMHC model is good to handle high-dimensional biomedical data and also helps in the development of new therapies. Not only disease diagnosis, but this model will also work successfully for research labs in hospitals. This AI-based Ml smart healthcare system uses the quality of clustering and classification algorithms to make use of these powerful techniques that are good to use in healthcare. As huge data of other diseases are also recognized, methods are generated and a proper algorithm is selected according to data need. There must be some security issues that arise or management problems are also solved properly as an alert system is active all the time.

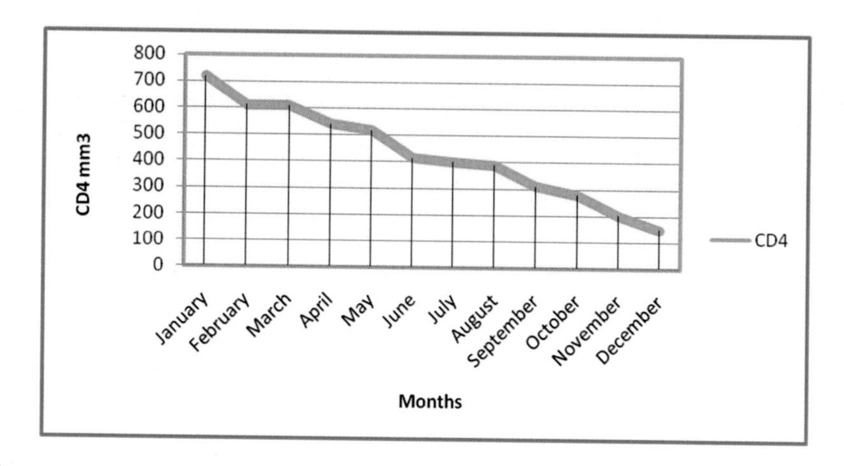

FIGURE 4.6

Visualization of CD4 count decreases from January 2020 to December 2020.

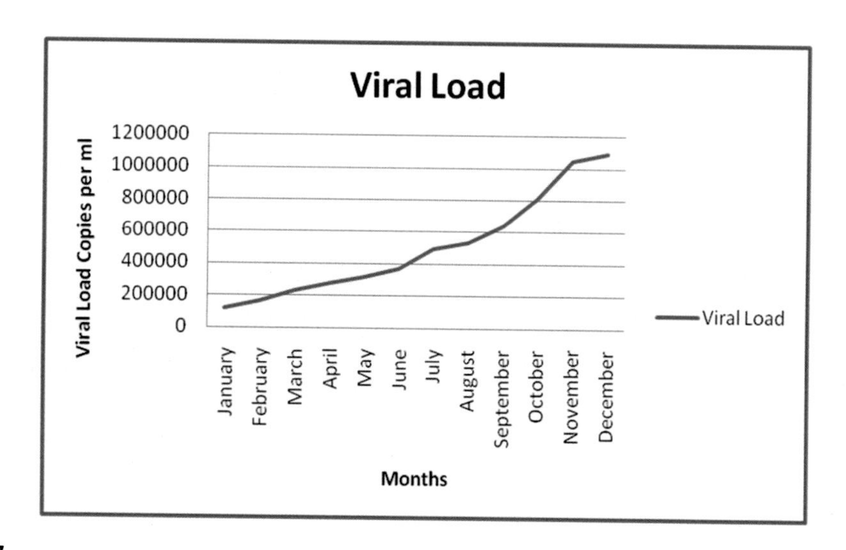

FIGURE 4.7

Visualization of Viral load increases from January 2020 to December 2020.

IoTM applications are used across industries or any other organization like automation, retail, healthcare, public sector, transportation, and many more sectors. Our SMHC model discussed four core applications:-

- **Analysis**: According to our SMHC model this property is to analyze when it is connected with the health care data. When it is accessed through an authorized user then also IoT analyzes the user's each action.

- **Efficiency improved:** In our SMHC model this property works very efficiently. Healthcare data is very well connected with the cloud through IoT. This increases the efficiency of the model.
- **Safety increased:** This IoT enhances the security of our SMHC model after cloud continuous monitoring and implemented Kerberos with SELinux. Safely connected with healthcare data no one can perform any malicious activity on the database.
- **Productivity optimized:** This feature enhances our SMHC model productivity when there is no malicious activity. When there is any access to data is executed IoT optimized the performance of our proposed model.

The current edge technologies like sensor technology, machine learning, and artificial intelligence had given a boost to make the use of IoTM practically possible.

- Sensor technology: IoTM makes it affordable and reliable sensors for more manufacturers. In our SMHC model, this technology is well connected with the cloud and healthcare database.
- Connectivity: A network on the Internet has made it easier to connect sensors to the cloud for efficient transfer of data. Our SMHC model healthcare data is cross-connected with in network and it is secured and safe.
- Cloud computing platforms: Cloud platforms enable availability for both businesses and consumers to access the infrastructure. IoTM is helpful and very well supported with this platform as in our SMHC proposed model.
- Machine learning: With advances in machine learning that is auto-mL and vast amounts of data stored in the cloud, are faster and more easily compatible with each other. With these technologies, IoTM works more efficiently as it is implemented in the SMHC proposed model.
- Artificial intelligence (AI).: This technique enables the SMHC model to use all the IoTM and machine learning applications independently

4.5.1 Analysis of security issues in the surveillance machine learning health care model

As the whole process of disease diagnosis to data management is under cloud supervision and IoT handled, there must be some security issues that need to be discussed are described. Machine learning (ML) consists of a set of laws, methods, or complex patterns to expect a behavior, which is an essential role in the field of security. Authors have tried different features to solve security-related tasks.

- Unauthorized access explains that access to the information in networks, systems, or data without permission can breach security. But this will not be possible in our proposed model (Sarker, Kayes, & Badsha, 2020). Only authorized persons can enter the network.
- Malware is said to be malicious software, which is purposely considered to damage a network or system. There are many types of malware such as computer viruses, worms, Trojan horses, adware, ransomware, and spyware. As in our proposed model, it's in the cloud environment where any kind of software cannot be installed.
- Denial-of-service it's a kind of attack that shuts down a system or network. The distributed denial-of-service attack utilizes multiple computers and Internet connections (Sarker et al., 2020).

Here we provided dual security with continuous monitoring to deactivate any attack in our proposed model.

- Phishing is a type of malicious activity that attempts to obtain sensitive information such as banking and credit card details, login credentials, and personal identification information. Any unauthorized person cannot modify any kind of information in our model (Sarker et al., 2020).
- A zero-day attack is an unidentified security liability. The application developers were unaware. Our proposed model is free from any kind of attack from attackers.

It is well said that the whole work of healthcare data management is under control in the SMART-AI-IOTM system which focuses on the method given and developed in the model. The cloud-based SMHC model is well suited for disease diagnosis and it is discussed already that well worked for data handling and maintenance.

4.5.2 Sustainability of work

The SMHC model development showed good accuracy to identify the disease. As this model is implemented and automated as the requirement occurs, will be great in the diagnosis of disease. As this model works well with HIV. It could be working well with other diseases as the availability of the medical data. Sustainability of the data management of diseases is very important. Present work also helps in maintaining the data on time and can be recognized as meeting the needs of research. Successful data management improves the knowledge, and awareness of the diseases, and disease epidemiology, and this data helps in deciding the treatment. Machine learning with automation (autoML) models helps to develop the uncover connections within millions of records. The importance of data management in hospitals involves:

1. It reduces the consequences of unmanaged medical data growth.
2. Medical data is protected at less cost.
3. Decreases the risk of data loss or theft as cloud security is available.
4. Every hospital is aware to maintain data governance.

This will provide the right approach to obtaining useful and accurate information. As fast medical emergencies are a need of today's world this effective SMHC model will work well as all automation processes of ML and data collection from various sources are secured and collected when required. This cloud-based system alerts and automatically sends notifications to the admin if any fraud/discrepancy is observed.

4.6 Pros and cons of model

1. Storage of data: Data storage is one of the most demanding tasks. The traditional method is unfit to manage the healthcare monitoring system. But in cloud storage, data from heterogeneous sources are managed as MongoDB and SQL databases, or a combined form of database is used as desired. Hence this method is scalable and has good storage capabilities.

Time-consuming: Here the method is task-based or designs a new method as per the requirement of the data. As automation is working with each and every step, the computational time is reduced.

2. In big medical data analysis, sometimes alerts are generated if any complications are found. This method indicates to prevent any health emergency from taking place. This process also assists medical scientists and clinicians in decision-making regarding any disease diagnosis which is beneficial for providing care on time.
3. As the clustering method is time taking and sometimes does not show the proper result, mainly classification algorithms are used. Clustering is performed only for the grouping of the data.
4. Quality of data: As electronic health record is fed to the system, they should be error-free for any of the application of auto-ML algorithm implementation. A little effort should be done by the auto-mL process itself to develop a good classifier for knowledge extraction for future use.

4.7 Application

This model is well suited for disease diagnosis and treatment with all the security in each and every step from electronic health record data to classifier development to the prediction of disease. Even the disease stage can be identified by the data added. All the treatment process is also secured. Without patient permission, no one can access the data. This is a highly secure model so no data breaching can happen. Care is taken while delivering any report and further tracked by the doctor's intervention. It is said that this model could be implemented in precision medicine in near future for diseases like Cancer. Deploying autoML is a difficult task but continuous research work will enhance the application power of the system. It is the joint effort of IT professionals and biomedical scientists that make this process a good working model and beneficial for doctors and patients. Machine learning is in its advanced form generates knowledge for further and future use. Not only patient data but reports generated during different tests are also secured and many hospitals have their own research center so the model developed is also helpful to secure all the data as such. If any data breaching happens or if any doubt of data misleading, the system. This model will work better in point-of-care diagnosis and treatment if required for further extension of the system.

4.8 Conclusion

This chapter discusses the proper healthcare data monitoring system which is developed with the help of IoTM-based smart auto WEKA. Big data processing approaches and recent technologies are applied to effectively make the Surveillance Machine Learning Healthcare SMHC model work. This proposed model is designed in such a way as to handle data in real-time with autoML approaches. As in WEKA some of the limitations are found while handling big data.

- WEKA is good to learn but the GUI is not well documented for handling big data.
- The WEKA-GUI provides several built-in "Visualization" panels, but these are very limited. It is basically based on AI algorithms, so maintaining a dataset is difficult sometimes.

Since the WEKA can handle only small datasets if a dataset is bigger than a few megabytes then an out-of-memory error occurs. Most importantly WEKA is a complete suite of data-mining/AI algorithms and knowledge recovery. Hence autoML is the need for handling big medical data. This cloud-based system is developed with Kerberos authentication and SE Linux which provides dual security in data management and all the working is possible in a secure environment. This system can work in data migration if there is a need to reuse any of the health data with patient and doctor permission for further treatment or experimental analysis.

4.9 Future scope

The partnership of IoT, Cloud, and machine learning advancing the development of the SMHC model. These techniques advance the model to work efficiently and effectively. In the future, this model will work more effectively in the following ways:

1. Data migration and recovery: Data migration is a planned strategy to relocate the data in a safer way. This is a little tedious, costly, and sometimes very time taking. So in near future, this proposed model is enhanced with a cloud migration strategy. Recovery of data is another challenge and it is tried to protect data during natural calamities. At the time of the disaster, the data is recovered through the approach so that data can protect from loss or damage. In the method of data recovery, protection and security are the main aspects of the cloud environment.
2. Linking of databases from other sources: All the databases of diseases are also linked with other sources to predict and add the new information gained during the analysis of the data. There are numerous databases like MySQL, NoSQL, DynamoDB, DB2, MongoDB, and OpenStack which can be used to manage databases.
3. Blockchain-based security: The proposed SMHC model also works best with blockchain implementation. AutoML with blockchain makes services faster than before, efficient, and reports generate on time. It also restricts data breaching, although the model is protected from any of the breaching attacks. With blockchain triple security layer is generated and any fraudulence is avoided (Casino et al., 2018).

References

Amit, K., Chinmay, C., & Wilson, J. (2021). Intelligent healthcare data segregation using fog computing with Internet of things and machine learning. *International Journal of Engineering Systems Modelling and Simulation*. Available from https://doi.org/10.1504/IJESMS.2021.10036745.

Amit, K., Chinmay, C., Wilson, J., Kishor, A., Chakraborty, C., & Jeberson, W. (2020). A novel fog computing approach for minimization of latency in healthcare using machine learning. *International Journal of Interactive Multimedia and Artificial Intelligence*, 1–11. Available from https://doi.org/10.9781/ijimai.2020.12.004.

Barreiro, E., Munteanu, C. R., Cruzmonteagudo, M., et al. (2018). Net-net auto machine learning (Auto ML) Prediction of complex Ecosystems. *Scientific Reports*, *8*. Available from https://doi.org/10.1038/s41598-018-30637-w.

Baydin, A. G., Pearlmutter, B. A., Radul, A., & Siskind, J. M. (2017). Automatic differentiation in machine learning: A survey. *Journal of Machine Learning Research, 18*, 1–43.

Bergstra, J., & Bengio, Y. (2012). Random searches for hyper-parameter optimization. *Journal of Machine Learning Research, 13*, 281–305.

Chinmay, C. (2017). Chronic wound image analysis by particle swarm optimization technique for tele-wound network, Springer *International Journal of Wireless Personal Communications, 96*(3), 3655–3671.

Chkirbene, Z., Erbad, A. & Hamila, R. (2019). A combined decision for secure cloud computing based on machine learning and past information. In: *IEEE wireless communications and networking conference (WCNC)* (pp. 1–6). Marrakesh, Morocco.

Chkirbene Z., Erbad A., Hamila R., Gouissem A., Mohamed A. & Hamdi M. (2020), *Machine learning based cloud computing anomalies detection*, IEEE Network, (34, pp. 178–183), no. 6.

Cirilio, D., & Vakancia, A. (2019). Big data analytics for pesonaized medicine. *Current opinion in Biotechnology, 58*, 161–167. Available from https://doi.org/10.1016/j.copbio.2019.03.004.

Cox, M., & Ellsworth, D. (1997). Application-controlled demand paging for out-of-core visualization. *IEEE Vis*, 235–244.

Dubey, A., Pant, B., & Adlakha, N. (2010). SVM model for amino acid composition based classification of HIV-1 groups. *IEEE EXPLORE, 2010 International Conference on Bioinformatics and Biomedical Technology*. Available from https://doi.org/10.1109/ICBBT.2010.5478996.

Ellen, F., & Kostas, T. (2016). Introduction to apache flink: Stream processing for real time and beyond, *inc*. O'Reilly Media.

Evans, R. (2015). Apache storm, a hands on tutorial. In: *Proceedings of the IEEE international conference on cloud engineering (IC2E)* (pp. 2–2).

Feurer, F., & Hutter, F. (2018). *Hyperparameter optimization* [Online]. Available from https://www.mL4aad.org/wp-content/uploads/2018/09/chapter1-hpo.pdf.

García, A.L., Lucas J.M., Antonai, M., David, M., et al. (2020). *A cloud-based framework for machine learning workloads and applications*, IEEE Access (8, pp. 18681–18692).

Hojabri, M., & Rao, K. V. (2013). Innovation in cloud computing: Implementation of Kerberos version5in cloud computing in order to enhance the security issues. In: *2013 International conference on information communication and embedded systems (ICICES)* (pp. 452–456). Available from https://doi.org/10.1109/ICICES.2013.6508293. Chennai, India.

Hows, D., Membrey, P., Plugge, E., & Hawkins, T. (2015). *Introduction to MongoDB*. Avaialble from https://www.mL4aad.org/wp-content/uploads/2018/09/chapter2-metalearning.pdf.

Guyon, A., Buback, A. B. L., & Knittel, B. (2016). Mobile based nutrition and child health monitoring to inform program development, an experience from Liberia. *Global Health Science and Practice, 4*(4), 661–670.

Jiang, F., Jiang, Y., Zhi, H., et al. (2017). Artificial intelligence in healthcare: Past, present and future. *Stroke and Vascular Neurology, 2*, e000101. Available from https://doi.org/10.1136/svn-2017-000101.

Kanter, J. M., & Veeramachaneni, K. (2015). Deep feature synthesis: Towards automating data science endeavors. In: *IEEE international conference on data science and advanced analytics* (pp. 1–10).

Katz, G., Shin, E. C. R., & Song, D. (2016). Explorekit: Automatic feature generation and selection. In: *International conference on data mining* (pp. 979–984).

Kotthoff, L., Thornton, C., Hoos, H., Hutter, F., & Leyton-Brown, K. (2017). Auto-WEKA 2.0: Automatic model selection and hyperparameter optimization in WEKA. *Journal of Machine Learning Research, 18* (1), 826–830.

Kuo, B. C., & Golnaraghi, F. (1995). *Automatic control systems. Englewood Cliffs, NJ: Prentice- Hall.*

Lemke, C., Budka, M., & Gabrys, B. (2015). Metalearning: A survey of trends and technologies. *Artificial Intelligence Review, 44*(1), 117–130.

Li, K., Gibson, C., Ho, D., Zhou Q, Kim, J., et al. (2013). Assessment of machine learning algorithms in cloud computing frameworks. In: *IEEE systems and information engineering design symposium* (pp. 98–103). Charlottesville, VA, USA.

Mandler, B., Marquez-Barja, J., Campista, M. E. M., Cagáňová, D., Chaouchi, H., Zeadally, S., … Somov, A. (2016). Internet of Things. IoT infrastructures: Second international summit, IoT 360°. Springer Revised Selected Papers, 27–29, October 2015. Rome, Italy.

Manogaran, G., Varatharajan, R., & Lopeza, D. (2018). *A new architecture of internet of things and big data ecosystem for secured smart healthcare monitoring and alerting system*. Elsevier.

Margolis, R., Derr, L., Dunn, M., Huerta, M., Larkin, J., Sheehan, J., … Green, E. D. (2014). The national institutes of health's big data to knowledge (BD2K) initiative: Capitalizing on biomedical big data. *Journal of the American Medical Informatics Association: JAMIA*, *21*, 957–958.

Mohr, F., Wever, M., & Hüllermeier, E. (2018). ML-plan: Automated machine learning via hierarchical planning [J]. *Machine Learning*, *107*(8–10), 1495–1515.

Muhammad, L. J., Ebrahem, A. A., Sani, S. U., Abdulkadir, A., Chinmay, C., & Mohammed, I. A. (2021). *Supervised machine learning models for prediction of COVID-19 infection using epidemiology Datase*t. *SN Computer Science*, *2*(11), 1–13. Available from https://doi.org/10.1007/s42979-020-00394-7.

Naoual, E. A., & Benhamila, L. (2018). Big data management for healthcare systems: architecture, requirements and implementation. *Advances in Bioinformatics*, *1*, 1–10.

Pan, S. J., & Yang, Q. (2009). A survey on transfer learning. *IEEE Transactions on Knowledge and Data Engineering*, *10*, 1345–1359.

Phillips C.L., & Habor R. D., (1995). *Feedback control systems. Simon & Schuster. Prediction of complex ecosystems [J]*. Scientific Reports (8).

Quinn, T. C., Kline, R., Moss, M. W., Livingston, R. A., & Hutton, N. (1993). Acid dissociation of immune complexes improves diagnostic utility of p24 antigen detection in perinatally acquired human immunodeficiency virus infection. *The Journal of Infectious Diseases*, *167*, 1193–1196.

Rich, J. D., Merriman, N. A., Mylonakis, E., et al. (1999). Misdiagnosis of HIV infection by HIV-1 plasma viral load testing: A case series. *Annals of Internal Medicine*, *130*, 37–39.

Rovnyagin, M. M., Timofeev, K. V., Elenkin, A. A., & Shipugin, V. A. (2019). Cloud computing architecture for high-volume ML-based solutions. In: *IEEE conference of Russian young researchers in electrical and electronic engineering (EIConRus)* (pp. 315–318). Saint Petersburg and Moscow, Russia.

Salvador, M. M., Budka, M., & Gabrys, B. (2018). Automatic composition and optimization of multicomponent predictive systems with an extended auto-weka. *IEEE Transactions on Automation Science and Engineering*, 1–14.

Sarker, I. H., Kayes, A. S. M., Badsha, S., et al. (2020). Cybersecurity data science: An overview from machine learning perspective. *Journal of Big Data*, *7*, 41. Available from https://doi.org/10.1186/s40537-020-00318-5.

Saxena, A., & Dubey, A. (2019). Kerberos authentication model for data security in cloud computing using honey-pot. *Global Journal of Engineering Science and Researches(GJESR)*, *6*(8), 35–44.

Sniezynski, B., Nawrocki, P., Wilk, M., et al. (2019). VM reservation plan adaptation using machine learning in cloud computing. *Journal of Grid Computing*, *17*, 797–812.

Sutradhar, M. R., Sultana, N., Dey, H., & Arif, H. (2018). A new version of kerberos authentication protocol using ECC and threshold cryptography for cloud security. In *2018 Joint 7th international conference on informatics, electronics & vision (ICIEV) and 2018 2nd international conference on imaging, vision & pattern recognition (icIVPR)* (pp. 239–244). Available from https://doi.org/10.1109/ICIEV.2018.8641010. Kitakyushu, Japan.

Vanschoren, J. (2019). Meta-Learning. *Automated machine learning, The springer series on Challenges in Machine Learning book series (SSCML).*

Vilalta, R., & Drissi, Y. (2002). A perspective view and survey of metalearning. *Artificial Intelligence Review*, *18*(2), 77–95.

Wang, Y., Kung, L., & Byrd, T. A. (2018). Big data analytics: Understanding its capabilities and potential benefits for healthcare organizations. *Technological Forecasting and Social Change*, *126*, 3–13.

Wolpert, D. H., & Macready, W. G. (1997). No free lunch theorems for optimization. *IEEE Transactions on Evolutionary Computation*, *1*(1).

Wolpert, D. H. (1996). The lack of a priori distinctions between learning algorithms. *Neural Computation*, *8* (7), 1341–1390.

Zaharia, R., Xin, S., Wendell, P., et al. (2016). Apache spark: A unified engine for big data processing. *Communications of the ACM*, *59*(11), 56–65.

Zhu, Z, & Fan, P. (2019). Machine learning based prediction and classification of computational jobs in cloud computing centers. In: *15th International wireless communications & mobile computing conference (IWCMC)* (pp. 1482–1487). Tangier, Morocco.

CHAPTER

5 A predictive method for emotional sentiment analysis by machine learning from electroencephalography of brainwave data

Pijush Dutta[1], Shobhandeb Paul[2], Korhan Cengiz[3], Rishabh Anand[4] and Madhurima Majumder[5]

[1]Department of Electronics and Communication Engineering, Greater Kolkata College of Engineerting & Management, West Bengal, India [2]Guru Nanak Institute of Technology, West Bengal, India [3]Department of Telecommunication, Trakya University, Edirne, Turkey [4]Service Deliver Manager, HCL Technologies Limited, New Delhi, India [5]Department of Computer Science Engineering, Global Institute of Management and Technology, Krishnagar, West Bengal, India

5.1 Introduction

In very recent times, artificial intelligence and machine learning are common techniques utilized in clinical diagnostics. Artificial intelligence-based frameworks frequently help a doctor in diagnosing the illness of a patient. Previously a doctor affirmed the analysis after getting the symptoms insights into the patient's body. Hence, demonstrative exactness is exceptionally reliant on a doctor's insight. In such cases, machine learning is used to procure and store a huge amount of datasets and send the automatic clinical decision in a practical way to deal with helping doctors quickly and precisely analyze patients' conditions (Martino, Samamé, & Strejilevich, 2016). Some of the statistical strategies have significant drawbacks which are applied in clinical classification (Petrantonakis & Hadjileontiadis, 2009). Sometimes the characteristics of the dataset are troublesome and not feasible. In such cases, soft computing methodologies are less reliant on such information.

Emotion plays a significant function in individuals' everyday life and discernment. At present research, emotion acknowledgment has become an intriguing issue in the fields of brain-computer interface (BCI), artificial intelligence, and clinical wellbeing, particularly for the exploration and treatment of the instrument and seizure law of sicknesses, for example, dysfunctional behavior and mental issues (Martino et al., 2016). However, emotion acknowledgment dependent on the EEG signals is as yet a notable test. Investigation of passionate states improves human-computer interaction, yet additionally, it is a significant issue from a clinical application perspective (Brave & Nass, 2007). Besides, considering inborn parts of feelings can likewise help psychiatrists in the treatment of mental problems, for example, autism spectrum disorders (Witwer & Lecavalier, 2010), attention deficit hyperactivity disorder (Rowland, Lesesne, & Abramowitz, 2002), and anxiety disorders (Celano, Daunis, Lokko, Campbell, & Huffman, 2016). In recent years, physiological signals, for example, electrocardiography, electromyography, galvanic skin reaction (Pourmohammadi & Maleki, 2020),

Implementation of Smart Healthcare Systems using AI, IoT, and Blockchain. DOI: https://doi.org/10.1016/B978-0-323-91916-6.00008-4

respiration rate (Y. Lin et al., 2018), and, especially, electroencephalography (EEG) have been utilized to arrange emotional states (Das, Khasnobish, & Tibarewala, 2016).

There are several techniques which include frequency domain, time domain, nonlinear analysis, mixed time and frequency domain, etc. are used for emotion acknowledgment on EEG signals (Frantzidis et al., 2010). Besides this there are several techniques also used to highlight the emotional acknowledgment like Differential asymmetry, rational asymmetry and power spectral density (Lin et al., 2010), higher-order crossing features (Petrantonakis & Hadjileontiadis, 2009), a fractal measurement (FD) concerning the evaluation of fundamental feelings (Liu, Sourina, & Nguyen, 2010) entropy-based measurements (Hosseini & Naghibi-Sistani, 2011), discrete wavelet change (DWT) (Murugappan, Rizon, Nagarajan, & Yaacob, 2010), etc.

To explore the biological sign several promising advancements in EEG signals also be examined by the researchers like the multivariate-multiscale approach dependent on the utilization of the Fourier-Bessel series development based experimental wavelet change (FBSE-EWT) and temporal entropies (Bhattacharyya, Tripathy, Garg, & Pachori, 2020), A multiscale sample entropy for unconscious emotion (Shi, Zheng, & Li, 2018), based upon error and a correlation loss functions (Atmaja & Akagi, 2021), Hyperchaotic encryption algorithm (Xue, Du, Li, & Ma, 2018), Lyapunov spectrum (Valenza, Citi, Scilingo, & Barbieri, 2017) distinctive entropy assessors for the computerized diagnostics of epileptic subjects (Acharya, Fujita, Sudarshan, Bhat, & Koh, 2015; Kannathal, Choo, Acharya, & Sadasivan, 2005) and the emotional acknowledgment approach was introduced on the base of test entropy (Jie, Cao, & Li, 2014).

Researchers have applied AI for emotional acknowledgment by utilizing a dataset for emotional investigation utilizing the Muse headband which is an economically accessible EEG recording gadget (Asif, Majid, & Anwar, 2019). To make a moderate gadget that shows these patients' feelings with straightforward and understandable pictures or charts so they can be perceived by the individuals who deal with them (Tian, 2018), the chance of utilizing EEG signals as a sign of misdirection or untruths (Youssef, Ouda, & Azab, 2018), the value in estimating EEG sign of Muse is amazingly helpful in test work (Przegalinska, Ciechanowski, Magnuski, & Gloor, 2018). This arrangement collects 2549 datasets based on time-frequency domain statistical features where a subset of 640 datasets chosen by their symmetrical uncertainty was discovered to be best when utilized with three different classifiers Random Forest (RF), XG Boost, and Decision Tree (DT) for emotion detection. In this major contribution is here we used three prescient models were created utilizing linear regression, DT, and RF separately

The rest of this article is presented as follows: After the brief introduction in Section 5.1, Section 5.2 gives an investigation of the cutting-edge works identified, quickly presenting the most pertinent ideas applied to deal with AI with electroencephalographic information. Section 5.3 portrays the techniques used to play out the investigations performed, feature selections, and machine learning algorithms. The consequences of the tests, including graphical portrayals of results, are introduced in Section 5.4. Section 5.5 subtleties the ends separated from the trials and the proposed future work.

5.2 Review of literatures

In previous studies, researchers communicated their endeavors in finding the best model for foreseeing the detection of emotion. Meanwhile, different investigations give just a brief look into anticipating emotion detection utilizing AI procedures. This segment investigates the exploration works that are identified with the proposed approach (Table 5.1).

In contrast to the previously mentioned works, it was seen that during the identification of a significant pattern of a dataset from a large dataset it becomes a very complicated one as noise in the information additionally sabotages the capacities of the cutting-edge classification algorithms.

Table 5.1 Notable machine learning techniques (ML) for emotion detection.

Algorithm used	Disease	Outcomes
Bayesian networks, support vector machines (SVMs), and random forests classifier (Bird, Ekárt, Buckingham, & Faria, 2019)	Human-machine interaction (HMI) is based on mental state recognization	In this research 2100 datasets with 44 features were used for mental state recognization. The maximum accuracy obtained by the algorithm is 87%
Random Forest (RF) classifier (Edla, Mangalorekar, Dhavalikar, & Dodia, 2018)	HMI is based on mental state recognization with the help of data collected from 40 subjects (33 male and 7 female)	The proposed model shows an accuracy of 75%. The proposed model is further applied to the IoT-based home automation and offers faithful operation
Support vector machine, Optimum path Forest, Multilayer perception, k-nearest neighbor, and Naïve Bayes (Rodrigues et al., 2019)	Wavelet Packet Decomposition and machine learning classification were done for alcoholic electroencephalographic signals	In this research wavelet packet decomposition and five Machine learning algorithms are used. Among all the algorithms Naïve Bayes offers the maximum accuracy of 99.87%
Sparse discriminative ensemble learning algorithm (Ullah et al., 2019)	Emotion recognition by EEG	Maximum Computational efficiency and classification accuracy are obtained by kNN and SVM are about 82.1% by reducing the amount of data size
Sequential minimal optimization (SMO), stochastic decent gradient, logistic regression (LR), and multilayer perceptron (MLP) (Asif et al., 2019)	Classify the human stress with the effect of music tracks in English and Urdu language	To perform this study the datasets were collected from 27 persons (including 14 male and 13 female) with ages ranging from 20 to 35 years, Among all these four classifier algorithms LR provides the maximum accuracy of 98.76
Support vector machine (SVM), Universum SVM, Twin SVM and Universum twin SVM (Richhariya and Tanveer, 2018)	Diagnosis of neurological disorders	Universum SVM classifier has achieved the maximum accuracy of 99% after extraction of healthy and seizure signals
MLP, a Logistic Regression, and a Support Vector Machine-based classifier (Ieracitano et al., 2019)	Classify the patient affected by classifying the patient Alzheimer's disease (AD) Mild Cognitive Impairment (MCI) and Healthy Control	From the result analysis, it is seen that single hidden layer MLP achieved the highest accuracy rate up to 95.76% ± 0.0045 for AD versus HC and 86.84% ± 0.0098 for AD versus MCI respectively

(Continued)

Table 5.1 Notable machine learning techniques (ML) for emotion detection. *Continued*

Algorithm used	Disease	Outcomes
Naïve Bayes (NB), SVM (Dabas, Sethi, Dua, Dalawat, & Sethia, 2018)	A 3D emotional state is classified from a set of videos based on various parameters such as arousal, valence, and dominance	NB classifier gave the highest accuracy about 78.06% from the 3D Emotional Model which comprised 8 different emotional states namely relaxed, peaceful, bored, disgusted, nervous, sad, surprised, and excited
SVM, k-Nearest Neighbor, and artificial neural network (Islam, Sajol, Huang, & Ou, 2016)	To build the model for the different cognitive tasks by the movement of the eyes	The highest accuracy provides by the kNN is about 98.59%
SVM with RBF kernel (Y. Zhang et al., 2017)	The strategy of feature extraction from EEG signals	The proposed model shows that the combination strategy of the AR model effectively improves the classification

Hence proper feature selection takes an important role to classify the features with a higher degree of accuracy. In this research symmetrical uncertainty, feature selection is performed on the dataset taken from short temporal lapses of EEG information and finally classifies the mental state of the patient using three different classification machine learning algorithms.

5.3 Materials and methods

A BCI is an immediate correspondence pathway between a brain and an outside gadget (Kumar et al., 2020). The purpose of BCI is that it is Assisting, enlarging, or fixing psychological or sensory-motor functions, Gaming, creating advancements for individuals with handicaps, help daze individuals to imagine outer pictures, help deadened individuals to work outside gadgets without actual development, disentangle data put away on the human cerebrum.

There are three sorts of Human BCI EEG securing measures. They are

1. Invasive: In this process, anodes are legitimately embedded into the dark matter of the cerebrum during neurosurgery.
2. Partially invasive: BCI gadgets embed inside the skull yet rest outside the brain
3. Noninvasive: In this BCI system cathodes are set externally in the skull. Among these strategies, the Noninvasive technique is exceptionally utilized in research. Attractive Resonance Imaging (MRI) and EEG are non-obtrusive and they have cathodes put from Scalp. Electroencephalogram (EEG) is utilized in estimating the electrical movement of the cerebrum which is produced by billions of nerve cells called neurons. This EEG movement is recorded through terminals put on the scalp (Sameer, Gupta, Chakraborty, & Gupta, 2019). EEG signals are ordered into various rhythm frequencies going from 0.5 to 30 Hz. The characteristics of four different EEG waves are named Delta, Theta, Alpha, Beta, and Gamma waves are shown as

Table 5.2 Different brain waves with ranges of frequency and emotions.

Serial no.	Brainwave	Ranges of frequency (Hz)	Emotions
1	Delta	0.5–4	Deep sleep, coma
2	Theta	4–8	Trance, dreams
3	Alpha	8–13	Relaxation with eye closed
4	Beta	13–20 for beta 1and 20–30 for beta 2	Thinking, tension, anxiety, excitement
5	Gamma	26–70 for gamma	High stress and disturbance

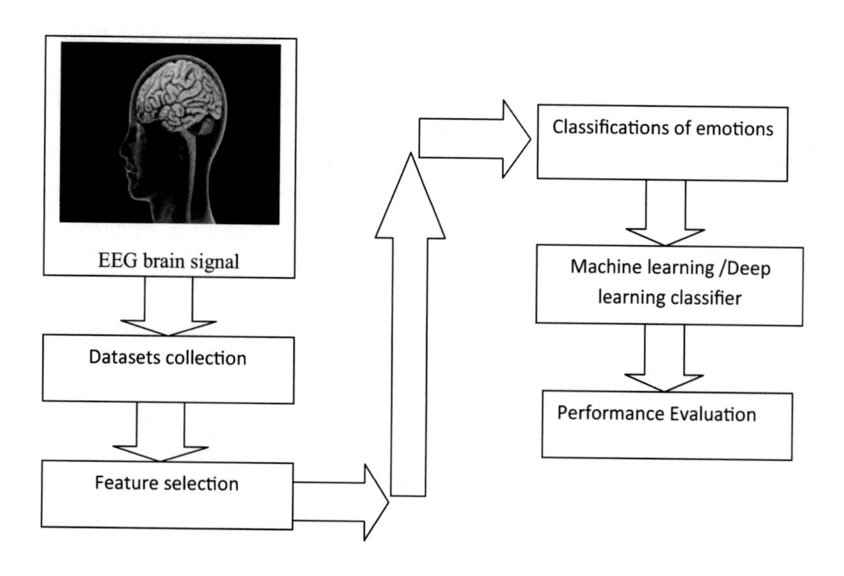

FIGURE 5.1

The overall framework of the proposed model.

appeared in Table 5.2. After appropriate training, subjects create cerebrum signals (Features) and a component of the consistent EEG yield is utilized that the client can dependably produce (waves), one can likewise inspire an EEG reaction with an outer improvement (evoked potential).

Fig. 5.1 shows the proposed model to perform research on EEG brainwave signals collected from the muse band, send them to one of the best feature selection processes and segment different types of emotions and finally apply the machine learning algorithm to calculate the performance index.

5.3.1 Muse headband

The Muse headband is an EEG recording gadget with four anodes put on the TP9, AF7, AF8, and TP10 positions dependent on the worldwide EEG arrangement framework (Jasper, 1958). Since the signs are very feeble, signal commotion is a significant issue because it viably covers the helpful data (Symeonidou, Nordin, Hairston, & Ferris, 2018). The EEG headband utilizes different antique detachment methods to best hold the brainwave information and dispose of undesirable commotion (Oliveira, Schlink, Hairston, König, & Ferris, 2016). Already, the headband has been utilized alongside AI techniques to quantify various degrees of client projects, regarding it as an inclination much like in opinion examination ventures, scientists effectively figured out how to gauge various degrees of a client's delight (Abujelala, Abellanoza, Sharma, & Makedon, 2016). Dream headbands are also used in neuroscience research extends because of their ease and simplicity of sending, just as their viability as far as order and exactness (Krigolson, Williams, Norton, Hassall, & Colino, 2017).

5.3.2 Features selection

Highlight unsheathing and characterization of EEG is a central issue in the field of the BCI domain. The signs are viewed as fixed uniquely inside short spans that are the reason the best practice is to apply a brief timeframe windowing method to meet this prerequisite. Feature selection (Dwivedi, Dey, Chakraborty, & Tiwari, 2021) can be performed by several statistical techniques like max-min features in temporal sequence, log-covariance, Shannon entropy, Time-frequency based Fourier Transform, etc. All highlights proposed to arrange the psychological states are registered as far as the worldly dissemination of the sign in a given time window. This sliding window is characterized as a time of 1 s at 250 Hz to identify the Beta band where the shortest period and the minimum number of samples that need to be detected would be $1/30 = 33.33$ Ms and 17 samples for a short beta period. A cover of 1 s is utilized while moving each of the four windows. Another significant highlight figure is that the EEG Muse headband offered five kinds of sign frequencies $\{\alpha, \beta, \theta, \delta, \gamma\}$.

Statistical Features: To have a minimal portrayal of the sensor information in a given time range, several measurable features are demonstrated.

1. Mean value

$$\mu = \frac{1}{N}\sum_{i}^{N} x_i \tag{5.1}$$

2. The standard deviation

$$\sigma = \sqrt{\frac{1}{N}} \sum_{i}^{N} (x_i - \mu)^2 \tag{5.2}$$

3. Skewness and kurtosis

$$\text{Skewness: } y = \frac{\mu^k}{\sigma^k} \tag{5.3}$$

and

$$\text{kurtosis: } \mu^k = \frac{1}{N}\sum_{i}^{N} (x_i - \mu)^k \tag{5.4}$$

Log-covariance features: To attain a 12×12 square matrix to process the log-covariance from 150 temporal features, we disposed of 6 features

$$\text{LcM} = (\log m\,(\text{cov}(M))) \tag{5.5}$$

where LcM is an upper triangular element of the matrix

The Covariance matrix is given by

$$(M) = covij = \frac{1}{N}\sum (x_{ik} - \mu_i)\left(x_{kj} - \mu_j\right) \tag{5.6}$$

Shannon entropy and log-energy entropy: Shannon entropy is one type of nonlinear analysis widely applicable in the field of signal processing and time series. In the data hypothesis, the Shannon entropy is given by:

$$h = -\Sigma_j S_j \times \log(S_j) \tag{5.7}$$

Where h is a feature computed in every time window of 1 s. and Sj is each element (normalized) of this temporal window.

The log-energy entropy can be given by:

$$\log e = \Sigma_i \log(s_i^2) + \Sigma_j \log(s_j^2) \tag{5.8}$$

where i *and j* represent an index for the elements of the first and second sub-window (0–0.5 s).

Frequency domain: *It is a* frequency domain method to analyze the time series spectrum can be represented by:

$$\mathrm{X(k)} = \Sigma_{(n=0)}^{(N-1)} x(n).e^{\frac{-j(2\pi)*nk}{N}} \quad \text{where, } \mathrm{k} = 0, 1, 2, 3\ldots(\mathrm{N}-1) \tag{5.9}$$

5.3.3 Datasets

5.3.3.1 Feature selection algorithms

Highlight choice intends to eliminate information that has no valuable application and just serves to unneeded increment the interest for assets (Witten & Frank, 2002). In modern research there five different feature selection algorithms were utilized to diminish the number of characteristics, chosen by the calculation. The evaluators utilized were as per the following:

1. **One R**: It helps to calculate the error rate of every forecast and selects the lowest risk classification (Faria, Premebida, Manso, Ribeiro, & Núñez, 2018).
2. **Information gain**: Estimating the class (Jadhav, He, & Jenkins, 2018).
3. **Correlation**: Measures concerning with the help of attribute and class utilizing their *Pearson's* coefficient (Jain, Jain, & Jain, 2018).
4. **Symmetrical uncertainty:** Measures concerning the class (Saikhu, Arifin, & Fatichah, 2019).
5. **Evolutionary algorithm:** Measure concerning subsets and position their viability with a fitness capacity to gauge their prescient capacity of the class (Li, He, Liang, & Quan, 2019).

5.3.3.1.1 Symmetric uncertainty

When a lot of information is amassed then it becomes a daunting task to find the significant patterns. Noise in the information additionally sabotages the capacities of the main cutting-edge classification methods. In such a manner, feature determination assumes a fundamental job. This study has presented an extremely productive technique for feature selection. The straightforwardness and simplicity of execution alongside the prescient precision of the selected features are the solid inspiration for the proposed technique. As referenced above for a filter-based feature selection mechanism their requirements to have some proxy classifier measure that gauges the value of a chosen highlight subset (Ali & Shahzad, 2012). The expression for symmetric uncertainty is given by

$$\mathrm{SU}(X,Y)=2\left[\frac{\mathrm{IG}\left(\frac{X}{Y}\right)}{H(x)+H(Y)}\right] \tag{5.10}$$

where IG $\left(\frac{X}{Y}\right)$ is the information gain of feature X, concerning independent attributes Y. $H(X)$ and $H(Y)$, is the entropy of feature X and Y.

5.3.4 Artificial intelligence process

As a benchmark, every dataset is set to run by a symmetric Uncertainty classifier. This oversimplified classifier picks one single class to apply to the entirety of the information to diminish off base arrangements, it is normal that the exactness of 33% is accomplished with a reasonable appropriation of the three mental states AI alludes to the utilization of computational algorithms that can figure out how to perform given errands from model information without the need directions (Dutta, Paul, & Majumder, 2021; Dutta, Paul, Shaw, Sen, & Majumder, 2022). AI applied advanced statistical methods to remove prescient or unfair samples to play out the most precise forecasts of new information (Sen, Saha, Chaki, Saha, & Dutta, 2021; Dutta et al., 2021). Here we proposed ML algorithms as an AI tool in the field of cardiovascular diagnosis for a non-expert audience. In this research under the python platform, distinctive prescient calculations were picked to assemble the model, namely: RF, DT, and XG Boost. Accordingly, we chose the most ideal model to accomplish the best on clinical datasets and permit viable data classification (Dutta, Paul, & Kumar, 2021; Dutta, Paul, Obaid, Pal, & Mukhopadhyay, 2021).

The dataset was taken and sent for analysis; it was found that there were a total of 2132 records present in the dataset. On analyzing the label column of the dataset it was identified as a classification problem and the classification algorithms were used to train the models and get the model accuracy.

5.3.4.1 Contribution of the present research

This arrangement collects 2549 datasets based on time-frequency domain statistical features where a subset of 640 datasets chosen by their symmetrical uncertainty was discovered to be best when utilized with three different classifiers RF, XG Boost, and DT for emotion detection. In this major contribution is here we used three prescient models created utilizing linear regression, DT, and RF separately. We assess the presentation of the models utilizing distinctive famous metrics from the literature, specifically, exactness, recall, accuracy, F-score, Receiver operating curve (ROC), (AUPR), and (RMSE). Subtleties of these measurements and important notions have been discussed in the review of machine learning. All through the tests, the 10 cross-approval results are the normal outcomes. All these metrics are described in the following subsections.

5.3.4.2 XG Boost

XG Boost is an improved gradient- enhance machine learning algorithm, which can successfully build improved trees and run equal processing (Jia et al., 2019; XingFen, Xiangbin, & Yangchun, 2018). This algorithm is applicable on 2nd order Taylor expansion on loss function and offers a

	# mean_0_a	mean_1_a	mean_2_a	mean_3_a	mean_4_a	mean_d_0_a	mean_d_1_a	mean_d_2_a	mean_d_3_a	mean_d_4_a	...	fft_741_b	fft_742_b	fft_
0	4.62	30.3	-356.0	15.6	26.3	1.070	0.411	-15.70	2.06	3.15	...	23.5	20.3	
1	28.80	33.1	32.0	25.8	22.8	6.550	1.680	2.88	3.83	-4.82	...	-23.3	-21.8	
2	8.90	29.4	-416.0	16.7	23.7	79.900	3.360	90.20	89.90	2.03	...	462.0	-233.0	
3	14.90	31.6	-143.0	19.8	24.3	-0.584	-0.284	8.82	2.30	-1.97	...	299.0	-243.0	
4	28.30	31.3	45.2	27.3	24.5	34.800	-5.790	3.06	41.40	5.52	...	12.0	38.1	

5 rows × 2549 columns

FIGURE 5.2

Actual data collection from the electroencephalography wave.

higher degree of proficiency (Suo, Song, Dou, & Cui, 2019) other than the first-order derivative performed by gradient boosting DT. Actual data obtained from EEG wave shown in Fig. 5.2.

(1) Simpler to locate the ideal arrangement; (2) deal with scanty and missing data; (3) Generate a choice base basic score; (4) The split hub calculation runs fast. The fundamental hindrance is When the measure of information is large, the strategy is tedious (Chen et al., 2018; Rahman et al., 2020). This improved algorithm is applied in different field of research some of them is Crude oil pricing (Gumus & Kiran, 2017), image classification (Ren, Guo, Li, Wang, & Li, 2017), renewable energy (D. Zhang et al., 2018), disease diagnosis (Ogunleye & Wang, 2019).

Pseudocode algorithm

Input D: Datasets of EEG

N: The number of Trees in XGBOOST

Output: Discrete data

For i = 0; i < D; i++ do

Preprocessing the data

Text participle

End for

While LDA doesn't converge **do**

Extracting the text subjects characteristics

End while

Extracting the user characteristics U, constructing the union feature C

For i = 0; i < C; i++ **do**

End for

5.3.4.3 Random forest

The RF classifier consists of a combination of tree classifiers where each classifier is generated using a random vector sampled independently from the input vector, and each tree casts a unit vote for the most popular class to classify an input vector (Dutta et al., 2021; Dutta, Shaw, et al., 2021). The RF classifier used for this study consists of using randomly selected features or a combination of features at each node to grow a tree. Bagging, a method to generate a training data set by randomly drawing with replacement *N* examples, where *N* is the size of the original training set (Dutta, Paul, Obaid, et al., 2021), was used for each feature/feature combination selected. There are many features in the RF classifier: (1) each time a tree is grown to the maximum depth on new training data using a combination of features. These full-grown trees are not pruned. (2) As the number of trees increases, the generalization error always converges even without pruning the tree, and overfitting is not a problem because of the strong law of large numbers.

5.3.4.4 Decision tree

DTs are a type of Supervised Machine Learning where the data is continuously split according to a certain parameter (Dutta et al., 2021; Dutta, Shaw, et al., 2021). The tree can be explained by two entities, namely decision nodes, and leaves. The leaves are the decisions or the outcomes. And the decision nodes are where the data is split. DT algorithms have the following advantages: (1) they required less effort for data preparation during the preprocessing stage. (2) It can operate on a non-normalized dataset (3) missing data does not affect the performance of the algorithms (4) does not require any scaling data.

5.4 Result analysis and discussion

Statistical features of the datasets shown in Fig. 5.3. To prove the validity of ML, results must be broken down from two viewpoints: measurable legitimacy, the rightness of factual qualities, reproducibility with various cohorts, and intra-legitimacy, concerning the clinical and genuine ramifications of the calculations consistently (i.e., clinical viability). The accompanying sub-segments will

	# mean_0_a	mean_1_a	mean_2_a	mean_3_a	mean_4_a	mean_d_0_a	mean_d_1_a	mean_d_2_a	mean_d_3_a	mean_d_4_a	...	fft_74(
count	2132.000000	2132.000000	2132.000000	2132.000000	2132.000000	2132.000000	2132.000000	2132.000000	2132.000000	2132.000000	...	2132.000(
mean	15.256914	27.012462	-104.975629	13.605898	24.150483	0.025378	0.052282	0.301655	0.036793	0.083567	...	-22.938!
std	15.284621	9.265141	206.271960	16.874676	14.187340	17.981796	8.509174	68.098894	17.010031	18.935378	...	298.034:
min	-61.300000	-114.000000	-970.000000	-137.000000	-217.000000	-218.000000	-255.000000	-1360.000000	-203.000000	-553.000000	...	-1180.000(
25%	6.577500	26.075000	-195.000000	4.857500	23.600000	-3.105000	-1.340000	-4.002500	-2.905000	-2.622500	...	-106.500(
50%	14.100000	30.000000	14.950000	15.400000	25.200000	-0.044600	0.132000	0.957500	-0.099750	0.146500	...	83.850(
75%	27.700000	31.400000	29.600000	26.500000	26.800000	2.920000	1.540000	6.735000	2.535000	2.870000	...	154.000(
max	304.000000	42.300000	661.000000	206.000000	213.000000	402.000000	257.000000	1150.000000	349.000000	444.000000	...	1070.000(

8 rows × 2548 columns

FIGURE 5.3

Descriptive statistics of the datasets.

depict how the measurements and the clinical viability are thought of. An associate is arranged unmistakably for ML purposes. To play out all these three algorithms we utilized the following features processor and platform: Intel i3, sixth era processor, OS: Ubuntu 20.04 and RAM +8 Gb, python 3.7.6, and Jupyter journal 6.03.

For the legitimacy of the algorithm, an entire informational index ought to be part into 3 unique subgroups, called the preparing set, approval set, and testing set, separately. These gatherings are regularly chosen in such a manner that subgroups share segment disseminations, for example, age or sex, to speak to a certifiable situation.

5.4.1 Confusion matrix

Accuracy measures the level of the calculation arranging the information accurately (Dutta et al., 2021; Lever, Krzywinski, & Altman, 2016). It is a basic measure utilized in numerous logical situations if there is no class awkwardness. One of the disadvantages of utilizing precision as the measurement is that there is an information misfortune when estimating False Negative and False Positive. To survey the exhibition of a calculation and to comprehend where there may be a miss-grouping issue, a table report called Confusion Matrix is used (Ohsaki et al., 2017; Ruuska et al., 2018; Xu, Zhang, & Miao, 2020).

Figs. 5.4 and 5.5 represent the confusion matrix and its classification of the RF algorithm. Fig. 5.4 represents the accuracy of the RF model which is about 96.25% while other classification parameters like precision, recall, and F1 score is about 96%. So from the confusion matrix, it concludes that the RF model offers a better agreement of classification of sentiment analysis.

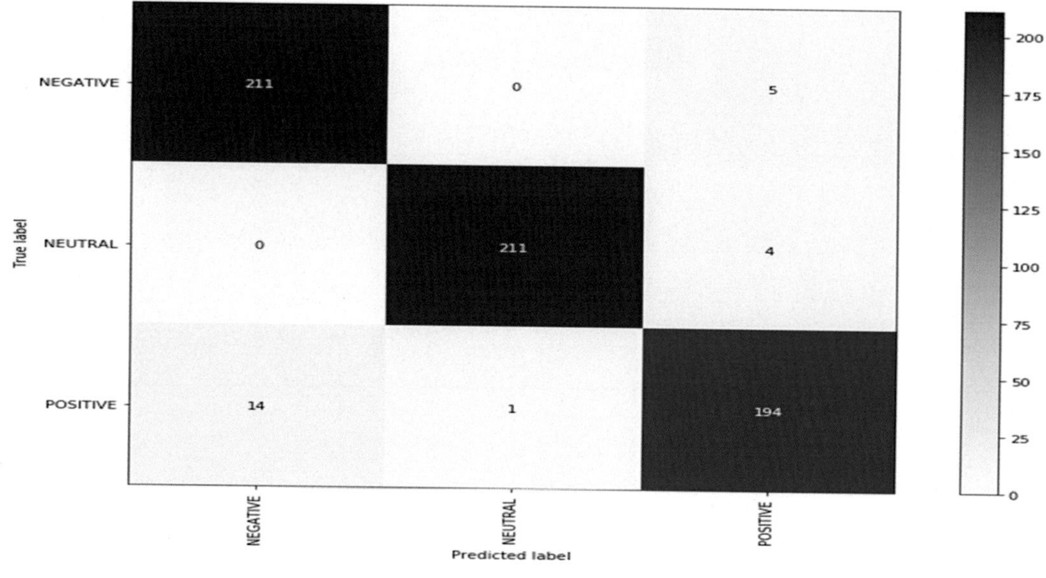

FIGURE 5.4

Confusion matrix for random forest.

Classification Report

	precision	recall	f1-score	support
NEGATIVE	0.94	0.98	0.96	216
NEUTRAL	1.00	0.98	0.99	215
POSITIVE	0.96	0.93	0.94	209
accuracy			0.96	640
macro avg	0.96	0.96	0.96	640
weighted avg	0.96	0.96	0.96	640

accuracy:

0.9625

FIGURE 5.5

Confusion matrix classification report of random forest.

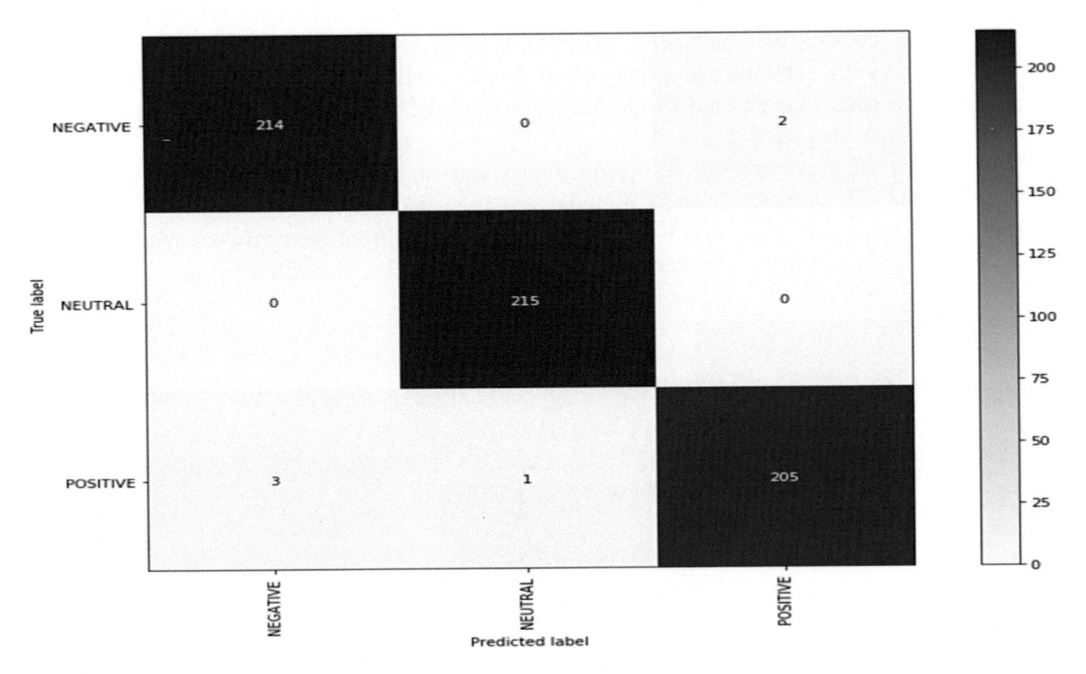

FIGURE 5.6

Confusion matrix of XG Boost.

Figs. 5.6 and 5.7 represent the confusion matrix and its classification of the RF algorithm. From Fig. 5.7 it is seen that the accuracy of the RF model is about 99.06% while other classification parameters like precision, recall, and F1 score is about 99%. So from the confusion matrix, it concludes that the RF model offers a better agreement of classification of sentiment analysis.

Figs. 5.8 and 5.9 represent the confusion matrix and its classification of the RF algorithm. From Fig. 5.9 it is seen that the Accuracy of the RF model is about 95.78% while other classification

Classification Report

	precision	recall	f1-score	support
NEGATIVE	0.99	0.99	0.99	216
NEUTRAL	1.00	1.00	1.00	215
POSITIVE	0.99	0.98	0.99	209
accuracy			0.99	640
macro avg	0.99	0.99	0.99	640
weighted avg	0.99	0.99	0.99	640

accuracy:

0.990625

FIGURE 5.7

Confusion matrix classification report of XG Boost.

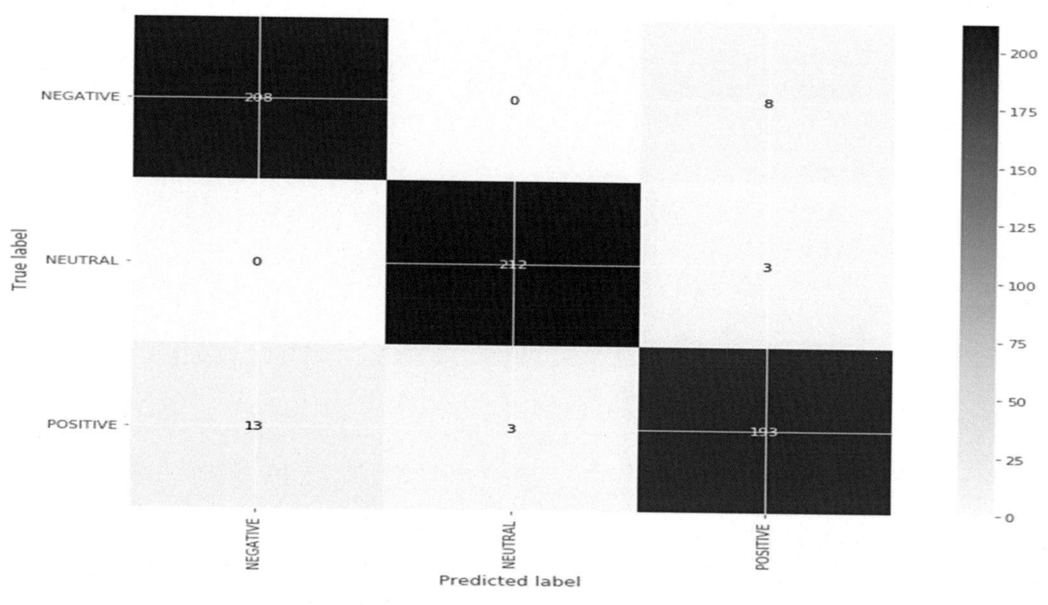

FIGURE 5.8

Confusion matrix of random forest.

parameters like precision, recall, and F1 score is about 96%. So from the confusion matrix, it concludes that the RF model offers a better agreement of classification of sentiment analysis.

5.4.2 Execution time

It is the time taken to execute the testing datasets after symmetrical analysis of the 20% brainwave datasets

Random Forest

```
Classification Report
              precision    recall  f1-score   support

    NEGATIVE       0.94      0.96      0.95       216
     NEUTRAL       0.99      0.99      0.99       215
    POSITIVE       0.95      0.92      0.93       209

    accuracy                           0.96       640
   macro avg       0.96      0.96      0.96       640
weighted avg       0.96      0.96      0.96       640

accuracy:

0.9578125
```

FIGURE 5.9

Confusion matrix classification report of decision tree.

```
CPU times: user 102 µs, sys: 20 µs, total: 122 µs
Wall time: 136 µs
```

XG Boost

```
CPU times: user 20.8 s, sys: 212 ms, total: 21.1 s
Wall time: 9.83 s
```

Decision Tree

```
CPU times: user 14 µs, sys: 0 ns, total: 14 µs
Wall time: 17.6 µs
```

5.4.3 Misclassified samples

Random Forest

```
Misclassified samples: 24
Accuracy: 0.96
```

XG Boost

```
Misclassified samples: 6
Accuracy: 0.99
```

Decision Tree

```
Misclassified samples: 27
Accuracy: 0.96
```

5.4.4 Receiver operating curve

From the confusion matrix, specificity, and sensitivity we can extricate a presentation plot called the receiver operating (ROC) curve (Dutta et al., 2021, 2022) with the help True positive rate (TP rate) against the False positive rate (FP rate). In ML, the genuine positive rate is called the probability of correctness, and the genuine negative shows the error probability. The zone under the ROC bend (AUC) is another estimation used to measure the performance effectiveness of the algorithm. It is observable that AUC can be gotten from choice limits got by ML models despite the way that it is prepared with discrete yields. At the point when a prepared model is approached to make a forecast, the likelihood can be figured out and used to create a ROC plot. Fig. 5.10 shows the ROC for all three machine learning algorithms like RF, XG Boost, and DT. From the graph, it is seen that among these entire machine learning classifier algorithms RF occupied the maximum area under the curve of ROC (AUC). Moreover, all the algorithms have an AUC of more than 0.85 which indicates the classification of emotional identification done by the above-mentioned algorithm quite satisfactorily (Fig. 5.11).

Table 5.3 is used to represent the comparative study of these three algorithms (RF, XGB, and DT) employing parameters of the confusion matrix (already shown in Figs. 5.4 to 5.9 individually) which concludes XG Boost algorithm is comparatively better than the other two algorithms RF and DT. Figs. 5.1, 5.12 and Table 5.4 represents the comparative study on Training time, Accuracy, AUC, and many misclassified datasets. From Table 5.4 it is seen that DT has the least training

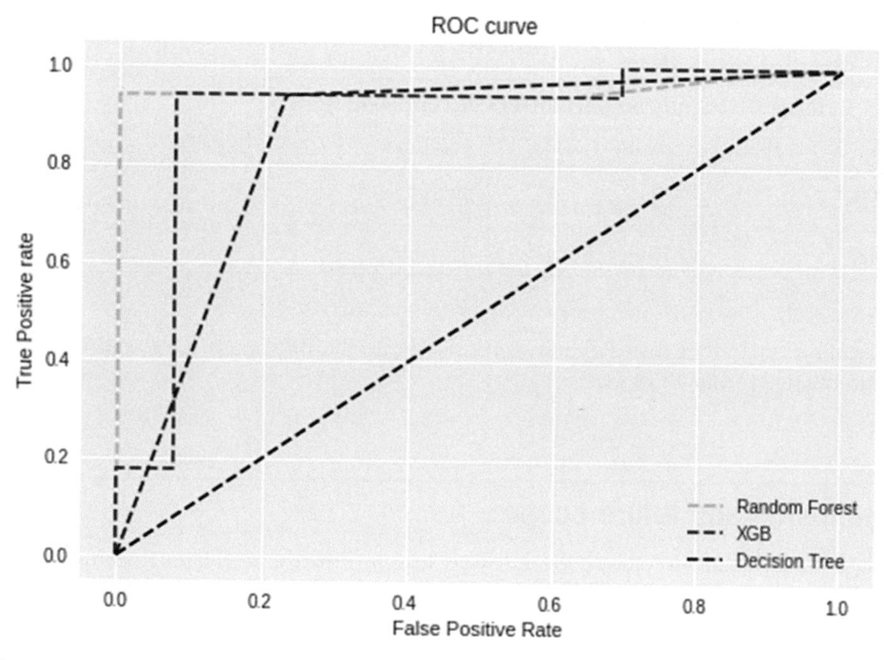

FIGURE 5.10

Comparative study for receiver operating curve graph of three machine learning algorithms.

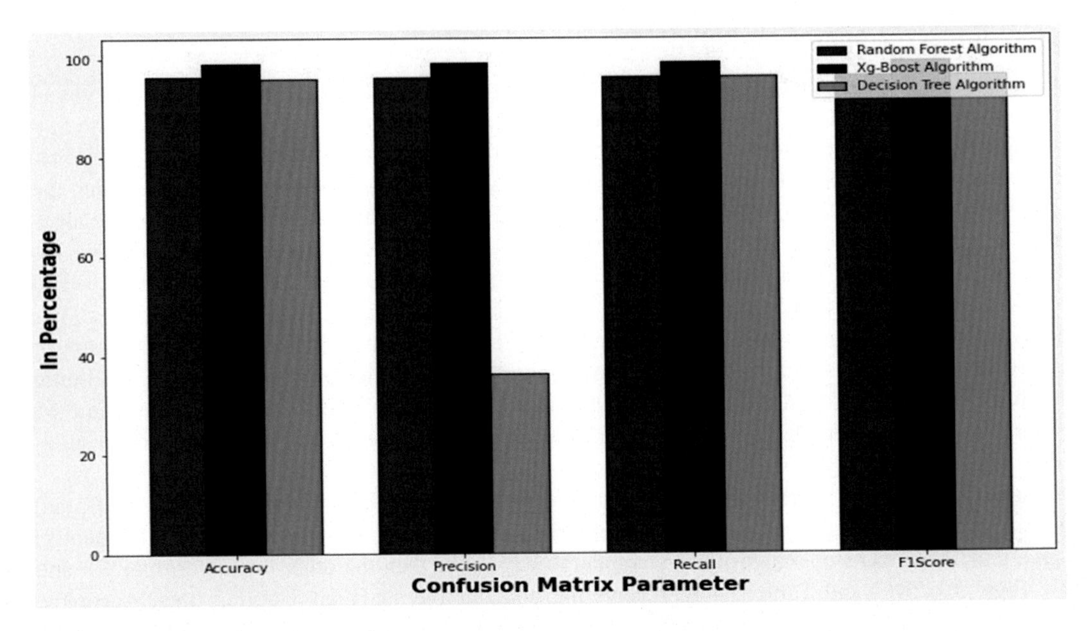

FIGURE 5.11

Comparative study on confusion matrix parameters of three algorithms.

Table 5.3 Comparative study on parameters of confusion matrix.

Algorithm	Accuracy	Precision	Recall	F1 score
Random forest	96.25	96	96	96
XG boost	99.06	99	99	99
Decision tree	95.75	96	96	96

dataset execution time other than RF and XGBoost while the maximum AUC and least misclassified data are obtained from XGBoost.

5.5 Conclusion and future scope

Clinical data classification is one of the intricate testing errands in clinical informatics. Because of its intricate nature, different strategies have been proposed in the literature. Like other clinical informatics, Emotion assumes a significant function in individuals' day-by-day life and insight. Recently, emotion acknowledgment has become an interesting issue in the fields of BCI, AI, and clinical health, particularly for the exploration and treatment of the instrument and seizure law of sicknesses, for example, dysfunctional behavior and mental issues.

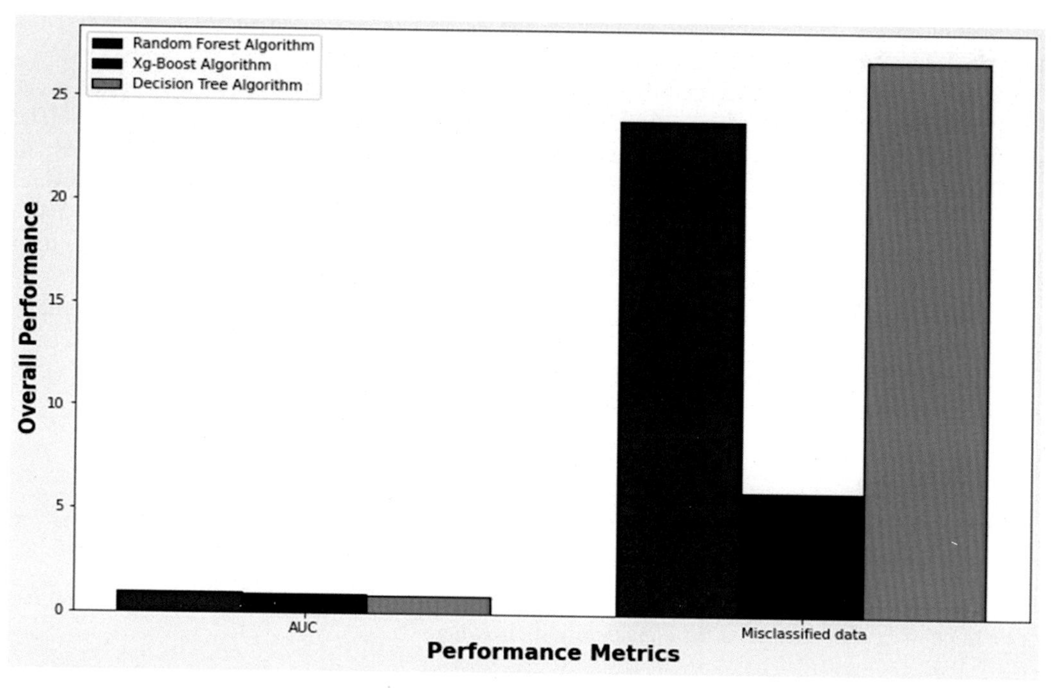

FIGURE 5.12

Comparative study on AUC and misclassified data of three algorithms.

Table 5.4 Comparative study on the overall performance index.

Algorithm	Training time	Accuracy	AUC	Miss classified datasets
Random Forest	122 μs	96.25	0.9547	24
XG Boost	21.1 μsec	99.06	0.9904	6
Decision tree	14 μs	95.75	0.8552	27

In this exploration, we have revisited this research and have put forth an attempt to introduce generalized methods for this classification task. Electroencephalogram signals are considered the best Non Invasive feeling acknowledgment-based gadget where EEG sensor classes three distinct states: neutral, relaxed, and concentrated. This course of action gathers 2549 datasets dependent on time-frequency domain statistical features taken (EEG Brainwave Dataset: Feeling Emotions Kaggle, 2019).

In this research, the overall work is performed in two stages. At the initial stage, a subset of 640 datasets was chosen by the symmetrical uncertainty feature selection to be best for further utilization of classification. In the final stage, emotion detection classification was performed by the three different machine learning algorithms such as RF, XG Boost, and DT. During the classification of the dataset using Ml techniques, the number of misclassified datasets observed is shown in Table 5.4.

The best classification model identifies based on some performance index parameters like accuracy (taken from confusion matrix), training time, AUC, and many miss classified datasets. The detailed result is described in Tables 5.3 and 5.4. From the result analysis, it is seen that among all these three algorithms XG Boost performs better by employing accuracy, the AUC, and several miss classified datasets. But the major drawbacks of the XG Boost are to train the model takes maximum time than the other two algorithms. The least training time was taken by the RF to train the model. The Accuracy of RF and DT is comparatively the same.

In future work will be centered around contrasting our best results with hybrid machine learning, Deep learning strategies in real-time applications like detecting positive and negative moods useful for applications in mental health care, emotions detection in the person's brain from music or video data, utilization of other techniques to perform better on noise removal or artifact expulsion.

5.6 Ethical statement

The authors confirm that this manuscript reports the original research works of the authors and has not been published or submitted for consideration elsewhere.

5.7 Conflicts of interest

None declared.

References

Abujelala, M., Abellanoza, C., Sharma, A., & Makedon, F. (2016). Brain-ee: Brain enjoyment evaluation using commercial EEG headband. *Proceedings of the 9th Acm International Conference on Pervasive Technologies Related to Assistive Environments*, 1–5.

Acharya, U. R., Fujita, H., Sudarshan, V. K., Bhat, S., & Koh, J. E. (2015). Application of entropies for automated diagnosis of epilepsy using EEG signals: A review. *Knowledge-Based Systems*, *88*, 85–96.

Ali, S. I., & Shahzad, W. (2012). A feature subset selection method based on symmetric uncertainty and ant colony optimization. In *2012 International conference on emerging technologies* (pp. 1–6).

Asif, A., Majid, M., & Anwar, S. M. (2019). Human stress classification using EEG signals in response to music tracks. *Computers in Biology and Medicine*, *107*, 182–196.

Atmaja, B. T., & Akagi, M. (2021). Evaluation of error-and correlation-based loss functions for multitask learning dimensional speech emotion recognition. *Journal of Physics: Conference Series*, *1896*(1), 012004.

Bhattacharyya, A., Tripathy, R. K., Garg, L., & Pachori, R. B. (2020). A novel multivariate-multiscale approach for computing EEG spectral and temporal complexity for human emotion recognition. *IEEE Sensors Journal*, *21*(3), 3579–3591.

Bird, J. J., Ekárt, A., Buckingham, C. D., & Faria, D. R. (2019). High resolution sentiment analysis by ensemble classification. In K. Arai, R. Bhatia, & S. Kapoor (Eds.), *Intelligent computing* (pp. 593–606). Springer International Publishing. Available from https://doi.org/10.1007/978-3-030-22871-2_40.

Brave, S., & Nass, C. (2007). *Emotion in human-computer interaction. The human-computer interaction handbook* (2nd ed.). CRC Press.

Celano, C. M., Daunis, D. J., Lokko, H. N., Campbell, K. A., & Huffman, J. C. (2016). Anxiety disorders and cardiovascular disease. *Current Psychiatry Reports*, *18*(11), 1–11.

Chen, Z., Jiang, F., Cheng, Y., Gu, X., Liu, W., & Peng, J. (2018). XGBoost classifier for DDoS attack detection and analysis in SDN-based cloud. In: *2018 IEEE international conference on big data and smart computing (Bigcomp)* (pp. 251–256).

Dabas, H., Sethi, C., Dua, C., Dalawat, M., & Sethia, D. (2018). Emotion classification using EEG signals. In: *Proceedings of the 2018 2nd international conference on computer science and artificial intelligence* (pp. 380–384).

Das, P., Khasnobish, A., & Tibarewala, D. N. (2016). Emotion recognition employing ECG and GSR signals as markers of ANS. In: *2016 conference on advances in signal processing (CASP)* (pp. 37–42).

Dutta, P., Paul, S., & Kumar, A. (2021). *Comparative analysis of various supervised machine learning techniques for diagnosis of COVID-19. Electronic devices, circuits, and systems for biomedical applications* (pp. 521–540). Elsevier.

Dutta, P., Paul, S., & Majumder, M. (2021). *An efficient SMOTE based machine learning classification for prediction and detection of PCOS* [Preprint]. *Review*. Available from https://doi.org/10.21203/rs.3.rs-1043852/v1.

Dutta, P., Paul, S., Obaid, A. J., Pal, S., & Mukhopadhyay, K. (2021). Feature selection based artificial intelligence techniques for the prediction of COVID like diseases. *Journal of Physics: Conference Series*, *1963* (1), 012167. Available from https://doi.org/10.1088/1742-6596/1963/1/012167.

Dutta, P., Paul, S., Shaw, N., Sen, S., & Majumder, M. (2022). *Heart disease prediction: A comparative study based on a machine-learning approach. Artificial intelligence and cybersecurity*. CRC Press.

Dutta, P., Shaw, N., Das, K., & Ghosh, L. (2021). Early and accurate forecasting of mid term wind energy based on PCA empowered supervised regression model. *2*(2), 12.

Dwivedi, R., Dey, S., Chakraborty, C., & Tiwari, S. (2021). Grape disease detection network based on multi-task learning and attention features. *IEEE Sensors Journal*.

Edla, D. R., Mangalorekar, K., Dhavalikar, G., & Dodia, S. (2018). Classification of EEG data for human mental state analysis using random forest classifier. *Procedia Computer Science*, *132*, 1523–1532.

EEG Brainwave Dataset: Feeling Emotions | Kaggle. (2019). https://www.kaggle.com/birdy654/eeg-brainwave-dataset-feeling-emotions

Faria, D. R., Premebida, C., Manso, L. J., Ribeiro, E. P., & Núñez, P. (2018). *Multimodal bayesian network for artificial perception. Bayesian networks-advances and novel applications*. IntechOpen.

Frantzidis, C. A., Bratsas, C., Papadelis, C. L., Konstantinidis, E., Pappas, C., & Bamidis, P. D. (2010). Toward emotion aware computing: An integrated approach using multichannel neurophysiological recordings and affective visual stimuli. *IEEE Transactions on Information Technology in Biomedicine*, *14*(3), 589–597.

Gumus, M., & Kiran, M. S. (2017). Crude oil price forecasting using XGBoost. In: *2017 International conference on computer science and engineering (UBMK)* (pp. 1100–1103).

Hosseini, S. A., & Naghibi-Sistani, M. B. (2011). Emotion recognition method using entropy analysis of EEG signals. *International Journal of Image. Graphics and Signal Processing*, *3*(5), 30.

Ieracitano, C., Mammone, N., Bramanti, A., Marino, S., Hussain, A., & Morabito, F. C. (2019). A time-frequency based machine learning system for brain states classification via EEG signal processing. In: *2019 International joint conference on neural networks (IJCNN)*, (pp. 1–8).

Islam, S. M. R., Sajol, A., Huang, X., & Ou, K. L. (2016). Feature extraction and classification of EEG signal for different brain control machine. In: *2016 3rd international conference on electrical engineering and information communication technology (ICEEICT)*, (pp. 1–6).

Jadhav, S., He, H., & Jenkins, K. (2018). Information gain directed genetic algorithm wrapper feature selection for credit rating. *Applied Soft Computing*, *69*, 541–553.

Jain, I., Jain, V. K., & Jain, R. (2018). Correlation feature selection based improved-binary particle swarm optimization for gene selection and cancer classification. *Applied Soft Computing*, *62*, 203–215.

Jasper, H. H. (1958). The ten-twenty electrode system of the international federation. *Electroencephalography and Clinical Neurophysiology*, *10*, 370–375.

Jia, Y., Jin, S., Savi, P., Gao, Y., Tang, J., Chen, Y., ... Li, W. (2019). GNSS-R soil moisture retrieval based on a XGboost machine learning aided method: Performance and validation. *Remote Sensing*, *11*(14), 1655.

Jie, X., Cao, R., & Li, L. (2014). Emotion recognition based on the sample entropy of EEG. *Bio-medical Materials and Engineering*, *24*(1), 1185–1192.

Kannathal, N., Choo, M. L., Acharya, U. R., & Sadasivan, P. K. (2005). Entropies for detection of epilepsy in EEG. *Computer Methods and Programs in Biomedicine*, *80*(3), 187–194. Available from https://doi.org/10.1016/j.cmpb.2005.06.012.

Krigolson, O. E., Williams, C. C., Norton, A., Hassall, C. D., & Colino, F. L. (2017). Choosing MUSE: Validation of a low-cost, portable EEG system for ERP research. *Frontiers in Neuroscience*, *11*, 109.

Kumar, R. S., Misritha, K., Gupta, B., Peddi, A., Srinivas, K. K., & Chakraborty, C. (2020). A survey on recent trends in brain computer interface classification and applications. *Journal of Critical Reviews*, *7* (11), 650–658.

Lever, J., Krzywinski, M., & Altman, N. (2016). Points of significance: Model selection and overfitting. *Nature Methods*, *13*(9), 703–705.

Li, H., He, F., Liang, Y., & Quan, Q. (2019). A dividing-based many-objective evolutionary algorithm for large-scale feature selection. *Soft Computing*, 1–20.

Lin, Y., Wang, L., Xiao, Y., Urman, R. D., Dutton, R., & Ramsay, M. (2018). Objective pain measurement based on physiological signals. *Proceedings of the International Symposium on Human Factors and Ergonomics in Health Care*, *7*(1), 240–247.

Lin, Y.-P., Wang, C.-H., Jung, T.-P., Wu, T.-L., Jeng, S.-K., Duann, J.-R., ... Chen, J.-H. (2010). EEG-based emotion recognition in music listening. *IEEE Transactions on Biomedical Engineering*, *57*(7), 1798–1806.

Liu, Y., Sourina, O., & Nguyen, M. K. (2010). Real-time EEG-based human emotion recognition and visualization. In: *2010 International conference on cyberworlds* (pp. 262–269).

Martino, D. J., Samamé, C., & Strejilevich, S. A. (2016). Stability of facial emotion recognition performance in bipolar disorder. *Psychiatry Research*, *243*, 182–184.

Murugappan, M., Rizon, M., Nagarajan, R., & Yaacob, S. (2010). Inferring of human emotional states using multichannel EEG. *European Journal of Scientific Research*, *48*(2), 281–299.

Ogunleye, A., & Wang, Q.-G. (2019). XGBoost model for chronic kidney disease diagnosis. *IEEE/ACM Transactions on Computational Biology and Bioinformatics*, *17*(6), 2131–2140.

Ohsaki, M., Wang, P., Matsuda, K., Katagiri, S., Watanabe, H., & Ralescu, A. (2017). Confusion-matrix-based kernel logistic regression for imbalanced data classification. *IEEE Transactions on Knowledge and Data Engineering*, *29*(9), 1806–1819.

Oliveira, A. S., Schlink, B. R., Hairston, W. D., König, P., & Ferris, D. P. (2016). Induction and separation of motion artifacts in EEG data using a mobile phantom head device. *Journal of Neural Engineering*, *13*(3), 036014.

Petrantonakis, P. C., & Hadjileontiadis, L. J. (2009). Emotion recognition from EEG using higher order crossings. *IEEE Transactions on Information Technology in Biomedicine*, *14*(2), 186–197.

Pourmohammadi, S., & Maleki, A. (2020). Stress detection using ECG and EMG signals: A comprehensive study. *Computer Methods and Programs in Biomedicine*, *193*, 105482.

Przegalinska, A., Ciechanowski, L., Magnuski, M., & Gloor, P. (2018). *Muse headband: Measuring tool or a collaborative gadget? Collaborative innovation networks* (pp. 93–101). Springer.

Rahman, S., Irfan, M., Raza, M., Moyeezullah Ghori, K., Yaqoob, S., & Awais, M. (2020). Performance analysis of boosting classifiers in recognizing activities of daily living. *International Journal of Environmental Research and Public Health*, *17*(3), 1082.

Ren, X., Guo, H., Li, S., Wang, S., & Li, J. (2017). A novel image classification method with CNN-XGBoost model. *International Workshop on Digital Watermarking*, 378–390.

Richhariya, B., & Tanveer, M. (2018). EEG signal classification using universum support vector machine. *Expert Systems with Applications*, *106*, 169–182. Available from https://doi.org/10.1016/j.eswa.2018.03.053.

Rodrigues, J., Das, C., Rebouças Filho, P. P., Peixoto, E., Kumar, A., & de Albuquerque, V. H. C. (2019). Classification of EEG signals to detect alcoholism using machine learning techniques. *Pattern Recognition Letters*, *125*, 140–149.

Rowland, A. S., Lesesne, C. A., & Abramowitz, A. J. (2002). The epidemiology of attention-deficit/hyperactivity disorder (ADHD): A public health view. *Mental Retardation and Developmental Disabilities Research Reviews*, *8*(3), 162–170. Available from https://doi.org/10.1002/mrdd.10036.

Ruuska, S., Hämäläinen, W., Kajava, S., Mughal, M., Matilainen, P., & Mononen, J. (2018). Evaluation of the confusion matrix method in the validation of an automated system for measuring feeding behaviour of cattle. *Behavioural Processes*, *148*, 56–62.

Saikhu, A., Arifin, A. Z., & Fatichah, C. (2019). Correlation and symmetrical uncertainty-based feature selection for multivariate time series classification. *International Journal of Intelligent Systems*, *12*(3), 129–137.

Sameer, M., Gupta, A. K., Chakraborty, C., & Gupta, B. (2019). Epileptical seizure detection: Performance analysis of gamma band in EEG signal using short-time Fourier transform. In: *2019 22nd international symposium on wireless personal multimedia communications (WPMC)* (pp. 1–6).

Shi, Y., Zheng, X., & Li, T. (2018). Unconscious emotion recognition based on multi-scale sample entropy. In: *2018 IEEE international conference on bioinformatics and biomedicine (BIBM)* (pp. 1221–1226).

Suo, G., Song, L., Dou, Y., & Cui, Z. (2019). Multi-dimensional short-term load Forecasting based on XGBoost and fireworks algorithm. In: *2019 18th international symposium on distributed computing and applications for business engineering and science (DCABES)* (pp. 245–248).

Sen, S., Saha, S., Chaki, S., Saha, P., & Dutta, P. (2021). Analysis of PCA based AdaBoost machine learning model for predict mid-term weather forecasting. *Computational Intelligence and Machine Learning*, *2*(2), 41–52.

Symeonidou, E.-R., Nordin, A. D., Hairston, W. D., & Ferris, D. P. (2018). Effects of cable sway, electrode surface area, and electrode mass on electroencephalography signal quality during motion. *Sensors*, *18*(4), 1073.

Tian, K. (2018). Muse headband: Potential communication tool for locked-in people. *Mechanical Engineering Research*, *8*(1), 16–22.

Ullah, H., Uzair, M., Mahmood, A., Ullah, M., Khan, S. D., & Cheikh, F. A. (2019). Internal emotion classification using EEG signal with sparse discriminative ensemble. *IEEE Access*, *7*, 40144–40153.

Valenza, G., Citi, L., Scilingo, E. P., & Barbieri, R. (2017). *Time-varying cardiovascular complexity with focus on entropy and lyapunov exponents. Complexity and nonlinearity in cardiovascular signals* (pp. 233–256). Springer.

Witten, I. H., & Frank, E. (2002). Data mining: Practical machine learning tools and techniques with Java implementations. *Acm Sigmod Record*, *31*(1), 76–77.

Witwer, A. N., & Lecavalier, L. (2010). Validity of comorbid psychiatric disorders in youngsters with autism spectrum disorders. *Journal of Developmental and Physical Disabilities*, *22*(4), 367–380. Available from https://doi.org/10.1007/s10882-010-9194-0.

XingFen, W., Xiangbin, Y., & Yangchun, M. (2018). Research on user consumption behavior prediction based on improved XGBoost algorithm. In: *2018 IEEE international conference on big data (Big Data)* (pp. 4169–4175).

Xu, J., Zhang, Y., & Miao, D. (2020). Three-way confusion matrix for classification: A measure driven view. *Information Sciences*, *507*, 772–794. Available from https://doi.org/10.1016/j.ins.2019.06.064.

Xue, H., Du, J., Li, S., & Ma, W. (2018). Region of interest encryption for color images based on a hyperchaotic system with three positive Lyapunov exponets. *Optics and Laser Technology*, *106*, 506–516. Available from https://doi.org/10.1016/j.optlastec.2018.04.030.

Youssef, A. E., Ouda, H. T., & Azab, M. (2018). MUSE: A portable cost-efficient lie detector. In: *2018 IEEE 9th annual information technology, electronics and mobile communication conference (IEMCON)* (pp. 242–246). Available from https://doi.org/10.1109/IEMCON.2018.8614795.

Zhang, D., Qian, L., Mao, B., Huang, C., Huang, B., & Si, Y. (2018). A data-driven design for fault detection of wind turbines using random forests and XGboost. *IEEE Access*, *6*, 21020–21031. Available from https://doi.org/10.1109/ACCESS.2018.2818678.

Zhang, Y., Ji, X., Liu, B., Huang, D., Xie, F., & Zhang, Y. (2017). Combined feature extraction method for classification of EEG signals. *Neural Computing and Applications*, *28*(11), 3153–3161. Available from https://doi.org/10.1007/s00521-016-2230-y.

CHAPTER

6 Role of artificial intelligence and internet of things based medical diagnostics smart health care system for a post-COVID-19 world

Sanjay Kumar Sinha
Department of Physics, Birla Institute of Technology, Mesra, Patna Campus, Patna, Bihar, India

6.1 Introduction

Post Corona Virus period will have a large number of people who have recovered from the infection. As per a report (Li et al., 2020), the virus infection makes the immune system weak and it has many after-recovery side effects on the allied body parts apart from the lungs. Lungs are the first most affected body part but have a damaging effect on cardiac, nephron, and other body organs. Those persons who have already suffered from an autoimmune disease such as diabetes, lupus, rheumatoid arthritis, etc. are in the most health altercation position after recovering from corona diseases. Autoimmune diseases are the disorder of the immune system, in this disease, the immune responses are already been suppressed by the medication, thus these patients have a chance to develop serious illnesses after recovery from corona disease (Li et al., 2020). A report says that most of the casualty due to infection of coronavirus was reported in those persons whose immune system was already compromised.

The present chapter is focused on the latest available technique and technology helpful in monitoring a large number of people having after corona disease effect. The most favorable way of monitoring a large number of people together can be possible only through the online wireless monitoring system. Artificial intelligence (AI) and machine learning (ML) technique-based systems can only handle this kind of post COVID scenario, as it is quick, accurate and many a time automatic. Thus present book chapter is focused on the review of the present latest AI/ML-based health monitoring systems. Separate sub-topics on cardiac, nephrology, diabetics, etc. have been taken elaborately. The health monitoring system shall be capable of monitoring diseases such as cardiac, nephrology, and diabetes.

Healthcare is an essential area in which IOTM technology can improve the quality of services beyond imagination. According to a report (Petrlík et al., 2020), IOTM based diagnostic devices in the healthcare trade will be more than \$117 Billion by 2020 and As per another report (Markopoulou, Papakonstantinou, & de Hert, 2019), there may be a yearly increase speed of 38% in the IOTM (Internet of Things) based health monitoring system in next five years. IoT has a high possibility to make the masses independent to understand the diseases; Doctors have also found it

Implementation of Smart Healthcare Systems using AI, IoT, and Blockchain. DOI: https://doi.org/10.1016/B978-0-323-91916-6.00006-0

useful to make a decision, especially in complicated cases. IOTM wearable devices (medical sensors) are useful for recording various body parameters of the patient like comprehensive pressure, fever, physics activity, heart rate (HR), etc. A real-time IOTM-based system is capable to deliver the data to caregiving medical centers, doctors, or family members for proper treatment (Dwivedi, Dey, Chakraborty, & Tiwari, 2021).

Integration of Medical transducers available as a Wearable with other AI-based IOTM-systems are completely changing the work culture of the Healthcare sector. Now even a non-medical person can monitor the well-being state of a patient remotely by using this equipment. These wearable sensors are available in attractive forms such as wrist bands, garments, goggles, caps, footwear, and other devices such as mobile phones, hearing devices, bands, etc. Medical-related monitoring systems are categorized based on their use, for example—to monitor sensors for physiological behavior that is electrocardiogram (ECG), EMG, EEG, there are physical sensors, For sensing sweat, glucose, saliva- chemical sensors, for optical sensing such as oximetry, tissue properties, etc. (Mugisha et al., 2019).

Medical Sensors as a Wearable form with IOTM Applications and ML technique is now becoming an integral part of the health industry. It is granting a better quality of life to the patient. Real-time monitoring is saving lives and making treatment costs affordable. IOTM, ML, and big data-based technologies make it possible to do an error-free accurate diagnosis (De Hert, Detraux, Van Winkel, Yu, & Correll, 2012; Mebazaa et al., 2010; Mugisha, De Hert, Stubbs, Basangwa, & Vancampfort, 2017; Sinha, Kumari, & Chaudhary, 2019a, 2019b) IoT-based medical sensors system can record accurately and can send data related to comprehensive pressure, fever, physics activity and HR, blood sugar, etc.

Major contribution in this chapter includes the useful discussion on the challenges and probable solutions on the topics related to the internet of thing (IOTM), artificial neural networks (ANN), and ML in the field of the heart monitoring system. 10-fold cross-validation techniques responsible for high accuracy in heart disease detection is become possible due to the introduction of ANN in the heart monitoring system. The second major contribution in the chapter is the discussion on the proposal of fog computing in the place of cloud computing to improve the efficiency of the continuous glucose monitoring (CGM) system. The discussion on a non-invasive kidney monitoring system is also a major contribution to this chapter, especially the introduction of AI and ML-based voice recognition in kidney monitoring is going to make the system highly useful.

The chapter is organized into various subsections for example, smart cardiac monitoring system, mobile ML model, ANN based diagnostics, smart glucose monitoring system, mid 20 models for integrated monitoring system, smart kidney monitoring system, real-time monitoring of GFR, contrast-enhanced ultrasound and other technique, recent challenges in AI & ML system and the future work.

6.2 Challenges

This new emerging monitory system has a few challenges. IoT-based patient monitoring has a few drawbacks related to the error in analysis and acceptability among the medical fraternity is another problem. A few others such as Security & Privacy Issues, since these devices, capture, private

Table 6.1 Challenges in artificial intelligence and machine learning-based monitoring systems.

Sr. number	Challenges
1.	In renal GFR, a better detector to monitor the renal glomerular filtration rate is required.
2.	There is an issue related to the transfer of data from microservice computing to serverless computing in an AI-based monitoring system.
3.	Handling multisensory data and sharing information in the monitoring system requires a better security system to avoid any misuse of the data.
4.	The compatibility of monitoring devices due to the use of different communication protocols by various manufacturers needs a proper solution.
5.	The present monitoring systems require better high storage capacity and they should have better-elevated cell capability.

health-related information, and these data are highly vulnerable as being in the public domain through the internet. Thus it may attract unethical people for misuse (Queralta, Gia, Tenhunen, & Westerlund, 2019).

Interpretability or comprehensibility of the data is a big issue for the last many years in the field of ML. The involvement of deep learning techniques in ML makes it very difficult to explain the logic behind the decision in simple man terms, which becomes mandatory after the law imposed by European Union. Decades of research and investment by scientists and technologists are now in a situation where it is in the dark future of implementation of outcome. This era of application of revolutionary outcomes of information technology in the field of medical science is now an unknown future that requires the attention of lawmakers.

As mentioned above, as per so far reported research there is still research required to bring out a proper IOTM and AI-based renal monitoring system to help the kidney patient, especially for those who are on kidney dialysis and require continuous monitoring of GFR and kidney functioning (Aymé et al., 2017).

Table 6.1 is placed below to illustrate the recent challenges point-wise in the field of AI and ML-based monitoring systems.

6.3 Smart cardiac monitoring system

In the post period of the corona pandemic, it is predicted that due to side effects of Corona or due to the medication during the corona period the heart failure cases may go up and may become difficult to control. Hence in the present writing work, a brief account of the so far reported latest techniques and technology to monitor the heart health of the patient from a remote place. Detection of cardiovascular abnormalities at the initial stage and an efficient system of proper monitoring and counseling by expert Doctors would be a great help in saving the life of people.

At present days due to the introduction of Big Data application in the field of preserving the medical history of the patient, there is a useful implication of data in designing the predictive model for cardiovascular diseases. Machine learning technique or data mining is being used for analyzing big data and making it possible to decide on a particular case. Data mining is non-trivial and has many attractive features useful for the patient monitoring system. Data mining have the noble feature to discover the hidden patterns or similarity of data available through big data. Health care industries are generating a large amount of data on every disease and those are nowadays available online for further analysis. The machine learning algorithm is found most suitable for the prediction and monitoring of heart disease of a patient. However, monitoring systems require integrating the IOTM along with various physiological signal sensors as well as the microcontroller (Cai, Xu, Jiang, & Vasilakos, 2016; Lalit, Emeka, Nasser, Chinmay, & Garg, 2020).

IOTM has recently emerged as a rapidly expanding technology that handles multiple sensor data and shares information using communication systems such as private networks, Internet protocol (IP), or public networks. IOTM-based products having embedded technology are having the capability to exchange information among billions of people through smartphones and mobile communication. Sensors meant for heart health and microcontroller integrated system can collect the data after a specific interval of time, analyze it and pass the information among the concerning people related to the patient. The integrated system uses an intelligent cloud-based network for analysis, planning, and decision-making (Ganesan & Sivakumar, 2019).

Let us discuss the challenges and the probable solution in the latest available facility in the field of technology for the monitoring of heart patients.

6.3.1 Mobile machine learning model

There are challenges and shortcomings related to the late detection or the wrong detection of a cardiac problem in a patient by providing the provision for simulation of data from both the online as well as off-line obtained that is through various sensors and due to detected so far responsible for the error in results of the present cardiac detection system. These problems are related to lack in the nano sensors used for the measurement. The author of the paper has developed a few sensors having better piezo electric responses (Sinha et al., 2019a, 2019b). A mobile machining learning system known as Mobile Machine Learning Model for Monitoring Cardiovascular Disease (M4CVD) is reported. This is a quite friendly system helpful in establishing the live connection to the entire stakeholder related to the patient. Support vector machine is utilized to keep an eye on the collection of data from various wearable sensors and clinical databases (Boursalie, Samavi, & Doyle, 2015).

The reported proposed system is known as M4CVD that is mobile machine learning model (Muhammad et al., 2021) for monitoring of cardiovascular disease. In the reported system support vector machine (SVM) is also helpful in making the final decision. In the system, the decision output data is communicated to a server for storage purposes and it is also available for remote viewing. There are three components of the system architecture that is Input, Machine learning consisting of data processing as well as decision making, and output. Data processing, Machine learning, and decision making, all can be performed through mobile devices. The architecture of M4CVD redrawn based on the reported article is shown in Fig. 6.1.

In the reported paper, the input data were collected from two sources. The first source is a physiological signal measured through wearable sensors. The second is the clinical data from health records. Physiological sensors supply the signals such as ECG, blood pressure (BP), accelerometer, and stress level measurer. Stress level measurement is carried out by a galvanic skin response

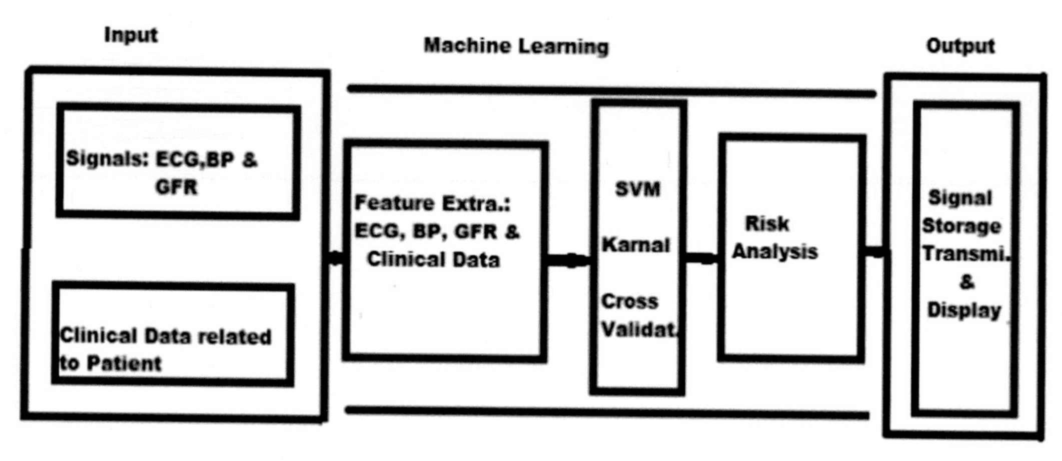

FIGURE 6.1

Architecture of mobile machine learning model for monitoring cardiovascular disease.

sensor. During accelerometer provides the data for the physical activity. Wearable sensors are required to wear by the patient; the output signals by various sensors are then collected by an application-specific integrated circuit or by a data acquisition microcontroller. Here database includes records of physiological readings such as ECG, BP, and SpO_2. Here provision for entry of the clinical data is also there. In this system, clinical data includes information such as age, gender, medication record, body mass indexed, and old records related to ECG, BP, SpO_2, etc.

M4CVD system has successfully solved the shortcomings related to the late detection or the wrong detection of a cardiac problem in a patient by providing the provision for simulation of data from both the online as well as off-line obtained that is through various sensors and due to detected so far responsible for the error in results. The first one is related to the data collected from sensors and the second one is data entered manually by the treatment team. Since there is difficulty in getting information if data are entered manually that is clinical data as it is not as fast as the feeding through the sensors. In this system dual accumulation of data sets and the matching from already recorded is carried out through the SVM.

Data processing is carried out by the SVM unit of the system. Initially, the input data have noise and environmental effects such as interference of already existing fields. These noises and disturbances make it difficult to do the extraction of the information. Hence use of a signal-to-noise ratio filter solves this problem.

Systems have another special feature related to the conversion of a continuous signal into a discrete one, discrete features provide better input to the SVM. Analysis of discrete data helps produce accurate output. In this system data from various sensors are analyzed together, they require various techniques of intelligence. For example, Systolic BP has a physiological range from 120 to 150 mmHg, BMI has a smaller range from 16 to 50 kg/m^2, and analyzing of heartbeat requires a different kind of intelligence. This problem is sorted out here by a normalization technique by assuming the threshold value such as the patient's BP is either above the level ($BP = 1$) or below the level ($BP = 0$). Similarly, the threshold is assumed in other measurements or analyses also.

BMI is represented by category as the highest risk, high, medium, or low. ECG signals are computed as time-domain discrete signals in which HR and R-R interval are normalized for normal or exaggerated.

6.3.2 Artificial neural network based diagnostics

One of the published reports (Learning, 2017) nicely presented the application of ANN in the diagnosis and communication of the result related to heart patients. Neurons are the processing units in this system; the architecture of the neural network comprises layers of neurons and connections between them. Layers of neurons are termed here as subgroups. ANN-based heart monitoring technique is reported to have high predictive power that is classifier property. ANN has another noteworthy accurate capacity related to fault tolerance and learning from the environment. Output result in ANN diagnostic system is unsupervised learning type, hence only the inputs are required for the unknown target. The required data set can be in the form of attributes and instances. An ANN is efficiently able to compute the back propagation learning algorithm on the available data. In this reported system target samples are divided as 60% training set, 20% validation set, and the remaining 20% for the test set. Tangent sigmoid for hidden layers and linear transfer function for the output layer is used in this system as the activation function. The reported ANN diagnostic system has 88% classification accuracy for heart disease and it has a mean square error equal to 0.1071.

In one of the reports (Learning, 2017), an interesting mixture of different classification techniques is used and tested for a high accuracy heart disease detection and monitoring system. Various classification techniques such as the J48 Decision tree, K- Nearest Neighbors (KNN), Naive Bayes (NB), and Sequential minimal optimization (SMO) have been used in the reported system. This system exhibits high accuracy of detection that is 83.732%. The gain ratio evaluation technique on dataset features for selection is used in this system to extract the important features. For the implementation of classical algorithms in this system WEKA software is employed. The high accuracy in the result in this system is said to be due to the 10-fold cross-validation technique.

6.4 Smart glucose monitoring system

Post Corona 19 pandemic eras and in general during the normal time, continuous diabetes monitoring and accordingly proper medication may save a precious life. Diabetic is considered the main cause of death by the medical fraternity. As per a report, there may be an increase in Diabetic patients to the tune of 330 million by 2025 in the world. A fast hike in the number of Diabetic patients is slowly creating an emergency sort of situation. Diabetes is a manageable disease if detection, care, and prevention measure are adopted at the proper stage. Manual testing of glucose levels is painful and requires skin puncture. The report says that there is a rapid change in glucose level in blood that is 0.125 mM/ minute, hence manual testing becomes not sufficient at the time of emergency. Careful control of glucose levels at the appropriate time is essential for diabetic patients. Type 1 diabetes mellitus (T1DM) may convert into hypoglycemia if the glucose level remains on the higher side for a long. Hence proper continuous monitoring by adopting an advanced wireless technology-based glucose monitoring system would have immense help for the patients.

6.4.1 Continuous glucose monitoring system

Continuous monitoring of glucose levels in the blood may become a lifesaver for the diabetic patient as in a report (Ding & Schumacher, 2016) it is told that uncontrolled blood sugar leads to loss of conciseness and heart failure. A neurological disorder is also reported in a diabetic patient if the elevated blood sugar remains unrestrained. Abnormal glucose levels are reported to be the main cause of blood vessel damage; it is also reported to be causing blindness and serious kidney disease. A 2–4 times increase in heart failure is reported due to high blood sugar. (Beier et al., 2011) Controlling the blood sugar properly may save the priceless life of the patients. Continuous monitoring and alarm system may be very effective in stopping the diabetic disease from a pandemic in the coming time. As per one report of the diabetes control and complication trial, the tight control of glucose levels (80 to 110 mg/dL) decreases the occurrence and harshness of long-term difficulties (Beier et al., 2011).

The flow diagram of a general diabetic monitoring system is shown in Fig. 6.2. It has a provision for three various types of data saving and transmission. The first decision is taken by the intelligent system itself if the blood glucose level is normal. The second decision by the system is communicated after the validation of the doctor. The third data save provision is related to the monitoring of the dose of insulin administration, thus the monitoring system has three various kinds of data of a patient. This continuous monitoring of glucose is highly helpful for a patient.

No doubt that automated continuous Glucose Monitors have many advantages over conventional blood sample testing techniques. The conventional one requires the active involvement of the patient as well as the caretaker, there is a risk of infection in this method also. Rapid fluctuation in blood glucose levels also makes the conventional method not appropriate for taking a decision or for proper medication. While the continuous glucose monitor has many expediencies in use and

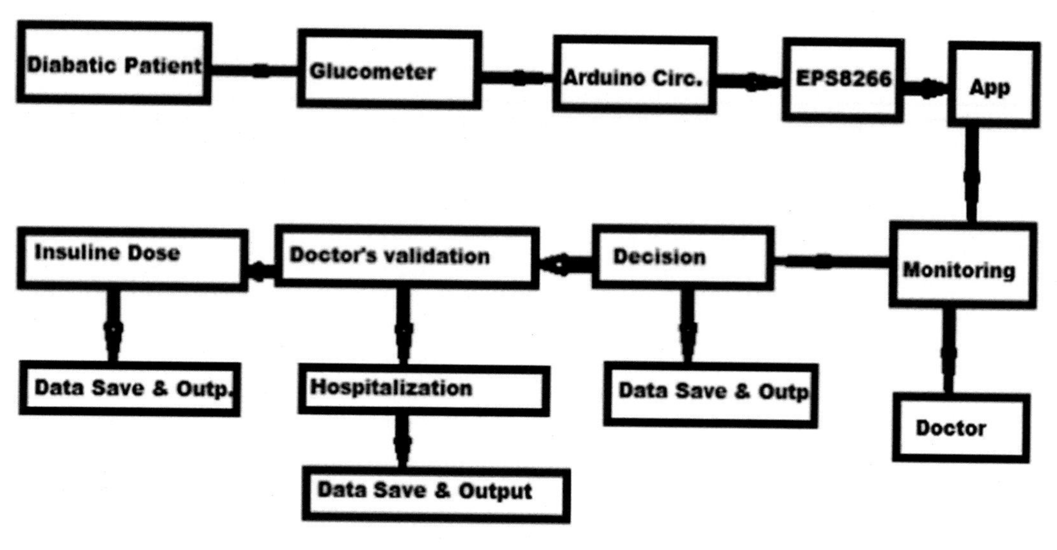

FIGURE 6.2

Flow diagram of a glucose monitoring system.

advantages. Being a small device having nanosensors to read the various parameters, the CGM system is patient-friendly; the patient has not had to go through the troublesome needle-based finger pricking conventional system. CGM grants the facility to do precise control over blood glucose levels by the patient as well as the connected health personnel or the family members. In diabetic patients, hyperglycemia (high blood glucose level) and hypoglycemia (low blood glucose level) are frequently been observed, in the whole day or even after taking insulin, the blood glucose goes to the lower side immediately. Thus, these types of fluctuations in blood sugar levels can be monitored and managed by CGM system. The first challenge related to keeping a CGM system is the high cost of the instrument. Another one is related to frequent calibration of the instrument to obtain error-free data. Calibration is done by using manual needle invasion on the skin. If suppose calibration is required to do twice or thrice in 24 h, then this kind of continuous monitoring system is not considered better than the old conventional one.

Apart from the above-mentioned points, the integration of CGM with IOTM (internet of thing) as well as the use of a remote cloud system to store the signals is a few other challenges. Effective use of AI in analyzing the data available in the cloud is also an open challenge. Once cloud computing comes into the scene or the online communication between various stakeholders comes into the picture, immediately the security of data becomes another challenge. Another challenge is related to keeping internet communication fast. What I mean to say is that if there is the involvement of a third party then delivery of information from one party (Doctor) to the patient (Party -2) becomes slow, which may obstruct proper medication and care.

However, above mentioned many challenges are effectively been addressed by various workers (Lin, Gal, Mayzel, Horman, & Bahartan, 2017; Van Enter & Von Hauff, 2018). The problem is assumed to solve by introducing fog computing in the place of cloud computing. To minimize the leniency, the communication capabilities are advised to keep close to the sensor nodes. Distribution and storage arrangements are also advised by a worker (Van Enter & Von Hauff, 2018) to keep the system safe and fast.

6.4.2 Mid 20 model

Recently in mid-2020 (Mid-20 model), an interesting and useful paper has been published by a group of workers (Rghioui, Lloret, Harane, & Oumnad, 2020). This system has a commonly accessible platform for patients and Doctors. Various data such as blood glucose level, body temperature, motion or movement data, etc. are recorded and provided by the system. This diabetic care surveillance system has an advantage such as it sends the data as fast as the wishes of the doctor or patient. The architecture of the system also has a provision for the smart analysis system, which not only collects the data but does a fast accurate analysis, and if there is an abnormality found then it is immediately been reported with an alarm to the concerning Doctor.

As per the Mid 20 model, it proposed that the architecture has an integrated system having various wearable Devices such as glucose sensors, motion sensors, and temperature sensors connected to an android based application. Continuous data is stored in a Data storage system accessible to Doctor as well as the patient. Any alarming data is immediately been conveyed to the doctor and patient through a real-time monitoring alert system. Every day's continuous data makes the patient aware of balancing the diet and workout exercise to keep the disease under control.

In Mid-20 diabetic system consists of three main components—sensor module, data acquisition module, and the database server. The database server is a mobile application installed on a Smartphone and this acts as a local server (Rghioui et al., 2020).

The sensor module has several sensors as shown in the above diagram. The glucose sensor, motion sensor, and temperature sensor are wearable devices in the form of an integrated wrist band. All three sensors are controlled by an Arduino Uno microcontroller. Arduino Uno is connected to a Smartphone wirelessly through Bluetooth. The received data that is glucose level, temperature range, and motion or workout related are sent to the database section through cellular network 4G.

The processing unit of the Mid 20 diabetic system consists of components such as a monitoring system and a Doctor's phone as components. The processing unit obtains the sensor data through a Smartphone by a 4G communication network. Data is processed and if any abnormality is detected then it is immediately informed to Doctor's phone. A provision to access the patient's data by family members is also being provided in the Mid-20 diabetic system. To prevent any unauthorized access to the data, a personal access code for family members can be generated, this makes this system safe and secure. A wearable integrated sensor device has been made flexible and very lightweight. It is easier to carry the wearable sensor while outside movement. Hence a real-time diabetic patient monitoring system is accessible to doctors as well as family members and it is priceless in case there is something serious issue with the patient. This system is a combination of AI and machine learning and it is making the health monitoring system smart by connecting all concerns and helping in taking decisions.

The hardware part of the Mid 20 system consists of a microcontroller integrated circuit with a Wi-Fi connection that is ESP8266, Glucose sensor, Pedometer, and Arduino Nano Board.

EPS 8266 can be programmed in various ways that is with the Arduino IDE3, in JavaScript or C language. Integrated circuit EPS 8266 is a popular component for the low-cost solution for the IoT. The EPS 8266 communicates the data by using Wi-Fi network technology. It has the capability of very precise location service; precision is in the order of 1 cm. An obstacle like a wall or any other object does no effect in communicating signals (Rghioui et al., 2020).

The glucose sensor produces the data related to the blood glucose concentration of the patient. This sensor is an integrated part of the CGM system. CGM is capable of measuring the glucose level in interstitial fluid that is the fluid surrounding the cell. The signal in the range of microvolt is picked up by the sensitive piezo sensors and that microvolt is further amplified for further processing. Calibration of the signal with the corresponding blood glucose range helps in determining the glucose level.

A pedometer is used here to monitor the physical activity of the patient. Physical activity is related to the burn of energy or calorie. Understanding the required burn rate of carbohydrates per day helps the patient to keep the glucose level under control. Continuous monitoring of the blood glucose level data makes the family members also alert mode and they can better decide what to serve the patient.

Running the complete program is done by Arduino Nano board. It helps in providing a suitable solution for controlling the electronics part of the system. Arduino is a small board capable of performing various tasks such as serial communication interface and input-output interface. Arduino works with a mini USB cable, which can also be used to transfer the data into a computer. The data collected from sensors are first sent to the microcontroller of the Arduino board and the data is processed, the glucose level is provided in the unit mg/dL (Rghioui et al., 2020).

6.5 Smart kidney monitoring system

The kidney being a vital organ of the body requires proper care during and post corona-19 period. A few reports on Corona infection reported the damage to the kidney and made it less efficient after the corona infection. (Li, 2020) The main body part affected by the coronavirus is the lungs. The lung is the main infected organ due to the coronavirus. However, the function of the kidney and lungs are correlated. Respiration and fluid metabolism is well established. The lungs regulate the water passage while the kidney controls the water. Thus the normal distribution of required water in the body is coordinated by both that is kidneys as well as lungs. Hence once the lungs of a person get infected then it usually reflects poor function or damage to the kidney also.

Smart renal function monitoring in the post-Corona-19 period is helpful, a non-invasive monitoring system capable of providing continuous data related to kidney health is the demand of time.

The kidney is small in size that is 10 cm (approx.) having a huge load of blood filtration per day. On average kidney filters 200 L of blood in 24 h intervals of time. Hence a little deficiency and inefficiency of kidney function can create a life threat.

It is well known that after the use of food for energy and self-repair body throws the waste into the blood. The most common waste product goes into the blood are urea and creatinine. The kidney has to filter out these toxins, but there are other toxins also to be filtered out from the blood. The kidney has the responsibility to filter properly the blood and returns the essential amino acid, vitamins, glucose, hormones, etc. in the blood flowing through it. The kidney also generates deficient hormones as well as balances the sodium and potassium electrolytes in the blood. Thus one can imagine the high workload of a kidney (Kumar, Sinha, & Sinha, 2017).

A few other essential functions of the kidney are the production of hormones-Renin and Erythropoietin. Improper production of Renin causes high or low BP. Similarly, Erythropoietin balances the formation of red blood corpuscles. Hence low production of RBC is also related to the kidney. The formation of vitamin D is also the duty of the kidney. Thus keeping the vital organ kidney health is essential for a healthy life.

6.5.1 Real-time monitoring of glomerular filtration rate

Non-invasive monitoring of renal function monitoring becomes helpful in the case of a critical condition of the patient. Real-time monitoring of renal GFR (glomerular filtration rate) is reported by workers (Rabito, Moore, BougaS, & Dragotakes, 1993). In this system cadmium-tellurium (Cd-Te) detector is used for non-invasive monitoring. Apart from the detector, a preamplifier and discriminator section constitutes the monitoring unit. In this real-time monitoring system, the filtration rate technique is used to determine the GFR. In one of the techniques, the clearance of the glomerular filtration rate is recorded with the help of a detector. The measurement of glomerular filtration is done for a very short interval of time to make the measurement approach a real-time continuous measurement. The normal glomerular filtration rate for a healthy man is 100 to 130 mL/min/1.73 m^2, while for a lady the range is 90 to 120 mL/min/1.73 m^2. This technique is having one issue related to the fluctuation of activity measurement as the renal glomerular filtration rate varies between ranges. This finally results in an error in the calculation of absolute renal function. There is a scope for improvement of this system by introducing a better-improved processing unit. The reported system does not have any IoT-based communication technique. There is a fair possibility of introducing the architecture of the system (Rabito et al., 1993).

6.5.2 Contrast-enhanced ultrasound and other techniques

One another report of real-time monitoring of renal blood flow is there by using a contrast-enhanced ultrasound technique (CEU). The CEU technique is a non-invasive imaging technique capable of monitoring the tissue blood flow. The rate of blood flow in this technique is determined by comparing the change in blood flow in normal conditions and after giving a high-protein food to the patient. High protein food accelerate the blood flow through the kidney at 2 h time interval. CEU technique is considered safe as there is no use of any radioactive kind of material in this technique as it was used in the earlier discussed method. Tissue blood flow measurement existing method involves radioactive tracer that is APH. In that method to clear the radioactive effect in the blood injection of a bolus dose is required. In the existing method, constant infusion and several timed urine and blood sample collection are needed to determine the renal blood flow rate. The nuclear medicine technique has required an alternative as the excretion of the injected radioisotopes clearance from blood and passing out by urine takes more than 48 h. A side effect of radioisotopes is also not been ruled out (Kalantarinia, Belcik, Patrie, & Wei, 2009).

Another report using computerized tomography, magnetic resonance imaging, and positron emission tomography to find out the renal blood flow is published recently. This imaging technique gives better high spatial resolution images easily. But due to the high cost of the instruments, toxicity involves with the contrast drug and many more restrictions make this technique not suitable for the determination of renal health (Andrews et al., 2014).

Recent most of work on real-time renal monitoring systems is focused on the problem of uncontrolled water or solute removal during dialysis. In the dialysis process, ultrafiltration (UF) and the dialyzed sodium profile should be controlled in such a way that there should be less formation of water or solute. However, as per the recent report published in the year 2020, (Jung et al., 2020), the present technology available for dialysis is not adequate to address the issue of production of a large quantity of water or solute. Uncontrolled production of water solute is the cause of most death during dialysis or making the patient senseless or patient goes down into a coma. Thus most of the recent work is on the development of a system having various biosensors to determine the important chemical/physical signal and accordingly able to do the adequate change in the dialysis process such that the solute level could be maintained and the problem of maintenance of required solute in the blood should be maintained during dialysis. However, the present problem during dialysis requires a monitoring system having smart biosensors capable of monitoring the solute such as sodium, potassium, vitamin D as well as essential hormones. It's not easy to monitor and control excess water generation during dialysis. But, the recent development and application of AI and ML grants confidence that this problem can be solved. Earlier contribution of AI and ML in the field of therapeutic image investigation, voice recognition, surgery, etc. indicates that a shortly effective solution for monitoring kidneys would also become possible (Cunningham, Danese, Olson, Klassen, & Chertow, 2005).

6.6 Result and discussion

Let us discuss the findings and the facts related to the above discussed various monitoring systems for example, heart monitoring system, glucose monitoring system, and Kidney monitoring system. As discussed in section 8.4.1 on heart monitoring system M4CVD system (Mobile machine learning model for monitoring of cardiovascular disease), BMI (Body mass index) is the basis for co-relation with BP and the HR. The problem of accurate correlation of BMI is sorted out here by adopting a customized kind of intelligence. Systolic BP has a physiological range from 120 to 150 mmHg, BMI has a smaller range from 16 to 50 kg/m^2, and analysis of heartbeat requires a different kind of intelligence. This problem is sorted out here by a normalization technique by assuming the threshold value such as the patient's BP is either above the level (BP = 1) or below the level (BP = 0). Similarly, the threshold is assumed in other measurements or analyses also.

In the case of the glucose monitoring system, as mentioned in above column 8.4.1 the rapid change in glucose level in type 1 diabetic patient normally changes very rapidly that is 0.125 mM/minute increases the chances of hypoglycemia. To avoid this grave situation, a CGM system having wireless technology is a good solution. In the above-discussed column, the proposed system having EPS 8266 can be programmed in various ways that is with the Adriano IDE3, in JavaScript or C language. Integrated circuit EPS 8266 is a popular component for the low-cost solution for the IoT. The EPS 8266 communicates the data by using Wi-Fi network technology. It has the capability of very precise monitoring.

The kidney function monitoring system is in the development stage, and the proper detector or transducer still requires further investigation and research. As discussed in above section 8.5 the monitoring of the dialysis process have many issues as there is the formation of a high amount of solute or water due to the comparability of proper continuous monitoring of sodium profile. Until and unless efficient sensors for monitoring solutes such as sodium, potassium, hormones, and vitamin D are not been available, it is difficult to propose a non-invasive wireless monitoring system to monitor kidney health. But, the recent development and application of AI and ML grants confidence that this problem can be solved shortly.

Comprehensibility of the data obtained by monitoring systems by applying AI and ML technique is a big issue for the last many years. The involvement of deep learning techniques in machine learning makes it hard to explain the logic behind the decision in common man expression, which becomes mandatory after the law imposed by European Union in the year 2018. Decades of research and investment by scientists and technologists are now in a situation where it is in the shady future of execution of the outcome. This important aspect requires the attention of lawmakers for an amicable solution.

The disease-specific monitoring system discussed in the above columns is summarized in Table 6.2.

Table 6.2 Disease-specific monitoring systems.

Specific disease	Monitoring system
Cardiovascular disease	Mobile Machine Learning Model for Monitoring Cardiovascular Disease—8.3.1
Heart disease	Artificial neural network (ANN) based heart disease detection—8.3.2
Diabetic disease	Continuous Glucose Monitoring system-8.4.1
Diabetic disease	Mid 20 diabetic monitoring system—8.4.2
Kidney disease	Real-time monitoring of GFR—8.5.1
Kidney disease	Contrast-Enhanced ultrasound technique—8.5.2

6.7 Future work for monitoring system on pranayama (breathing system)

During the Corona-19 pandemic, many research papers are published on the report of boosting the immune system by adopting regular breathing exercises (Pranayama). Breathing exercise makes the lungs strong. The report says a group of specific Pranayamas such as Nade Sodhan, Kapalbhate, and Vastrika in a particular specific interval of time per day capable of making the lungs strong; it also helps in curing the vital organ disease such as disease of heart, kidney, or related to mental health. A recently published paper on oxygen therapy on repairing of lungs or positive speedy effect on other organ diseases validates the findings related to Pranayama. As per one of the published reports Pranayama with simple Aasna such as Surya Namaskar (Sun Salutation) benefits muscles, ligaments, and joints and improves the flexibility of vertebral joints. Pranayamas improve the vagal tone that boosts the heart functioning and protects from any cardiovascular defect. Vagal tone also improves mental health. In the report, it is claimed that Bhramari Pranayama (Breathing in a low pitch in a way) increases nitric oxide in the body in the order of 15 times the normal position. Nitric oxide allows more blood to flow into the heart and other organs. Nitric oxide keeps the nerve cell energetic and healthy. It also boosts the immune system and helps in fighting against diseases. Research also suggested that nitric oxide is also responsible for long life and keeps away aging (Sinha, Deepak, & Gusain, 2013).

Development of a monitoring system based on the correlation of above mentioned various breathing exercises and their effect on various organs such as the heart, kidney functioning, or the improvement in mental health would be highly helpful for the prevention or cure of diseases.

6.8 Conclusion

Medical Sensors as a Wearable form with IoT Applications and machine learning technique is now becoming an integral part of the health industry. It is granting a better quality of life to the patient. Real-time monitoring is saving lives and making treatment costs affordable. IOTM, machine learning, and big data-based technologies make it possible to do an error-free accurate diagnosis. This system is helpful not only in taking quick decisions but it is capable of diagnosing diseases well in advance of the disease starts. Thus its prediction capability due to its ability to make a decision based on the data available as big data is made this system revolutionary. These days, our old age senior people face a common problem of loneliness, many of them have no one to take care of at the home. In this situation, IOTM based monitoring system is a boon. The monitoring system can send an alert signal to the health care team to save the elderly patient. The same fact is true for the post or present Corona patient. A mobile machining learning system for monitoring heart patients is reported (M4CVD). This is quite a friendly system helpful in establishing the live connection to the entire stakeholder related to the patient. A SVM is utilized to keep an eye on the collection of data from various wearable sensors and clinical databases. A mid-20 diabetic system is discussed in this chapter. The architecture of mid-20 models has various wearable Devices such as glucose sensors, motion sensors, and temperature sensors connected to an android based application. Continuous data is stored in a Data storage system accessible to Doctor as well as the patient. Any alarming data is immediately been conveyed to the doctor and patient through a real-time monitoring alert system. Every day's continuous data makes the patient aware of balancing the diet and workout exercise to keep the disease under control. Non-invasive monitoring of renal function monitoring becomes helpful in the case of critical condition patients. Real-time monitoring of renal GFR is discussed in detail in this chapter. In this system Cd-Te detector is used for non-invasive monitoring. Apart from the detector, the preamplifier and discriminator sections constitute the monitoring unit. In this real-time, monitoring system the filtration rate technique is used to determine the GFR.

These new emerging monitory systems have a few challenges. IOTM-based patient monitoring has a few drawbacks related to an error in analysis and acceptability among the medical fraternity is another problem. A few others such as Security and Privacy Issues, since this device captures private health-related information and these data, are highly vulnerable as being in the public domain through the internet. Thus it may attract unethical people for misuse. IOTM Based patient monitoring requires using a range of health check transducers or body contact signal material. Now monitoring system requires attaching to other gadgets for data communication as well as storage. Again the manufacturers are using different protocols for their devices for communication. Hence compatibility of the items manufactured by one with another is also one of the problems in this area.

As discussed in the above section the future scope of work is related to the development of a monitoring system based on the correlation of various breathing exercises and their effect on various organs such as the heart, kidney functioning, or the improvement in mental health. This future development would have immense help for the prevention or cure of diseases. There are good numbers of published papers related to a group of specific Pranayamas such as Nade Sodhan, Kapalbhate, and Vastrika in a particular specific interval of time per day and their capability of making the lungs strong; it also helps in curing the vital organ disease such as disease of heart, kidney or related to mental health.

References

Andrews, P. M., Wang, H. W., Wierwille, J., Gong, W., Verbesey, J., Cooper, M., & Chen, Y. (2014). Optical coherence tomography of the living human kidney. *Journal of Innovative Optical Health Sciences*, 7(02), 1350064.

Aymé, S., Bockenhauer, D., Day, S., Devuyst, O., Guay-Woodford, L.M., Ingelfinger, J.R., Schaefer F. (2017). Common elements in rare kidney diseases: Conclusions from a kidney disease: Improving global outcomes (KDIGO) controversies conference. Kidney International; 92(4): 796–808.

Beier, B., Musick, K., Matsumoto, A., Panitch, A., Nauman, E., & Irazoqui, P. (2011). Toward a continuous intravascular glucose monitoring system. *Sensors, 11*(1), 409–424.

Boursalie, O., Samavi, R., & Doyle, T. E. (2015). M4CVD: Mobile machine learning model for monitoring cardiovascular disease. *Procedia Computer Science* (63), 384–391.

Cai, H., Xu, B., Jiang, L., & Vasilakos, A. V. (2016). IoT-based big data storage systems in cloud computing: Perspectives and challenges. *IEEE Internet of Things Journal, 4*(1), 75–87.

Cunningham, J., Danese, M., Olson, K., Klassen, P., & Chertow, G. M. (2005). Effects of the calcimimetic cinacalcet HCl on cardiovascular disease, fracture, and health-related quality of life in secondary hyperparathyroidism. *Kidney International, 68*(4), 1793–1800.

De Hert, M., Detraux, J., Van Winkel, R., Yu, W., & Correll, C. U. (2012). Metabolic and cardiovascular adverse effects associated with antipsychotic drugs. *Nature Reviews Endocrinology, 8*(2), 114–126.

Ding, S., & Schumacher, M. (2016). Sensor monitoring of physical activity to improve glucose management in diabetic patients: a review. *Sensors, 16*(4), 589.

Dwivedi, R., Dey, S., Chakraborty, C., & Tiwari, S. (2021). Grape disease detection network based on multi-task learning and attention features. *IEEE Sensors Journal*, 1–8.

Ganesan, M., Sivakumar, N. (2019). IoT based heart disease prediction and diagnosis model for healthcare using machine learning models. In *2019 IEEE international conference on system, computation, automation and networking (ICSCAN) Mar 29* (pp. 1–5). IEEE.

Jung, H. Y., Jeon, Y., Seong, S. J., Seo, J. J., Choi, J. Y., Cho, J. H., ... Lee, J. S. (2020). ICT-based adherence monitoring in kidney transplant recipients: A randomized controlled trial. *BMC Medical Informatics and Decision Making, 20*(1), 1.

Kalantarinia, K., Belcik, J. T., Patrie, J. T., & Wei, K. (2009). Real-time measurement of renal blood flow in healthy subjects using contrast-enhanced ultrasound. *American Journal of Physiology-Renal Physiology, 297*(4), F1129–F1134.

Kumar, M., Sinha, R. K., & Sinha, S. K. (2017). Activation energy calculation for Li_2TiO_3 ceramics by java programming and validation by impedance analyzer. *International Journal on Future Revolution in Computer Science & Communication Engineering, 3*(9), 43–47.

Lalit, G., Emeka, C., Nasser, N., Chinmay, C., & Garg, G. (2020). Anonymity preserving IoT-based COVID-19 and other infectious disease contact tracing model. *IEEE Access, 8*, 159402–159414.

Learning, M. (2017). Heart disease diagnosis and prediction using machine learning and data mining techniques: A review. *Advances in Computational Sciences and Technology, 10*(7), 2137–2159.

Li, G., Fan, Y., Lai, Y., Han, T., Li, Z., Zhou, P., & Zhang, Q. (2020). Coronavirus infections and immune responses. *Journal of Medical Virology, 92*(4), 424–432.

Lin, T., Gal, A., Mayzel, Y., Horman, K., & Bahartan, K. (2017). Non-invasive glucose monitoring: A review of challenges and recent advances. *Current Trends in Biomedical Engineering & Biosciences, 6*(5), 1–8.

Markopoulou, D., Papakonstantinou, V., & de Hert, P. (2019). The new EU cybersecurity framework: The NIS directive, ENISA's role, and the general data protection regulation. *Computer Law & Security Review, 35*(6), 105336.

Mebazaa, A., Pitsis, A. A., Rudiger, A., Toller, W., Longrois, D., Ricksten, S. E., & von Segesser, L. K. (2010). Clinical review: Practical recommendations on the management of perioperative heart failure in cardiac surgery. *Critical Care, 14*(2), 1–4.

Mugisha, J., De Hert, M., Knizek, B. L., Kwiringira, J., Kinyanda, E., Byansi, W., & Vancampfort, D. (2019). Health care professionals' perspectives on physical activity within the Ugandan mental health care system. *Mental Health and Physical Activity* (16), 1–7.

Mugisha, J., De Hert, M., Stubbs, B., Basangwa, D., & Vancampfort, D. (2017). Physical health policies and metabolic screening in mental health care systems of sub-Saharan African countries: A systematic review. *International Journal of Mental Health Systems, 11*(1), 1–7.

Muhammad, L. J., Ebrahem, A. A., Sani, S. U., Abdulkadir, A., Chinmay, C., & Mohammed, I. A. (2021). Supervised machine learning models for prediction of COVID-19 infection using epidemiology dataset. *SN Computer Science*, *2*(11), 1–13.

Petrlík, M., Báča, T., Heřt, D., Vrba, M., Krajník, T., & Saska, M. (2020). A robust uav system for operations in a constrained environment. *IEEE Robotics and Automation Letters*, *5*(2), 2169–2176, 3.

Queralta, J.P., Gia, T.N., Tenhunen, H., Westerlund, T. (2019). Edge-AI in LoRa-based health monitoring: Fall detection system with fog computing and LSTM recurrent neural networks. In *2019 42nd international conference on telecommunications and signal processing (TSP)*, July 1 (pp. 601–604). IEEE.

Rabito, C. A., Moore, R. H., BougaS, C., & Dragotakes, S. C. (1993). Noninvasive, real-time monitoring of renal function: The ambulatory renal monitor. *Journal of Nuclear Medicine*, *34*(2), 199–207.

Rghioui, A., Lloret, J., Harane, M., & Oumnad, A. (2020). A smart glucose monitoring system for diabetic patient. *Electronics*, *9*(4), 678.

Sinha, A. N., Deepak, D., & Gusain, V. S. (2013). Assessment of the effects of pranayama/alternate nostril breathing on the parasympathetic nervous system in young adults. *Journal of Clinical and Diagnostic Research: JCDR*, *7*(5), 821.

Sinha, S. K., Kumari, S., & Chaudhary, R. K. (2019a). Dielectric and piezoelectric properties of $PbTi_{0.8-x}Se_{0.2}Sm_xO_3$ nanoceramics prepared by high energy ball milling. *Applied Physics A*, *125*(3), 171.

Sinha, S. K., Kumari, S., & Chaudhary, R. K. (2019b). Studies of dielectric and piezoelectric properties of $PbTi_{0.8-x}Te_{0}.2Gd_xO_3$ nanoceramics prepared by high energy ball milling. *Journal of Advanced Dielectrics*, *9*(02), 1950017.

Van Enter, B. J., & Von Hauff, E. (2018). Challenges and perspectives in continuous glucose monitoring. *Chemical Communications*, *54*(40), 5032–5045.

CHAPTER

7 Windowed modified discrete cosine transform based textural descriptor approach for voice disorder detection

Roohum Jegan and R. Jayagowri

Department of Electronics and Communication Engineering, BMS College of Engineering, Bengaluru, Karnataka, India

7.1 Introduction

Speech is the most powerful communication medium primarily used. Vocal fold disorders are present because different neurological, functional, and laryngeal diseases affect speech quality. In the present-day world, people are at a high risk of various pathological voice huddles including dysphonia and stuttering dysfluencies. The vocal cord shape affects due to the voice disease and resulting in an irregular spectral envelope of the person having voice disorder has a significant social and professional impact (Hegde, Shetty, Rai, & Dodderi, 2018).

Medical practitioners use laryngology procedures for vocal cord diagnosis and further treatment. The laryngoscopy performed by medical experts is an invasive method and it is expensive too (Sellam & Jagadeesan, 2014). Noninvasive approaches for automatic voice disorder detection based on signal processing techniques are becoming popular. These signal processing approaches can be employed as assisting tools by medical experts in the early stages of voice disorder detection. Voice pathology detection and classification is an important task useful in enhancing the algorithm performance and assisting the medical experts. The developed approach is applicable in a variety of clinical assistance systems. Moreover, the transform domain approaches are efficient in various bio-medical signal processing algorithm development.

Voice disorder detection algorithm using windowed modified discrete cosine transform (windowed MDCT) and texture features is proposed in this article. Normal and pathological speech utterance has different energy patterns and is exploited using windowed MDC transform. In this first step, a windowed MDCT coefficient image of healthy and pathological speech sample is generated. From this image, two types of textural descriptors are extracted: (1) completed local binary pattern (CLBP) and (2) local phase quantization (LPQ). To lower feature vector dimensionality and to select important features, two different feature selection approaches are explored: (1) Eigenvector centrality and (2) infinite selection method. Finally, a support vector machine (SVM) is used for the classification of voice samples of the Saarbrucken voice database (SVD).

The paper is organized as follows. Some recent techniques for speech disorder detection are discussed in Section 7.2. The proposed windowed MDC transform-based time-frequency texture descriptor feature approach is illustrated in Section 7.3. Details of windowed MDC transform,

Implementation of Smart Healthcare Systems using AI, IoT, and Blockchain. DOI: https://doi.org/10.1016/B978-0-323-91916-6.00007-2

different texture features, and feature selection techniques are briefly described in Section 7.4. Simulation results using the SVD database are presented in Section 7.5. Section 7.6 brings this article to an end.

7.2 Related works

Numerous automatic voice dysfluency detection methods have been proposed in the previous works to distinguish pathological voices from healthy ones using different feature extraction methods. In this article, a cancer detection algorithm based on time-frequency representation and textural descriptor is presented. A method is a noninvasive approach useful for speech therapists for prior identification of cancer. Early approaches were focused on the estimation of jitter for voice pathology detection as described in (Dárcio, Oliveira, & Andrea, 2009) with an average accuracy of 89%–96%. In (Lee, 2012), a two-stage method based on Gaussian mixture models (GMMs) and higher-order statistics is proposed for healthy and pathological voice classification. A 96.96% accuracy rate was achieved using this approach. A method (Vikram & Umarani, 2013) is presented using wavelet transform based Mel Frequency Cepstral Coefficients as features and fusion of Gaussian Model (GMM-UBM) classifier resulting in 93. 12% classification.

A wavelet packet decomposition of MFCCs is introduced to detect voice pathology with 91.54% detection accuracy obtained using the artificial neural network (ANN) classifier (Majidnezhad & Kheidorov, 2013). For feature reduction, PCA is used. SVM in combination with radial basis functional neural network classification strategy is employed (Sellam & Jagadeesan, 2014) to discriminate between healthy and pathological voices. Various time and frequency domain features were computed resulting in 91% classification accuracy.

Linear predictive coding (LPC) analysis is popularly employed as a feature extraction technique for voice pathology detection as presented in (Shakoor & Boostani, 2018). These proposed approaches are primarily developed for automatic health monitoring in smart city applications. Pathological speech detection algorithm using discrete wavelet transform energy features and ANN classifier is presented in (Shia & Jayasree, 2017). Nonlinear characterization of speech samples using time-delay embedded space is proposed by (Travieso et al., 2017) for the identification of various speech dysfluencies. Recently deep learning approaches are primarily developed for voice disorder detection and classification. From a 3-s sustained vowel, MFCC coefficients are obtained. A deep learning classifier in addition to the SVM and GMM with negative-fold cross-validation is used for performance evaluation with the MEEI database. Authors have shown that DNN outperforms the other two classifiers. Glottal flow parameters-based features are proposed for speech pathology detection (Ezzine & Frikha, 2018). Two different feature selection algorithms are used using SVD and MEEI databases. Detection is performed with ANN and SVM classification resulting in a 94% accuracy rate.

Parkinson's disease (PD) monitoring which measures the motor disorder and vocal impairments based on the tunable Q-factor wavelet transform (TQWT) is presented in (Sakar et al., 2019). TQWT has a higher resolution as compared to DWT which efficiently extracts features from sustained vowels. Different machine learning algorithms are compared during the evaluation stage. Machine learning algorithm-based PD algorithm is illustrated in (Mathur, Pathak, & Bandil, 2019) to improve the lifespan and early prediction of PD. The authors found that ANN performance is better as compared to the KNN classifier. In (Hammami, Salhi, & Labidi, 2019), voice dysfluency identification using a combination of empirical mode decomposition (EMD) and Discrete Wavelet Transform and Higher-order Statistic (HOS) features is introduced. Coefficient parameters obtained from Discrete wavelet transform are selected. They are evaluated in SVD. In the first step, the voice signal undergoes a process called EMD that results in intrinsic mode functions (IMFs). Amongst the functions, the best one is chosen based on the criteria of temporal energy. Then, the selected IMF is decomposed using the Discrete Wavelet Transform which results in the formation of two features vector; one consisting of six HOSs parameters, and another feature vector that has six DWT. SVM is used as a classifier. The proposed algorithm was trained using the SVD database and tested using the speech signals of speakers from a hospital. The new algorithm gave an accuracy of 98.26% and 94.1% using HOS and DWT parameters respectively validated on SVD.

Specific language impairment (SLI), is a type of speech dysfluency in kids that is characterized by difficulty in speaking and understanding the words described (Reddy & Alku, 2020). The authors proposed a novel method that analyses the glottal parameters obtained from input speech for picking out kids with this impairment. The algorithm makes use of both, time and frequency-domain glottal features, extracted from the speech sample employing glottal inverse filtering. In this method, Mel-frequency coefficients and open mile-based auditory parameters are also extricated from audio signals. SVM classifier and novel feed-forward neural network algorithms are individually validated for the cepstral coefficients, openSMILE- based auditory parameters, and laryngeal parameters. The classifiers are evaluated using the leave-fourteen-speakers-out cross-validation method and the evaluation is conducted on "audio compilation of SLI voice samples" formulated in the LANNA database. It was observed that laryngeal features consist of unique and delicate data to discriminate kids with SLI and from the experimental results it is observed that the glottal parameters along with MFC coefficients impart the best performance when used with FFNN classifier in the speaker-independent environment. It is being observed that many senior citizens and smokers are vulnerable to high risk of various forms of speech disorders. Hence voice pathological repair is of utmost importance. Earlier works on pathological speech repair focused only on the sustained vowel /a/. Simultaneous vowel repair is a demanding task because of the unsteady extrication of the pitch. Multiple vowels repair based on pitch extrication is proposed by (Zhang, Shao, Wu, Pang, & Liu, 2020). This widened the research in the area of voice reparation from individual vowel /a/ to assorted vowels. DNN is used as a classifier here and Wavelet Transform and Hilbert-Huang Transforms are used for extracting the pitch. The formant is reconstructed depending on the "Line Spectrum Pair" feature. The proposed method is tested on SVD and the improved accuracies obtained in the three features namely, Segmental Signal-to-Noise ratio, Line Spectrum Pair length, and MFC length, are 45.88%, 50.27%, and 15.46% respectively.

For effective clinical prognosis and accurate medication of various diseases in the nervous system such as Parkinson's syndrome, it is important to identify and discriminate the best parameters which can be used to successfully differentiate normal and pathological speech automatically. The dysphonic voice of PD patients is hoarse, partly airy, and is identified by unwanted pauses, staggering, and improper stammering. The property that spectro-temporal sparsity of dysphonic voice is different from the healthy speech is exploited here. In (Kodrasi & Bourlard, 2020), the authors discussed and proved the merits of spectro-temporal sparsity characterization for dysphonic speech detection. Experimental results were tested on a Spanish database which showed that evaluating the sparsity using the Gini index resulted in greater classifying accuracy of 84.3%. Narendra and Alku (2020) demonstrate the significance of source glottal details in the fully automatic classification of unhealthy speech signals by differentiating the general pipeline method from the end-to-end method. The former method has an extractor and a disparate classifier. The latter method consists of dual sets of glottal parameters along with the most commonly used openSMILE parameters and the SVM is used for classification. In this case, both input voice signals and original glottal flow waves are used to teach the two algorithms (1) a fusion of convolutional neural network and multilayer perceptron, and (2) a fusion of convolutional neural network and short-term long memory network. Experiments were carried out using three easily available databases, including the UA-Speech database, TORGO, and UPM databases. Based on the first approach, a combination of glottal features and openSMILE features, performance analysis of the detection system exhibited the best results.

Smart health is one of the most popular concepts in smart cities. It can be unique, easy, and fast support to the field of automatic voice pathology detection. A robust ML algorithm that can effectively distinguish pathological and healthy voice samples with increased accuracy is very much imperative to model an accurate and smart mobile health care system. In (Tuncer, Dogan, Özyurt, Belhaouari, & Bensmail, 2020), a multiclass-voice identification and discrimination model using a unique multileveled textural feature extractor with iterative selection is introduced. This method is an effective voice-based algorithm because it makes use of multicenter and multithreshold-based ternary designs.

In (Lauraitis & Maskeliunas, 2020), a Long and bidirectional Short-Term Memory neural network and Scatter wavelet Transform Fused SVM classifier are used to identify voice pathologies of people affected by disorders in the central neurological system. This research had 339 audio samples obtained from 15 people out of which 7 patients had early-stage Central nervous system disorders and the remaining 8 were healthy. Voice information was obtained using a simple voice recorder from a mobile application. Feature extraction was carried out from pitch outline and shapes, MFC coefficients and Gamma features, Gabor parameters, and audio spectrograms. Accuracy of 92.50% using BiLSTM and 97.3% using fusion WST-SVM was obtained for distinguishing normal and pathological audio signals. The model can be used to closely monitor the medical condition of patients having Central nervous system disorders. Fully automated computer-aided voice dysfluency classification tools can effectively diagnose voice dysfluencies in the initial stage that will be useful in delivering adequate, timely, and appropriate treatment.

In (Mohammed et al., 2020), an already-trained Convolutional Neural Network was given to a speech dysfluency dataset. The experimental outputs showed that the proposed model for voice dysfluency detection and classification achieved a higher accuracy of 94.41%. For F1-Score and Recall, it obtained an accuracy of 93.22% and 95.13% respectively. The proposed model proved to offer fast cum on-the-spot, fully automated identification and diagnosis and apt medication within 4 s to achieve the desired classification accuracy. Recent studies state that more than 10% of the general population suffer from some form of voice disorder. Fully automated diagnosis for such disorders using ML models has proved to minimize the expense and effort needed for diagnosis by clinicians, audiologists, and speech therapists. In (Chui, Lytras, & Vasant, 2020), a combination of the conditional regenerative network model and c-mean clustering model called COGAN-IFCCM is put-forth for the voice disorder discrimination and classification of three speech dysfluencies. The databases used are the SVD and the VOICE Database. the proposed COGAN-IFCCM performed better than traditional models.

In (Asmae, Abdelhadi, Bouchaib, & Sara, 2020), the authors have led a detailed study of dysphonia for the efficient detection of PD using various ML classifier models. For effective detection and to identify patients affected by Parkinson's syndrome from a group, ANNs along with the K Nearest Neighbors algorithm are used. Output proves that the ANN model performed better than the KNN model in terms of accuracy. The study consisted of 31 samples of which 24 people had Parkinson's disease and the remaining were healthy. The proposed model can discriminate between healthy individuals and people with PD from a huge group of people with greater accuracy of 95.67%.

Another automatic evaluation that helps the speech therapist make an efficient diagnosis of an affected larynx in a nonsurgical method is described in (Gidaye1, Nirmal, Ezzine, Shrivas, & Frikha, 2020). Voice dysfluencies affect the vocal cords and eventually vary the characteristics and dynamics of vocal cords. The glottal volume velocity waveform is calculated from the speech pressure signals of both normal and dysphonic subjects using a partially closed phase-glottal inverse filtering algorithm that captures the altered characteristics of the vocal cords. Various fusion-database approaches were explored to explain the merits of glottal features for voice dysfluency detection and classification and identification.

In (Gidaye, Nirmal, Ezzine, & Frikha, 2020), a wavelet model is proposed to evaluate speech dysfluencies, which does not dependent on databases. Stationary wavelet transform is used to break the speech signal and energy content and acoustic parameters are extracted from each band after multiple decompositions. The order of the feature vector depends on the level of decomposition. To reduce the dimensionality of the vector feature, the accuracy-based selection method selects the robust features and omits the unwanted ones. The final feature vector space is evaluated and classified using a SVM and ANN classifiers. German, English, Arabic, and Spanish speech databases are used here. In the initial phase of experiments, the raw audio signal is identified as healthy or nonhealthy. In the next phase, input speech samples are discriminated and classified into healthy, cyst, or polyp. Experiments prove that the energy of the signal and acoustic parameters can be used as a base for voice dysfluency analysis and the accuracy measures obtained are higher while considering other traditional and old methods.

The prime aim of (Lopes, Vieira, & Behlau, 2020) is to analyze various spectral and acoustic features in distinguishing people with and without speech dysfluencies. A database that has sustained vowel /ɛ/ from 484 aged people who underwent a laryngoscopy treatment in their larynx was developed. The laryngeal examination is carried out by a clinician or trained speech therapist. Once the results of the laryngeal examination are out, the auditory evaluation is done by an Audiologist or Pathologist. People were segregated into groups consisting of 58 people with voice disorder and a group of 434 people without voice dysfluencies. Four different measures of acoustic parameters were considered: state-of-the-art measures, cepstral measures, nonlinear measures, and recurrence quantification. Vowel /e/ was omitted from all voice recordings. The quadratic discriminant method was used as a classifier. Individual parameters achieved an above-average accuracy $\geq 75\%$. Combining the parameter measures enhanced the operation of the algorithm further. Children suffering from autism spectrum disorder (ASD) often exhibited few notable acoustic designs. Mohanta, Mukherjee, and Mittal (2020) show ways to discriminate autism speech. Previous research studies have given importance only to native English and American speakers. Therefore, an audio sample dataset of children who fall under the ASD spectrum and an audio sample dataset of healthy children were recorded and stored in English. For kids categorized under the ASD spectrum and healthy kids, the temporal and spectral parameters are analyzed and classified using various classifiers. The KNN classifier model obtained a peak accuracy of 94.5% in comparison to other traditional models. Jun and Kim (2018); Bahn, (2020) propose a new algorithm for fully automatic computer-aided detection of Pathological voice dysfluencies with state-of-the-art machine learning models and a few classifiers, including ANN, SVM, and Deep Convolutional Neural Network classifiers. The authors were able to transform the audio recordings into corresponding Mel spectrogram and the features extracted were classified using SVM, Dense-net, ANN, Recurrent Neural Network, and a few other feature-based classifiers. SVM achieved 75% accuracy, ANN achieved 40% of accuracy, RNN recorded a 35% accuracy rate, and the Random Forest approach, one of the commonly used feature-based classifier approaches achieved 69% of accuracy which is higher compared to the Deep learning-based classification approach. It was noted that major features that distinguish the dysfluency class were well distributed amongst the frequency domain, compared to the time domain (Jegan & Jayagowri, 2020).

The speech represents an intrinsic characteristic of human behavior. Any disturbances in the normal speech of a human being are called speech disorders. It affects communication and social integration. Such patients will also have psychological and emotional issues as a direct result of their voice disorder. These day-to-day problems may cause a deterioration of the quality of life of an affected person and this results in the person trying to isolate himself from the activities of his daily life. Hence early detection of speech disorders is of utmost importance. From the brief survey of state-of-the-art voice disorder detection and classification algorithms from recent years made, it was concluded that various conventional techniques for speech disorder detection were used earlier but extensive research has given rise to computer-based (noninvasive) methods of speech disorder detection. Although most of the multiple-voice disorder detection algorithms are the outcome of the last couple of years, the earlier research focus was only on neurological disorder detection. The computer-based techniques are easier to administer and are less expensive as compared to conventional methods. The algorithms were only divided based on the feature extraction techniques used for the detection task and not on any other criteria. Few databases were employed for the evaluation of the implemented approach. But the databases were repetitive and common in all works. Our work focuses on rare databases. One common procedure adopted by all papers are (1) extracting features from audio samples and (2) classifying abnormal voice from normal ones using a suitable classifier algorithm. It was seen that there is no concurrence between the audio feature and the classification algorithm that can provide the best accuracy in screening the pathological voice. It is observed that Deep learning techniques and a few fusion techniques for speech pathology detection have great potential and recent studies are focused on investigating deep learning architecture. Compared to the accuracies of other works, our algorithm yielded maximum accuracy and performed better compared to other state-of-the-art techniques. Comparing other machine learning techniques, Deep learning techniques for speech pathology detection have great potential and recent studies are focused on investigating more into deep learning architecture.

7.3 Proposed methodology

This article presents a voice pathology detection algorithm using windowed MDCT and texture features. Fig. 7.1 depicts the architecture of the proposed approach. The entire method is categorized

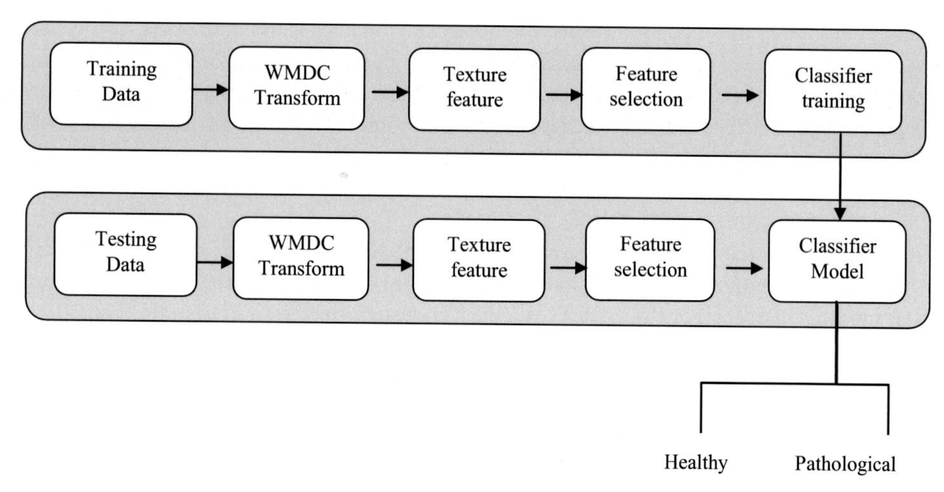

FIGURE 7.1

Windowed modified discrete cosine transform-based textural feature algorithm for voice disorder detection.

into the following parts: (1) training and (2) testing phase. In the training phase, a windowed MDCT coefficient image of healthy and pathological speech samples is generated. From this image, two types of textural descriptors are extracted: (a) CLBP and (b) LPQ. To lower the feature vector dimensionality and to select important features, two different feature selection approaches are explored: (a) Eigenvector centrality and (b) infinite feature selection technique. Finally, a SVM is used as a classifier for voice disorder detection using the SVD.

Similar steps are followed in the testing phase. 70% of the audio samples train the classifier and a balance of 30% is used for testing the model. Various evaluation parameters including detection accuracy, sensitivity, and specificity are computed. The result of different feature selection techniques is also analyzed in the experimental results section.

7.4 Feature extraction and selection

In this article, the first windowed modified DCT transform visual representation is obtained. From this image, two types of features are extracted: (1) completed LBP and (2) LPQ. The textural differences between healthy and cancer windowed MDCT images can be effectively captured by these two feature extraction techniques.

7.4.1 Windowed modified discrete cosine transform

Signal Processing is incomplete without Compression techniques. These techniques are of great importance because of the high amount of data transferred via the network's communication channel. Compressing the audio, video, image, and text data are imperative for effective and easy transmission of different types of information. Once the audio (speech) signals are compressed, transmission and reception become faster and easier, which in turn improves communication. Compression of speech signal deals with decreasing the number of data bits present in the input speech sample (Kamarul & Aini, 2008). Hence, the number of ways needed to store and transmit the input audio signal is considerably decreased. Signal compression, also called audio signal coding, involves various algorithms that help in distinguishing the lossy or lossless forms of Data Compression. The former audio compression method of speech signals is obtained using various compression methods such as FFT, DWT, DCT, etc. The fulfilment of these methods and the ease with which the raw voice sample is reconstructed from the compressed signal is still a hassle. Usually, the DCT technique is utilized and modified for better compression. Further research gave birth to the modified DCT algorithm. DCT was used to transform the time domain signal to its corresponding frequency domain. In this method, samples with energy levels less than 1 were eliminated, and speech signals with sample frequency lesser than the preceding sample frequency were concealed. A method called temporal auditory masking procedure eliminated voice signals that were concealed. MDCT is capable of operating on a large set of datasets with subsequent overlapping and hence called a lapsed transform. MDCT is derived from DCT type-IV with lapsed property and offers two major advantages: (1) energy compaction property and (2) overlapping. Hence it is used in various audio compression algorithms. The MDCT is further modified in (Chowdhury, Verma, & Stockwell, 2015). MDCT is 50% overlapped and because of the critical sampling, MDCT coefficients require larger space as compared to the raw input speech signal. Because of this overlapping property, MDCT is widely used in quantization applications that remove the features that are blocking which could pose serious problems during the reconstruction of the voice signals. It is a linear function that converts 2K real numbers to K numbers, as depicted in the following equation.

$$\mathrm{Pm} = \mathrm{Xk}\ \mathrm{Cos}\ ([(\pi/\mathrm{k})(\mathrm{k} + 0.5 + 0.5\ \mathrm{K})(\mathrm{m} + 0.5)]$$

where $m = 0, 1\ K - 1$.

In each band, the quantities $t(K)$ and $t\ (2K - 1)$ symbolize the latest input speech signals. Quantities $f\ (0)$ and $f\ (M - 1)$ symbolize the speech samples of the preceding band. The results in 55% overlapping (Musci, Feitosa, Velloso, & Novack, 2011). The above function depicts how MDCTs return the output sequence (F) that is half in number as compared to the number of inputs. Thus, the MDCT has the added merit of decreasing the number of distortion sources at the quantization phase.

The energy spectrum is different for healthy and pathological voice samples. This is shown by the time-frequency representation using the windowed MDC transform. Fig. 7.2 shows the time-frequency representation obtained using WMDCT. Four images from the first row illustrate healthy samples and the second row shows pathological samples. As it is seen from this figure, in pathological samples' time-frequency representation, energy content variations are prominent as compared to the healthy samples (Chowdhury, Borkar, & Birajdar, 2019).

7.4.2 Completed local binary pattern

Local binary pattern (LBP) is a unique, efficient textural operator that finds widespread application in the area of computers such as facial recognition and detection of targets (Ojala et al., 2002). Singh, Maurya, and Mittal (2013) introduced the local binary pattern operator that generates a

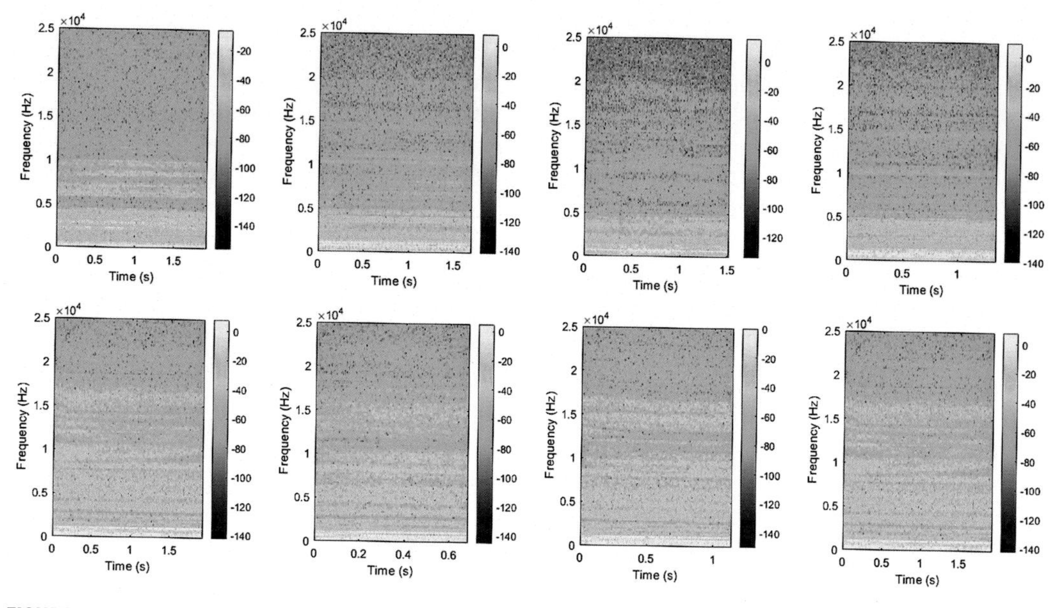

FIGURE 7.2

Windowed modified discrete cosine transform time-frequency representation for healthy and pathological samples.

binary code while comparing a neighboring pixel and its center patch gray unit. This operator assigns 0 if the neighbor pixel is smaller than the center value. Otherwise, it assigns a unit value.

LBP operates by assigning a threshold value of neighbor regarding the gray level of the center pixel as shown below:

$$LBP_{P,R} = \sum_{p=0}^{p-1} s(g_p - g_c)2^p$$

$$S(x) = \begin{cases} 1, x \geq 0 \\ 0, x < 0 \end{cases}$$

Where, l_c = gray value of center pixel and l_a ($a = 0,1,2,$ a-1) denotes the neighboring pixels taken along the circle having diameter *C/2* (*C/2* > 1) and the cumulative count of neighbors is denoted as *A* ($A > 1$). The $LBP_{A,B}$ value can be computed by introducing a binomial factor 2*a and assigning it* for each value of l_a. As LBP is a widely used method, researchers have been trying hard on ways to modify and improve this operator to widen the applications. One such improvised form of LBP is the CLBP.

Complete Local Binary Pattern is a hypothesized genre of LBP and it has proved to be effective in improving the robustness of the textural feature representation by including additional instructions that are not fulfilled by the original LBP operator. Guo et al. identified and proposed the concept of CLBP (Xu, Zhu, Wang, & Bao, 2005). Unlike LBP, the input signal is decomposed into the sign, magnitude, and different parts by CLBP which results in more accurate textural characterization. CLBP contains a local domain which is denoted by a center pixel and the difference in the magnitude information of the center and the adjacent gray value which is called the Local Difference Sign-Magnitude Transform that is omitted in a normal LBP texture descriptor. One of the major drawbacks of the LBP descriptor is that it fails to produce befitting binary codes and focuses entirely on the sign of the variance of the two gray values (Wang & Liu, 2005). It is made of three distinct parts; CLBP-SIG which implies the sign of the difference between the center pixel and the local pixel, CLBP-MAG implies the positive difference in magnitude between both the pixels and CLBP-CENT specified by the difference in the values of the local pixel and average central pixel. CLBP-SIG entails the normal LBP (Lu, 2016).

This proves that CLBP is favored over LBP for feature extraction (Ke, 2011). CLBP, unlike LBP, does not just assign 0 or 1 to l_a. Depending on the value of l_a-l_c, CLBP calculates the difference value between l_c and l_a to be $S_a = l_a\text{-}l_c$. This step is followed in LPQ also. But, CLBP can further break S_a into two parameters da and dm:

$$S_a = \text{da} * \text{dm}$$

where da is the sign of dp and the magnitude of dp is mp. CLBP_m can also be calculated where m is the mean value of dm for the entire image. CLBP_cent is obtained by transforming the center gray code to the corresponding binary code taking the threshold value as the reference. CLBP_Cent = f (Cep, ma) where Cep is marked as the center pixel and ma is the average of all the pixels' numbers. Three texture descriptors; the difference descriptor (denoted as CLBP-Sign), magnitude descriptor (CLBP-Magnit), and center pixel descriptor (CLBP-Cent), are put forward in the CLBP algorithm for textural analysis. The CLBP_magnit, CLBP_sign, and CLBP_cent denote the raw image, which is later conjoined to produce the CLBP feature mapping as output. CLBP_mag and CLBP_sig could be combined to CLBP_mag/sig by histogram method. CLBP texture features namely; CLBP_mag/sig and CLBP_mag/sig/cent, were the best results in Guo et al. Sun, Han, Sun, and Yuan (2006) too. The three descriptors, namely *CLBP_SIG*, *CLBP_MAG*, and *CLBP_CENT* are fused to construct a CLBP operator named *MSC* with the following equation.

$$\text{MSC} = \text{ks CLBP\_MAG} + \text{km CLBP\_SIG} + \text{kc CLBP\_CENT}$$

$$\text{ks} + \text{km} + \text{kc} = 1$$

where ks, km, and kc are weight coefficients.

The MSC texture descriptor describes the texture parameters of the pixels in the image individually. Robust models and high improvement depend on a few collaborations of CLBP designs such as CLBP_SIG/MAG/CENT.

7.4.3 Local phase quantization

Ahonen, Rahtu, Ojansivu, and Heikkila (2008) first came up with the idea of a texture descriptor called the LPQ operator (Cooley, Anderson, Felde, & Hoke, 2002). It utilizes individual local phase details extricated from the short-term FT calculated over a vicinity that has rectangular characteristics at the image's pixel position (Cheng et al., 2011). LPQ is a new technique that is not affected by blur. It is a textural classification method evaluated over the window of a local image (Wang, Fei, Xie, Liu, & Zhang, 2017). This quantization operator is used to identify different textures by analyzing them locally at each pixel location. The results are presented as codes as a histogram. The formation of the codes and their corresponding histograms are the same as that of the LBP method.

The equation for a blurring of a spatially nonvarying image in the Fourier domain, i(n) is given by,

$$A(t) = B(t)C(t)$$

where $A(t)$, $B(t)$, and $C(t)$ is the corresponding Fourier transforms of the blurred spatially nonvarying image $i(n)$, the original image $j(n)$, and the point spread function $q(n)$, respectively, and t is a vector of $[t;\ u]\ T$.

Equation 2 can be broken down into its corresponding magnitude and phase components as follows,

$$[A(t)] = [B(t)][C(t)]$$

$$A(t) = B(t) + C(t)$$

At centrally symmetric points of the blur $p(x)$, the output of FT is real-valued. Its phase is a function consisting of dual values (Ojala, Pietikäinen, & Harwood, 1996). LPQ method depends mainly on the above-mentioned properties of blur invariance. Many machine learning applications such as image analysis in various medical applications and surface inspection involve texture description and texture analysis. The applications extend to a series of textural identification and discrimination applications as well. Dedicated and exclusive research in the area of texture analysis over the last 35 years resulted in literature consisting of unique techniques for describing image textures. The blur in an image is the output of convolution between the image intensity and the point spread function of the image. If the raw image is represented by $i(t)$, and the observed image is given by $o(t)$, then the discrete model for a spatially invariant blurring of $o(t)$ can be expressed by a convolution:

$$o(t) = i(t) * p(t)$$

where $p(t)$ is the point spread function of the blur and it denotes the two-dimensional convolution and t is a vector. In the Fourier domain, this corresponds to $A(n) = b(n) *c(n)$ where $A(n)$, $B(n)$, and $C(n)$ is the corresponding discrete Fourier transforms of the blurred image $i(t)$, the original image o (t), and the PSF $p(t)$, respectively.

The magnitude and phase can be separated as follows,

$$|A(t)| = |B(t)|.|C(t)|$$

$$< A(t) = < B(t) + < C(t)$$

The blur point spread function $p(t)$ is symmetric with respect to the center and is represented by the equation $p(t) = p(-t)$. The corresponding Fourier transform is also real-valued and hence, its phase is only a dual-valued function.

In the case of linear phase quantization, the phase is evaluated in local neighborhoods La at each pixel position an of the image $i(a)$.

aϵ{a1,a2, aN} consists of a simple one-dimensional convolution of successive rows and columns. The local Fourier coefficients are calculated at the following points

$$\mathrm{v1} = [0, b]^T$$

$$v2 = [b, 0]^T$$
$$v3 = [b, b]^T$$
$$v4 = [b, -b]^T$$

where b is a scalar constant that satisfies the condition $B(\text{Ui}) > 0$. For each position of the pixel, a vector is generated given by: $A(b) = [A$ (U1, b) A (u2, b) A (u3, b) A (u4, b)]. Just by monitoring the signs of the real and imaginary parts of each component in $A(b)$, observation regarding phase can be obtained using the method of scalar quantization. From that information, the label image f LPQ (x) can be obtained easily.

7.5 Experimental results and discussions

This article presents a voice disorder detection technique using textural features extracted from the windowed MDC transform image. The algorithm is evaluated using the 123 normal and 125 pathological speech signals obtained from SVD. In the first step, a windowed MDC transform is applied over the input speech sample and a time-frequency visual image is generated. The sample length is set as 64 and the window length is set as 3. From these images, CLBP and LPQ texture descriptors are extracted. The radius is set as 8 while extracting these features.

Feature-length of CLBP is 59-D and with that of LPQ is 256. These features are then combined to generate a final feature vector. Two types of feature selection algorithms are evaluated to verify the importance of feature selection on accuracy. Finally, a SVM classifier is employed for detection purposes. Various evaluation parameters such as detection accuracy, sensitivity, and specificity are computed. 70% of the samples are used to train the classifier and 30% are used to validate the output.

The idea of SVM was first put forward by Vapnik leading to a greater interest in the field of machine learning. Numerous studies have outlined that, compared to other classifiers, the SVM is effective in providing greater performance with regard to classification accuracy. SVMs are mainly used for regression and categorization and it belongs to the class of supervised learning methods and the family of generalized linear classification (Guo, Zhang, & Zhang, 2010). One of the most commendable properties of SVM is that it curtails the classification error and boosts the geometric margin at the same time hence it is also called a Maximum Margin Classifier.

The first set of experiments involves the evaluation of detection rate, specificity, and sensitivity without a feature selection algorithm. RBF kernel function is used for SVM classification. A total of 125 pathological voice samples and 123 healthy samples are used during the evaluation of the presented algorithm. Table 7.1 shows the detection accuracy, sensitivity, and specificity of voice disorder detection using windowed MDC transform-based texture features without FS. From the table, it can be noted that compared to CLBP texture descriptors the detection rate is better in LPQ. Also, the combined rate is higher than the individual features.

Table 7.2 depicts the detection accuracy, sensitivity, and specificity of voice disorder detection using windowed MDC transform-based texture features using feature selection techniques. We have used two feature selection techniques: (1) Eigenvector centrality (Ojansivu & Heikkila, 2008) and (2) Infinite feature selection (Lowe, 2004). As illustrated in the table, the highest accuracy with

Table 7.1 Detection accuracy, sensitivity, and specificity of voice disorder detection using windowed modified discrete cosine transform-based texture features without feature selection.

Features	Sensitivity (%)	Specificity (%)	Detection accuracy (%)
CLBP	21. 62	79.49	50.55
LPQ	91.89	56.41	74.15
Combined	78.38	69.23	73.80

CLBP, completed local binary pattern; *LPQ*, local phase quantization.

Table 7.2 Detection accuracy, sensitivity, and specificity of voice disorder detection using windowed modified discrete cosine transform-based texture features using feature selection techniques.

Features	Sensitivity (%)	Specificity (%)	Detection accuracy (%)
Eigenvector centrality	97.30	93.33	95.32
Infinite	97.30	95.45	96.38

Table 7.3 Comparison with existing state-of-the art voice disorder detection algorithms.

Algorithm	Features	Classification accuracy (%)
(Dárcio et al., 2009)	Jitter	86
(Vasilakisa & Stylianou, 2009)	Short-term jitter	96
(Vikram & Umarani, 2013)	MFCC	93.30
(Majidnezhad & Kheidorov, 2013)	MFCC-Wavelet packets	91.54
(Sellam & Jagadeesan, 2014)	Spectral	91.70
(Ezzine & Frikha, 2018)	Glottal ow parameters	94.6
Proposed	Windowed modified discrete cosine transform	96.38

96.38% is obtained using the infinite feature selection approach. Additionally, a good detection rate is also obtained using the Eigenvector centrality algorithm (Dalal & Triggs, 2005), but it is less than the latter one. As both these feature selection approaches are based on the ranking technique (Manjunath & Ma, 1996), we have selected the top 20 features in each case for the evaluation purpose (Ojala, Pietikainen, & Maeenpaa, 2002).

The proposed WMDCT-based approach is compared with other voice disorder detection algorithms from the literature. A comparison with already available voice disorder discrimination algorithms is shown in Table 7.3. Methods based on jitter and MFCC resulted in an 86%–96% detection rate (Chan, Kittler, Poh, Ahonen, & Pietikäinen, 2009) as compared to the proposed technique which resulted in 96.38% detection accuracy. It is clear from Table 7.3 that the presented approach is better compared to other voice disorder techniques.

Many windowed modified DCT channels affect the algorithm performance significantly. To analyze the effect of a number of channels on different performance evaluation parameters, we have conducted several experiments using individual and combined feature sets. We have tested the proposed algorithm using four values of WMDCT channels: 128, 256, 512, and 1024.

Fig. 7.3 depicts sensitivity, specificity, precision, F1-measure, and average accuracy obtained using CLBP textural descriptors at different WMDCT channels. As illustrated four different channels corresponds to four figures ranging from 128 to 1024 channels. As evident from the figure, the specificity obtained using CLBP is highest as compared to other performance evaluation

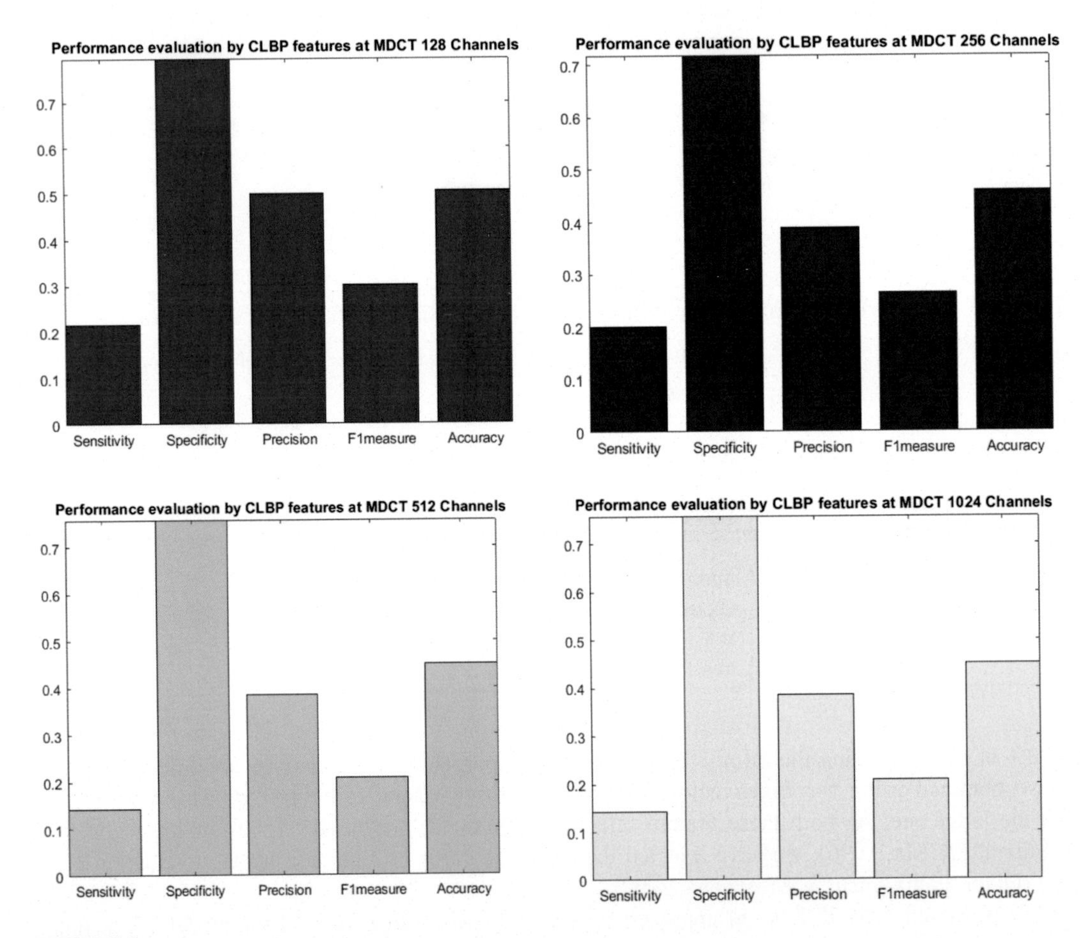

FIGURE 7.3

Sensitivity, specificity, precision, F1-measure, and accuracy obtained using completed local binary pattern textural descriptors at different windowed MDCT channels.

parameters. The proposed algorithm correctly evaluates true negatives using CLBP descriptors. However, overall accuracy attained using completed LBP features is lower.

Fig. 7.4 shows different performance metrics recorded using LPQ textural descriptors at four WMDCT channels. Similar to CLBP, LPQ specificity has achieved a maximum score compared to other parameters. Overall performance of the algorithm using LPQ is better than completed LBP and combined descriptors. Moreover, the best performance is attained on the WMDCT scale of 128. Hence, during the analysis, we selected 128 channels of windowed DMCT.

Finally, as shown in Fig. 7.5, sensitivity, specificity, precision, F1-measure, and accuracy obtained using combined textural descriptors at different WMDCT channels are evaluated. As it can be seen from the figure combined features achieved higher average accuracy and all

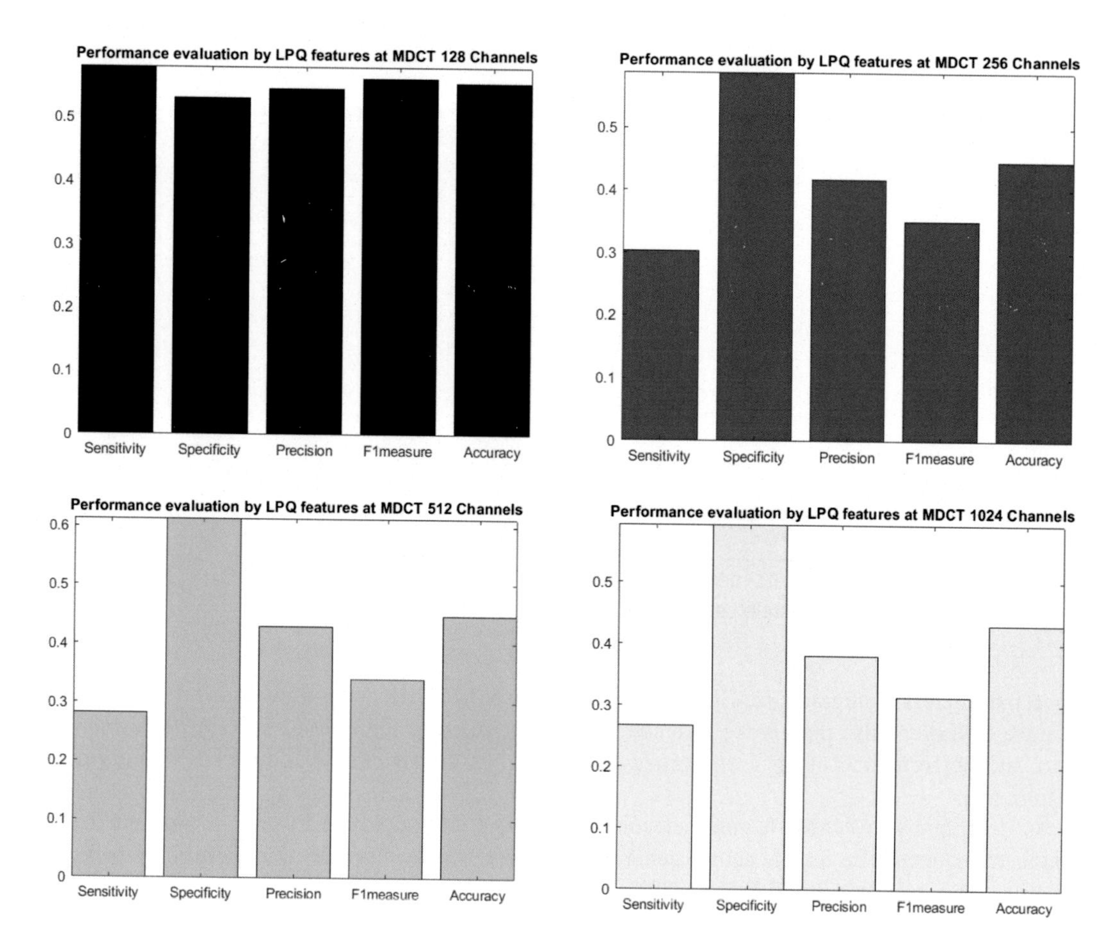

FIGURE 7.4

Sensitivity, specificity, precision, F1-measure, and accuracy obtained using LPQ textural descriptors at different WMDCT channels. *LPQ*, local phase quantization; *WMDCT*, windowed modified discrete cosine transform.

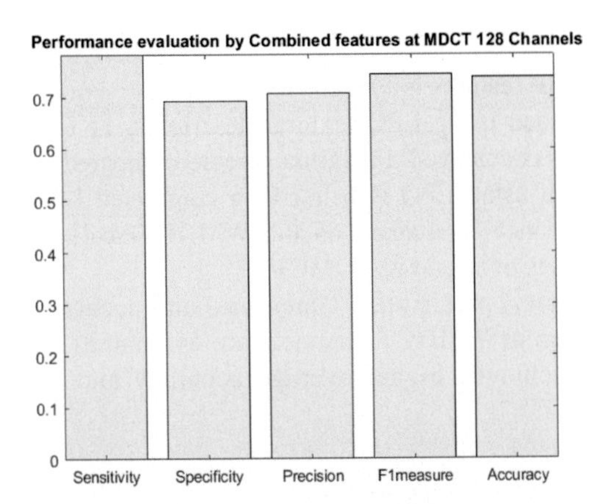

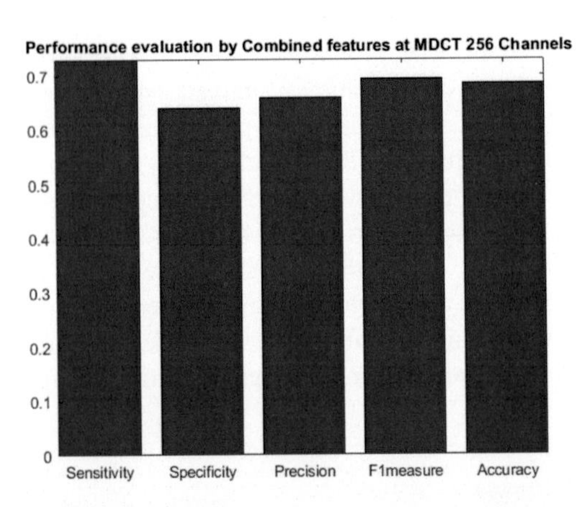

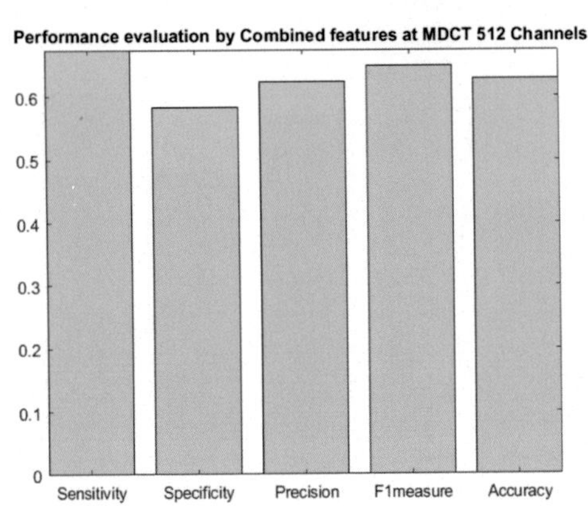

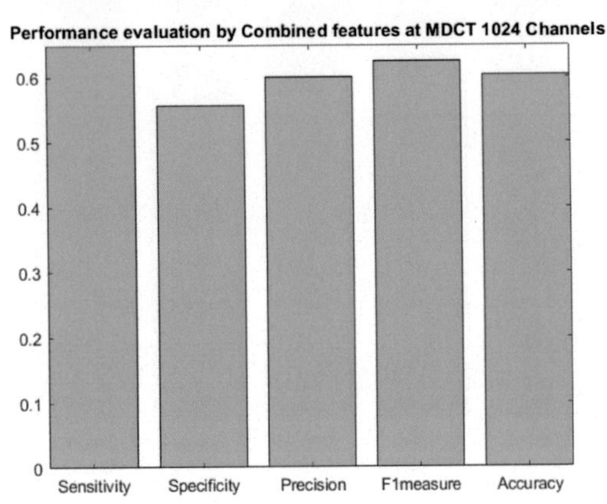

FIGURE 7.5

Sensitivity, specificity, precision, F1-measure, and accuracy obtained using combined textural descriptors at different windowed modified discrete cosine transform channels.

other parameters compared to CLBP but lower than LPQ descriptor evaluation. It may be because of larger false positives produced by CLBP affecting the overall system performance. Also, the performance using a combined feature set is consistent as compared to individual feature sets.

As described above, two feature selection techniques are employed to remove unwanted and redundant features. The feature subset generated by these two feature selection algorithms is found effective. Two selection algorithms are (1) eigenvector centrality and (2) infinite feature selection. To verify the effectiveness of these feature selection approaches, we have conducted several experiments.

The effect of eigenvector centrality and infinite feature selection approaches on overall accuracy using CLBP, LPQ, and combined feature set is depicted in Figs. 7.6 and 7.7. As it is clear from these figures that, the overall accuracy is enhanced by employing feature selection. The infinite

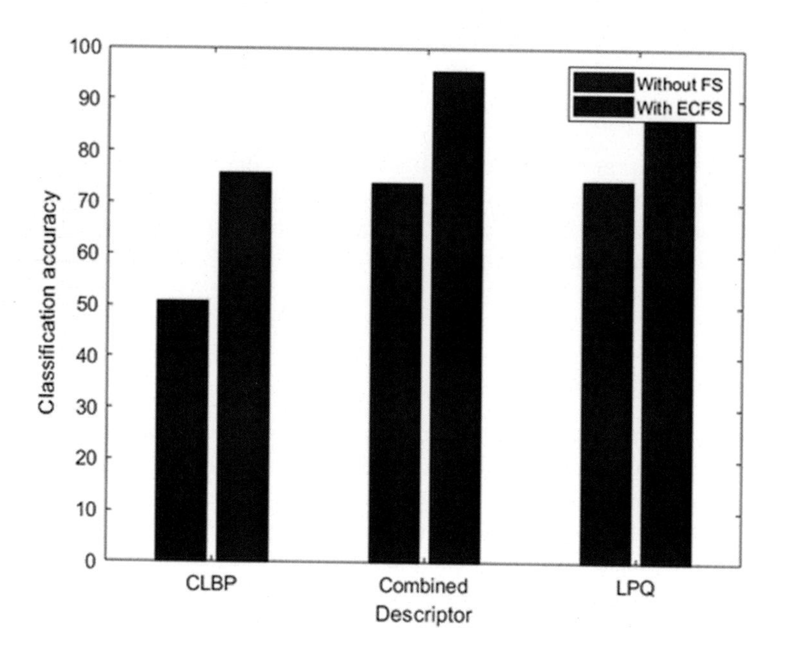

FIGURE 7.6

Effect of eigenvector centrality feature selection approach on overall accuracy using CLBP, LPQ and combined features. *CLBP*, completed local binary pattern; *LPQ*, local phase quantization.

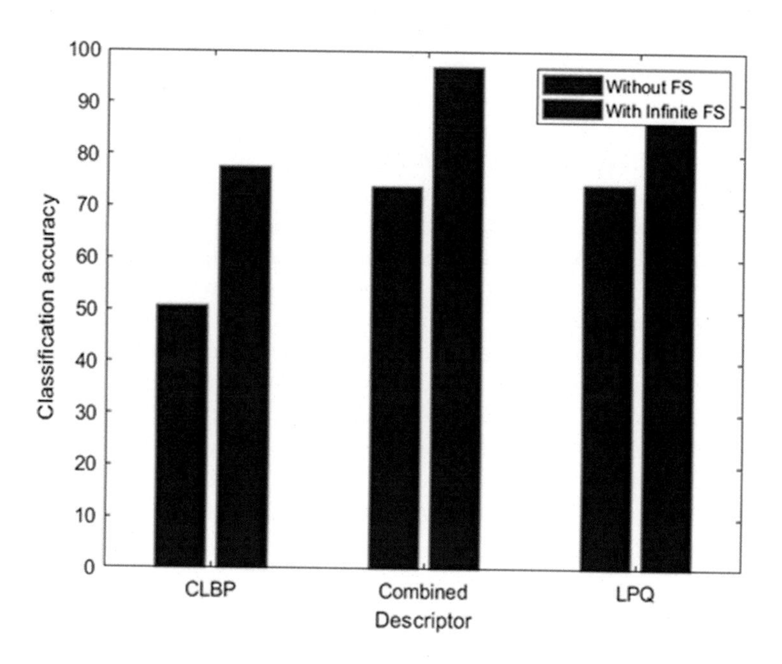

FIGURE 7.7

Effect of infinite feature selection approach on overall accuracy using CLBP, LPQ and combined features. *CLBP*, completed local binary pattern; *LPQ*, local phase quantization.

feature selection algorithm performance is slightly better compared to eigenvector centrality. Significant performance improvement is observed by employing these techniques. Hence, such approaches not only enhance the overall algorithm accuracy but also removes the nondiscriminable features lowering the feature vector dimension.

7.6 Current trends and future scope

In most of the works, it can be seen that voice signals are briefly classified into either normal or pathological. But few recent studies specially targeted people having PD and Alzheimer's disease. The research study happening in this particular area of speech processing has a very high scope and potential as they directly make a great impact on society. Considering this factor, it is the need of the hour to enhance more databases and explore new parameters in this area. More feature extraction techniques and learning algorithms must be modeled in the future for easier and smooth classification of normal and abnormal speech. Finally, it was observed that various Deep learning techniques were largely used to classify voice signals into normal and pathological ones.

7.7 Conclusion

From Table 7.1, it is observed that the detection accuracy, sensitivity, and specificity of voice disorder detection using windowed MDC transform-based texture features without feature selection is lesser than the detection accuracy, sensitivity, and specificity of voice disorder detection using windowed MDC transform-based texture features using feature selection techniques (eigenvector centrality and infinite features). This shows the importance of feature selection approaches. The detection accuracy was highest with the infinite feature selection method (96.38%) as compared to eigenvector centrality (95.32%). Comparison with existing state-of-the-art voice disorder detection algorithms shows the proposed methods outperformed them with a classification accuracy of 96.38%.

Windowed MDC transform time-frequency image representation-based texture descriptors are employed for voice pathology detection. The energy present in the normal (healthy) and pathological speech samples is different. These variable contents are extracted using CLBP and LPQ texture features and further its dimensionality is reduced using two ML algorithms. Different parameters are evaluated using an SVM classifier and 96.38% highest detection accuracy is achieved. The effect of feature selection is also analyzed and it is found that the infinite feature selection approach performs better as compared to the Eigenvector centrality method. Additionally, the proposed approach is compared with the conventional techniques including MFCC and various other methods and the performance of the proposed method is found to be better.

References

Ahonen, T., Rahtu, E., Ojansivu, V., & Heikkila, J. (2008). Recognition of blurred faces using local phase quantization. In *2008 19th international conference on pattern recognition*, pp. 1–4.

Asmae, O., Abdelhadi, R., Bouchaib, C., & Sara, S. (2020). Parkinson's disease identification using KNN and ANN algorithms based on voice disorder. *IEEE Xplore*.

Bahn, N. (2020). Critical evaluation of vocal disorder detection methods. Selected Computing Research Papers, Vol. 9.

Chan, C. H., Kittler, J., Poh, N., Ahonen, T., & Pietikäinen, M. (2009). (Multiscale) local phase quantization histogram discriminant analysis with score normalisation for robust face recognition. In *IEEE workshop on video-oriented object and event classification*, Kyoto, Japa, pp. 633–640.

Cheng, L., Youhua, M. A., Huang, Y., Zhi, X., Juan, Z. U., & Zhongwen, M. A. (2011). Comparison of atmospheric correction between ENVI FLAASH and ERDAS ATCOR2. *Agriculture Network Information, 12*, 007.

Chowdhury, A. A., Borkar, V. S., & Birajdar, G. K. (2019). Indian language identification using time-frequency image textural descriptors and gwo-based feature selection. *Journal of Experimental & Theoretical Artificial Intelligence*, 1–22.

Chowdhury, S., Verma, B., & Stockwell, D. (2015). A novel texture feature based multiple classifier technique for roadside vegetation classification. *Expert System Application, 42*, 5047–5055.

Chui, K. T., Lytras, M. D., & Vasant, P. (2020). Combined generative adversarial network and fuzzy C-means clustering for multi-class voice disorder detection with an imbalanced dataset. *Application Science*, 456–4571.

Cooley, T., Anderson, G. P., Felde, G. W., & Hoke, M. L. (2002). FLAASH, a MODTRAN4-based atmospheric correction algorithm, its application and validation. In *Proceedings of the 2002 IEEE international geoscience and remote sensing symposium (IGARSS '02)*, Toronto, ON, Canada, pp. 1414–1418.

Dalal, N., & Triggs, B. (2005). Histograms of oriented gradients for human detection. In *9th European conference on computer vision*, San Diego, CA.

Dárcio, S., Oliveira, L., & Andrea, M. (2009). Jitter estimation algorithms for detection of pathological voices. *EURASIP Journal on Advances in Signal Processing*, 567–575.

Ezzine, K., & Frikha, M. (2018). Investigation of glottal ow parameters for voice pathology detection on SVD and meei databases. In *2018 4th international conference on advanced technologies for signal and image processing (ATSIP)*, pp. 1–6.

Gidaye, G., Nirmal, J., Ezzine, K., Shrivas, A., & Frikha, M. (2020). Application of glottal flow descriptors for pathological voice diagnosis. *International Journal of Speech Technology, 23*, 205–222.

Gidaye, G., Nirmal, J., Ezzine, K., & Frikha, M. (2020). Wavelet sub-band features for voice disorder detection and classification. *Multimedia Tools and Applications, 79*, 28499–28523.

Guo, Z., Zhang, L., & Zhang, D. (2010). A completed modelling of local binary pattern operator for texture classification. *IEEE Transactions on Image Processing, 19*(6), 1657–1663, June.

Hammami, I., Salhi, L., & Labidi, S. (2019). Voice pathologies classification and detection using EMD-DWT analysis based on higher order statistic features. *Elsevier Masson SAS, IRBM*, 12–23.

Hegde, S., Shetty, S., Rai, S., & Dodderi, T. (2018). A survey on machine learning approaches for automatic detection of voice disorders. *Journal of Voice*, 1–23.

Jegan, R., & Jayagowri, R. (2020). Voice disorder detection and classification—A review. In *Second international conference on IoT, social, mobile, analytics & cloud in computational vision & bio-engineering (ISMAC-CVB)*.

Jun, T. J., & Kim, D. (2018). Pathological voice disorders classification from acoustic waveforms. Gct634, Kaist, Korea.

Kamarul, H. G., & Aini, H. (2008). Machine vision system for automatic weeding strategy using image processing technique. *American-Eurasian Journal of Agricultural & Environmental Sciences, 3*, 451–458.

Ke, C. Q. (2011). Analyzing coastal wetland change in the Yancheng national nature reserve, China. *Regional Environmental Change, 11*, 161–173, [CrossRef].

Kodrasi, I., & Bourlard, H. (2020). Spectro-temporal sparsity characterization for dysarthric speech detection. *IEEE/ACM Transactions on Audio, Speech, and Language Processing, 28.*

Lauraitis, A., & Maskeliunas, R. (2020). ROBERTAS DAMAŠEVIČIUS 2,3, (Member, IEEE), AND TOMAS KRILAVIČIUS: Detection of speech impairments using cepstrum, auditory spectrogram and wavelet time scattering domain features. *IEEE Special Section on Deep Learning Algorithms for Internet Of Medical Things, 8.*

Lee, J. Y. (2012). A two-stage approach using gaussian mixture models and higher-order statistics for a classification of normal and pathological voices. *EURASIP Journal on Advances in Signal Processing*, 234–252, Nov 2012.

Lopes, L., Vieira, V., & Behlau, M. (2020). Performance of different acoustic measures to discriminate individuals with and without voice disorders. *Journal of Voice*, 0892–1997, The Voice Foundation. Published by Elsevier.

Lowe, D. (2004). Distinctive image features from scale invariant key points. *International Journal of Computer Vision, 60*(2), 91–110.

Lu, Y. (2016). The role of local knowledge in Yancheng National Nature Reserve Management (Ph.D. Thesis). Dunedin, New Zealand: University of Otago.

Majidnezhad, V., & Kheidorov, I. (2013). An ANN-based method for detecting vocal fold pathology. *International Journal of Computer Applications, 62*(7), 113–124.

Manjunath, B., & Ma, W. (1996). Texture features for browsing and retrieval of image data. *IEEE Transactions on Pattern Analysis and Machine Intelligence, 18*(8), 837–842.

Mathur, R., Pathak, V., & Bandil, D. (2019). Parkinson disease prediction using machine learning algorithm, 363 In V. S. Rathore, M. Worring, D. K. Mishra, A. Joshi, & S. Maheshwari (Eds.), *Emerging Trends in Expert Applications and Security* (p. 357). Singapore: Springer Singapore.

Mohammed, M. A., Abdulkareem, K. H., Mostafa, S. A., Ghani, M. K. A., Maashi, M. S., Garcia-Zapirain, B., ... AL-Dhief, F. (2020). Voice pathology detection and classification using convolutional neural network model. *Application Science*, 3723–3743.

Mohanta, A., Mukherjee, P., & Mittal, V. K. (2020). Acoustic features characterization of autism speech for automated detection and classification. *IEEE Xplore*, 978-1-7281-5120-5/20/$31.00 ©.

Musci, M., Feitosa, R. Q., Velloso, M. L. F., & Novack, T. (2011). An evaluation of texture descriptors based on local binary patterns for classifications of remote sensing images. *Boletim de Ciencias Geodesicas, 17*, 549–570.

Narendra, N. P., & Alku, P. (2020). Glottal source information for pathological voice detection. *IEEE Access, 8*, 67745–67755, April 6.

Ojala, T., Pietikainen, M., & Maeenpaa, T. (2002). Multiresolution gray-scale and rotation invariant texture classification with local binary patterns. *IEEE Transactions on Pattern Analysis and Machine Intelligence, 24*(7), 971–987.

Ojala, T., Pietikäinen, M., & Harwood, D. (1996). A comparative study of texture measures with classification based on feature distributions. *Pattern Recognition, 29*, 51–59.

Ojansivu, V. & Heikkila, J. (2008). Blur insensitive texture classification using local phase quantization. In *ICISP*.

Reddy, M. K., & Alku, P. (2020). Fellow, IEEE), AND Krothapalli Sreenivasa Rao, (Senior Member, IEEE): Detection of specific language impairment in children using glottal source features. *IEEE Access, 8* (15273–15279), January.

Sakar, C. O., Serbes, G., Gunduz, A., Tunc, H. C., Nizam, H., Sakar, B. E., ... Apaydin, H. (2019). A comparative analysis of speech signal processing algorithms for parkinson's disease classification and the use of the tuneable q-factor wavelet transform. *Applied Soft Computing, 74*, 255–263.

Sellam, V., & Jagadeesan, J. (2014). Classification of normal and pathological voice using SVM and RBFNN. *Journal of Signal and Information Processing, 5*(1), 1–7.

Shakoor, M. H., & Boostani, R. (2018). A novel advanced local binary pattern for image based coral reef classification. *Multimedia Tools and Applications*, *77*(2), 2561.

Shia, S. E., & Jayasree, T. (2017) Detection of pathological voices using discrete wavelet transform and artificial neural networks. In *2017 IEEE international conference on intelligent techniques in control, optimization and signal processing (INCOS)*. pp. 1.

Singh, S., Maurya, R., & Mittal, A. (2012). Application of complete local binary pattern method for facial expression recognition. In *Proceedings of the international conference on intelligent human computer interaction*, Kharagpur, India, Vol. 2013, pp. 1–4.

Sun, J., Han, L. J., Sun, D. Y., & Yuan, Z. H. (2006). The studies on anti-drought of seaweed extracts. *Marine Science*, *30*, 40–45.

Travieso, C. M., Alonso, J. B., Orozco-Arroyave, J., Vargas-Bonilla, J., Noth, E., & Ravelo-Garc, A. G. (2017). Detection of different voice diseases based on the nonlinear characterization of speech signals. *Expert Systems with Applications*, *82*, 184–195.

Tuncer, T., Dogan, S., Özyurt, F., Belhaouari, S. B., & Bensmail, H. (2020). Novel multi centre and threshold ternary pattern based method for disease detection method using voice. *IEEE Engineering in Medicine and Biology Society Section*, *8*, 84532–84542.

Vasilakisa, M., & Stylianou, Y. (2009). Voice pathology detection based on short-term jitter estimations in running speech. *Folia Phoniatrica et Logopaedica: Official Organ of the International Association of Logopedics and Phoniatrics (IALP)*, *61*(3), 153–170, June.

Vikram, C. M., Umarani, K. (2013). A wavelet based MFCC approach for the phoneme independent pathological voice detection. In *2013 Third international conference on advances in computing and communications*, pp. 153–156.

Wang, M. Y., Fei, X. Y., Xie, H. Q., Liu, F., & Zhang, H. (2017). Study of fusion algorithms with high resolution remote sensing image for urban green space information extraction. *Bulletin of Survey and Mapping*, 36–40.

Wang, J., & Liu, Z. (2005). Protection and sustainable utilization for the biodiversity of Yancheng seashore. *Chinese Journal of Ecology*, *24*, 1090–1094.

Xu, H., Zhu, G. Q., Wang, L., & Bao, H. (2005). Design of nature reserve system for red-crowned crane in China. *Biodiversity and Conservation*, *14*, 2275–2289.

Zhang, T., Shao, Y., Wu, Y., Pang, Z., & Liu, G. (2020). Multiple vowels repair based on pitch extraction and line spectrum pair feature for voice disorder. *IEEE Journal of Biomedical And Health Informatics*, *24*(7), 2168–2194.

CHAPTER

8 Internet of medical things for abnormality detection in infants using mobile phone app with cry signal analysis

K. Sujatha[1], G. Nalinashini[2], A. Ganesan[3], A. Kalaivani[4], K. Sethil[5], Rajeswary Hari[6], F. Antony Xavier Bronson[6] and K. Bhaskar[7]

[1]Department of EEE, Dr. MGR Educational and Research Institute, Chennai, Tamil Nadu, India [2]Department of EIE, RMD Engineering College, Chennai, Tamil Nadu, India [3]Department of EEE, RRASE College of Engineering, Chennai, Tamil Nadu, India [4]Department of CSE, Saveetha School of Engineering, SIMATS, Chennai, Tamil Nadu, India [5]Department of Mechanical Engineering, Dr. MGR Educational and Research Institute, Chennai, Tamil Nadu, India [6]Department of Biotechnology, Dr. MGR Educational and Research Institute, Chennai, Tamil Nadu, India [7]Department of Automobile Engineering, Rajalakshmi Engineering College, Chennai, Tamil Nadu, India

8.1 Introduction

The primary objective behind this scheme is the instantaneous detection of earache, colic pain, cold, diaper rashes, or due to hunger from the baby's cry signal. To date, the diagnosis is done in consultation with the child's specialist by a direct diagnosis of the infant to detect the earache, colic pain, cold, diaper rashes, or hunger by clinical examination. This is a time-consuming procedure and also a challenging task. Current practice also has a disadvantage where there is a time gap involved in the analysis which increases the chance of risk due to infections. Telehealth monitoring using a smartphone app enables continuous, safe, remote, and online diagnosis from the baby's cry signal using intelligent signal processing algorithms. The proposed scheme is a non-invasive method that uses, signal processing and intelligent algorithms for the online detection of earache, colic pain, cold, diaper rashes, or due to hunger from the baby's cry signal. Thus, the Smartphone app once deployed will reduce the encumbrances faced by the young mothers who find it a challenge to immediately identify the reason for their baby's cry combined with the reduction in cost needed for the development of a separate device or instrument thereby preventing the usage of very difficult and time-consuming technologies. Also, this app can cater to the needs of a large number of infant's cry signals enabling to maintain social distancing during this pandemic COVID-19.

Routine recognition and categorization of audio signal events is a demanding investigation, related to computational intelligence (Yogesh & Chinmay, 2020). Due to the huge quantity of aural data gathered and stored in the current period, data annotation cannot be done manually, as it is a cumbersome process. Hence, there is a need to design an effective algorithm that can automatically identify and classify the data patterns. The stored database is very important for recognition and

Implementation of Smart Healthcare Systems using AI, IoT, and Blockchain. DOI: https://doi.org/10.1016/B978-0-323-91916-6.00012-6

classification problems. This work, in this chapter, focuses on detecting the various abnormal conditions existing among the infants, who try to express their discomforts by the way of crying whose sounds have incredible variation with respect to amplitude and frequency in the home environment. So, crying is an important signal from the babies, where the mothers need to closely monitor them to identify their distress and needs like hunger and urination. The mothers can be sent an alert signal if the babies are crying due to such discomforts. Also, this method serves as a non-invasive approach too closely monitoring the babies and providing them with the greatest care. Young mothers by using this smartphone app will be able to develop a close relationship with their newborn babies. Hence the monitoring process is very tedious which is time-consuming and requires a long time to gather useful data.

The research component in the analysis of a baby's cry signals has been a challenging task till today. The analysis is carried out to determine the researchers who provide sufficient information about the baby's cry signal and their related psychological conditions (Farsaie Alaie, Abou-Abbas, & Tadj, 2015; Barr, 2006; Golub, 1979; Soltis, 2004). History denotes that a substantial amount of work has been carried out in this domain which deals with newborn's cry signals. Orozco-García & Reyes-García (2003), Várallyay (2006), Hariharan, Sindhu, and Yaacob (2012), Reyes-Galaviz, Cano-Ortiz, and Reyes-García (2008) depicted an effective algorithm for identification and classification of the baby's cry signals and its associated reasons whose accuracy ranges from 88% to 100%. Nearly, 93.16% to 94% accuracy was achieved in classifying the healthy newborns and those suffering from suffocation due to deficit oxygen supply as discussed by Sahak, Lee, Mansor, Yassin, and Zabidi (2010) and Zabidi, Khuan, Mansor, Yassin, and Sahak (2010) respectively. Other conditions relating to infants like anger, pain, and fear can be detected from cry signals for which studies were conducted by Petroni, Malowany, Johnston, and Stevens (1995), with an identification rate of 90.4%.

Remote baby monitoring can be made automatic using audio signal recognition techniques. Some researchers try to correlate the crying pattern with the health conditions of infants. The network architectures with learning rules like convolutional neural network (CNN) to identify the baby's cry sound along with the talking sounds and other domestic sounds to build an automated home is proposed by the researchers. Mel-frequency cepstrum coefficients such as pitch and formants are used as features to train the CNN classifier (Sravanth et al., 2020). These results are also compared with the results from the conventional logistic regression classifier to state that the results from the CNN classifier are fruitful as compared with the conventional classifier. This mobile app once developed, finds a major application in the medical field to maintain the entire database of the signals corresponding to a baby's cry which forms a part of the internet of medical things (IoMT). Thus, to support young mothers in their child care, this smartphone-based app will be helpful. The next section enlightens the reader about the entire overview of this chapter.

8.1.1 Organization of the chapter

The entire chapter is organized as follows; Section 8.1 deals with the introduction to the analysis of a baby's cry signal using signal processing techniques. Section 8.2 describes the work carried out by various researchers. The objective to develop this smartphone app using intelligent image processing algorithms is discussed in Section 8.3 and the major contribution is discussed in

Section 8.4. Section 8.5 denotes the gaps identified with the existing technology followed by Materials and methods in Section 8.6. Section 8.7 throws light on how the baby's cry signal was gathered. Whereas, in Section 8.8, a clear explanation is given about the methodology of how to classify and detect the problems from a baby's cry signal. All the related results with respect to CNN to detect various conditions from a baby's cry signal are mentioned in Section 9. The conclusion and future scope are included in Sections 10 and 11, respectively.

8.1.2 Significance of the proposed technology

The proposed technology will find major applications in the following sectors

1. Safe, secure, and emergency home-based measurements for young mothers' utility
2. Biomedical instrumentation Industries and Mobile health monitoring vehicles
3. Hospitals and their associated institutions for Medical studies with the integration of the internet of things (IoT)
4. Research center and utility in COVID-19 zones
5. Medical testing laboratories where the testing of physiological parameters is carried out
6. Engineering and technology related institutions can conduct camps to promote the usage of the app for the young mothers
7. Diagnostic centers.

The purpose of this research is to reduce the feature set as well as the classification algorithm so that results with robustness can be achieved. Thus a methodical evaluation for choosing the features is carried out. Modeling of the proposed system can be done which is application-specific. Many attributes which are closely related to the system and describe the quality are identified to measure the classification performance in the case of audio signal processing applications.

8.2 Literature survey

The major parts during the analysis of the baby's cry sound are related to inhalation and exhalation phases with the production of voice which is audible. The key to competitiveness is to design a technique that is capable of locating the audio signal patterns for inspiration and expiration exactly. The challenge of segmenting the cry signal arises because it has two parts, one the voiced part and the other the nonvoiced part. The conventional modules associated with the baby's cry signal detection system will not be capable to resolve this issue because these cry signals may be corrupted by noise. This was already stated by the researchers like (Várallyay, 2006; Zabidi et al., 2015; Kuo, 2010; Ruiz et al., 2010). The cry signal detection system identifies the region of interest (RoI) in the acquired acoustic signal. Acoustic signals include noise, calmness, or also an alarm signal. The corrupted signals can be evaluated using the Peak Signal to Noise Ratio (PSNR). The voice audio recognition system comprises two subdivisions (1) feature extraction and (2) rules for decision-making logic. The feature extraction includes wavelet transform (WT), extraction of cepstrum coefficients, spectrum analysis, and entropy calculation. Decision-making logic uses a simple threshold to formulate the rules. This method had the disadvantages like fine-tuning the threshold value to get adapted to a noisy environment, segmentation of the cry signal to split the exhalation and inhalation parts, and distinguishing them with a clear demarcation.

Statistical approaches serve as an optimal solution to solve the problem involved in finalizing the dynamic threshold values. Statistical model-based approaches are introduced and discussed by (Abou-Abbas, Alaie, & Tadj, 2015a; Abou-Abbas, FersaieAlaie, & Tadj, 2015b; Abou-Abbas, Montazeri, Gargour, & Tadj, 2015c); Abou-Abbas et al., for automatically adjusting the threshold value to yield optimal results (Petroni, Bono, Marconi, & Stellingwerf, 2003).

The challenge lies in splitting the cry signal which has a silent phase and a respiratory phase in it (Várallyay et al., 2010). The noisy signals getting mixed up with the original cry signal is a major drawback of this system. Researchers have used Hidden Markov models (HMMs) to automatically slice the exhalation and inhalation portions from the baby's cry signal. The drawback of this method is the lack of a database relating to the baby's cry sound during the inspiration and expiration phases. Researchers have also tried support vector machines and Gaussian mixture models (GMMs) for classification purposes in this domain (Aucouturier, Nonaka, Katahira, & Okanoya, 2011). Kim, Kim, Hong, and Kim (2013) and Yamamoto, Yoshitomi, Tabuse, Kushida, and Asada (2013) have also contributed to the domain of a baby's cry signal. They have used segmented 2-DLinear Frequency Cepstral Coefficients (STDLFCCs). The Linear Frequency Cepstral Coefficients (LFCCs) form the basis for the STDLFCC algorithm. Only the low-frequency components are capable of offering a distinct variation among the baby's cry and non-cry signal segments. The inspiration was to detect the low and high-frequency values which are bounded by upper and lower limits for cry and non-cry segments of the signal. The deviation is nearly 4.42%.

Hence from the literature survey, it is understood that a smartphone-based app with an effective diagnosis system is required to facilitate the young mothers in their child care.

8.3 Objectives

1. The proposed technology is a smartphone app for the detection of earache, colic pain, cold, diaper rashes, fever, respiratory problems, or hunger from the cry signal of the infants.
2. This smartphone app will use intelligent image processing algorithms like WT and CNN for the classification of the audio signal captured from the baby's cry to identify earache, colic pain, cold, diaper rashes, fever, respiratory problem or hunger.
3. The time delay will be reduced, and results will be obtained no matter of time.
4. It will facilitate remote, non-invasive, online, and continuous monitoring for detection of earache, colic pain, cold, diaper rashes, fever, respiratory problems, or hunger and will be able to analyze at least 70,000 samples within no matter of time.
5. These indigenous intelligent signal processing algorithms and a smartphone-based app when developed will be able to cater to the need of many users thereby enabling social distancing during this pandemic COVID-19. The entire system also will have a cloud set that will facilitate data retrieval at any time.
6. To develop a strategy for structured Medical Internet of Things.

8.4 Major contribution

The novelty of the proposed online-based smartphone app lies in the detection of earache, colic pain, cold, diaper rashes, fever, respiratory problems, or hunger from the cry signal of the infants. This painless method enables user-friendly diagnosis for young mothers as a part of baby care by embedding intelligent signal processing algorithms which will process the baby's cry signal captured by the voice recorder in the smartphone. This smartphone app extracts the details and approximations using WT from the baby's cry signals that are captured and will classify the earache, colic pain, cold, diaper rashes, fever, respiratory problem, or hunger from the cry signal of the infants with specific and accurate measurements instantly. On the other dimension, this novel technology will place an end to the challenge involved, thereby offering a contactless diagnosis system during this pandemic COVID-19 situation.

8.5 Gaps identified

Presently, the smartphone-based app is used to capture the baby's cry signal and infer the reasons for the baby's cry which may be due to earache, colic pain, cold, diaper rashes, fever, respiratory problems, or hunger. The researchers have not developed the technology to infer the reason for the baby's cry by analyzing the audio signal. Hence from our side, we have focused to develop a robust smartphone app which will use intelligent signal processing algorithms to detect the reasons like earache, colic pain, cold, diaper rashes, fever, respiratory problems, or hunger from the baby's cry signal.

It is understood that very less quantum of work has been reported in this area, which serves as strong support to carry over this work. This proposed technology will be a smartphone app that will be installed on the mobile phone so that even a young mother, when capturing the baby's cry signal using the audio recorder on a mobile phone will be able to determine the reasons like earache, colic pain, cold, diaper rashes, fever, respiratory problem or hunger from the baby's cry signal.

The proposed smartphone app will infer the reasons like earache, colic pain, cold, diaper rashes, fever, respiratory problems, or hunger from the baby's cry signal which is an indigenous technique to relieve the anxiety of the mothers when the babies are crying continuously from pain. Also, this technology will reduce the time delay involved in contacting the child's specialist, thereby offering an immediate first aid to console the babies for time being. Moreover, the proposed technology is a layman's approach to the young and inexperienced mothers, who face the challenge to identify the exact reason for the baby's cry which will also avoid the direct contact despite COVID-19 enabling social distancing.

8.6 Materials and methods

Artificial intelligence (AI) research domain makes the machine perform tasks similar to human intelligence. Machine learning is used to obtain skills without any human interference.

Sophisticated intelligent machines can be deployed in AI Systems which use artificial intelligence, machine learning and deep learning algorithms. Better decisions can be automatically obtained using Machine Learning algorithms (Yamamoto, Bayat, Bellen, & Tan, 2013). A wide range of algorithms, on an iteration basis, learns the data patterns and provides direction to forecast optimal outputs. Hence, the patterns are analyzed to identify suitable algorithms to detect the required output. The branch of Machine Learning deals with deep learning algorithms which are shown in the correlation diagram depicted in Fig. 8.1.

The subset of machine learning is the Deep Learning neural networks. Optimal outcomes occur with human interference to achieve the desired output. They analyze the data in a logical structure to conclude.

It is a part of machine learning which trains the computer about the basic instincts of a human being. The wider applications include virtual assistants, facial recognition, driverless cars, etc. Supervised, semisupervised, and unsupervised learning algorithms are used to train the application-related data. The following are the points to be considered for deep learning algorithms

1. It mimics the activities of the human brain in decision making
2. Patterns of data are varied, diversified, and huge in number
3. Decision making can be made efficiently.

Applying deep neural networks with multiple layers on a large volume of data than traditional Machine Learning algorithms is effective. The performance of deep learning algorithms is directly proportional to the amount and variety of data as depicted in Fig. 8.2. Systems that learn to take actions based on examples, without an explicitly specific program.

Artificial neural network (ANN) architecture is made of three layers- input, output, and one hidden layer. Deep neural networks are ANNs with multiple layers between input and output layers which has multiple hidden layers. The major deep learning algorithms are deep neural network, deep belief network, recurrent neural network, and CNN. These algorithms are applied for different

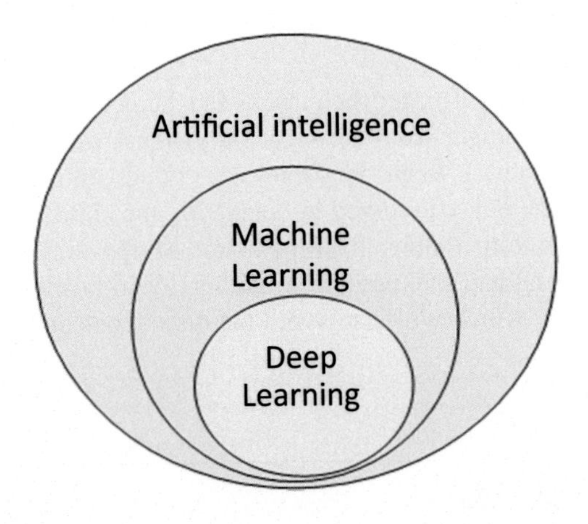

FIGURE 8.1

Correlation diagram for illustrating the association of various intelligent techniques.

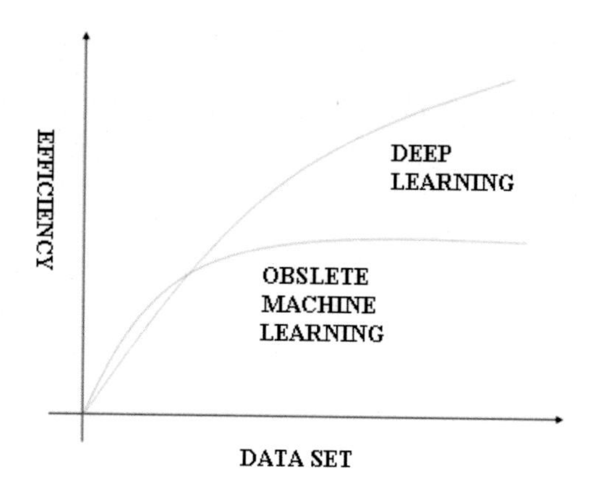

FIGURE 8.2

Machine learning and deep learning performance.

applications based on the requirement and performance of different types of data (Ntalampiras et al., 2013).

8.6.1 Classification of deep learning algorithms

Deep neural networks are not easy to train with backpropagation due to the problem of vanishing gradient which impacts the time taken for training and reduces accuracy. The cost function is calculated based on the net difference between the Neural Network's predicted output and actual output in the training data. Based on the cost, weights and biases are altered after each process until the cost is small. The gradient is the rate at which cost will change based on weights and biases. Deep learning algorithms perform best with problems with huge data sets. Fig. 8.3 shows the classification of deep learning algorithms.

8.6.2 Convolutional neural network

Convolutional neural network consists of three layers as shown in Fig. 8.4. They are, Convolution, Pooling, and fully connected layers. The convolution function is the integral product of the two functions after one is reversed and shifted (Reyes-Galaviz et al., 2008). The convolution layer is the first layer and it extracts the features from an input image. In the convolutional layer, an activation function is applied which helps in getting non-linearity and to get an output based on the input (Abou-Abbas, Tadj, Gargour, & Montazeri, 2017). The activation functions may be ReLU, Sigmoid, or tan h. The most commonly used in CNN is ReLU.

CNN can be used in Image Analysis, Optical Character Recognition to convert a handwritten document into digital text. The structural design for CNN is diagrammatically represented in Fig. 8.4. Deep Learning has the following benefits

1. Ability to generate new features from the limited available training data sets.
2. Capable of parallelism.

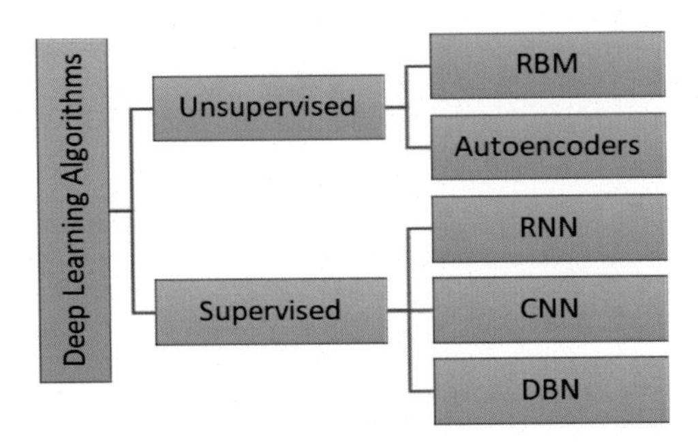

FIGURE 8.3

Deep learning algorithms classifications.

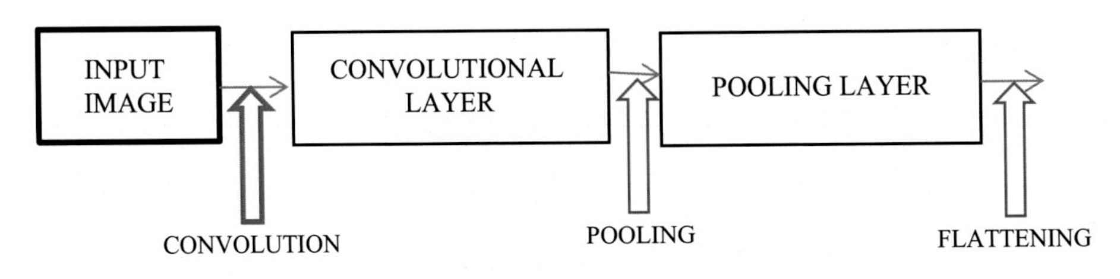

FIGURE 8.4

General architecture of convolution neural network.

3. Time complexity is reduced during feature analysis.
4. The model is robust.
5. One to one mapping is possible.

There are also some drawbacks associated with the CNN

1. Training the network using wide data set enhances the data flow leading to accurate output.
2. As the data set increases, the computational complexity also increases.
3. It is easily prone to faults whose identification and rectification is a challenge.
4. Higher-end processers with supercomputing facilities are required which is costly.

The Fourier Transform captures only global frequency information. This signal decomposition may not serve all applications, as in the case of certain signals relating to biomedical applications. Wavelets are obtained when the signal is decomposed using the standard set of functions.

Scale (or dilation) defines how "stretched" or "squished" (Fig. 8.5) a wavelet is which is related to frequency. The spatial and temporal coordinates of the wavelets are identified and represented graphically.

The factor "a" is defined as the scaling factor. The waveform gets squished for the decrease in the scaling factor and tracks the higher frequency components. Waveform for the wavelet is stretched, if the scaling factor is increased and tracks the low-frequency information.

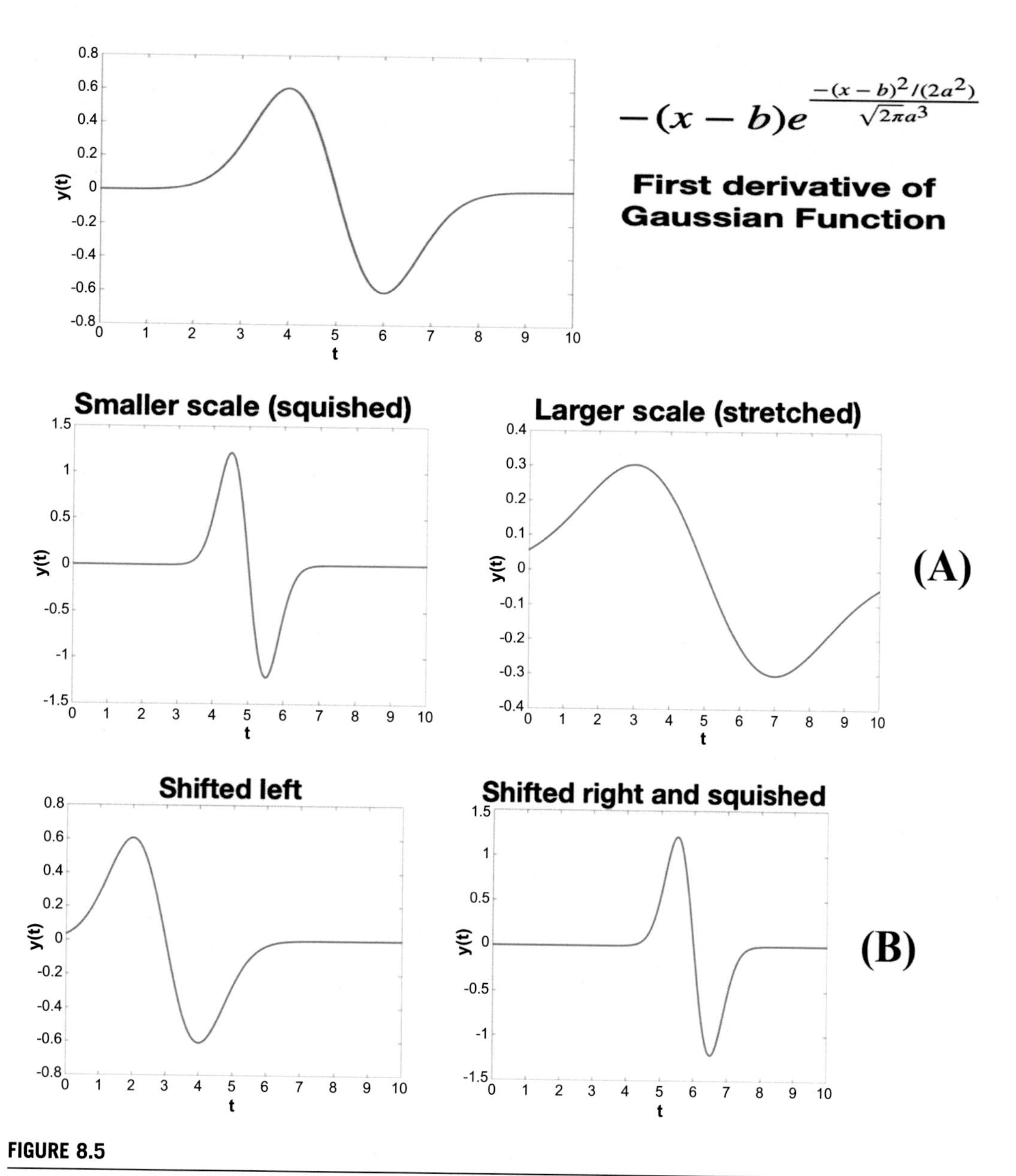

FIGURE 8.5

Graphical representation for scaled and stretched wavelets.

The factor "b" denotes the position of the wavelet. The wavelet is shifted to its left if the value of "b" is decreased and vice-versa. Like the continuous and discrete signals, the WTs are also classified as continuous and discrete transforms. The scaling factor and the position are countless in the case of continuous wavelet transform. Conversely, for discrete wavelet transform, the scaling factor

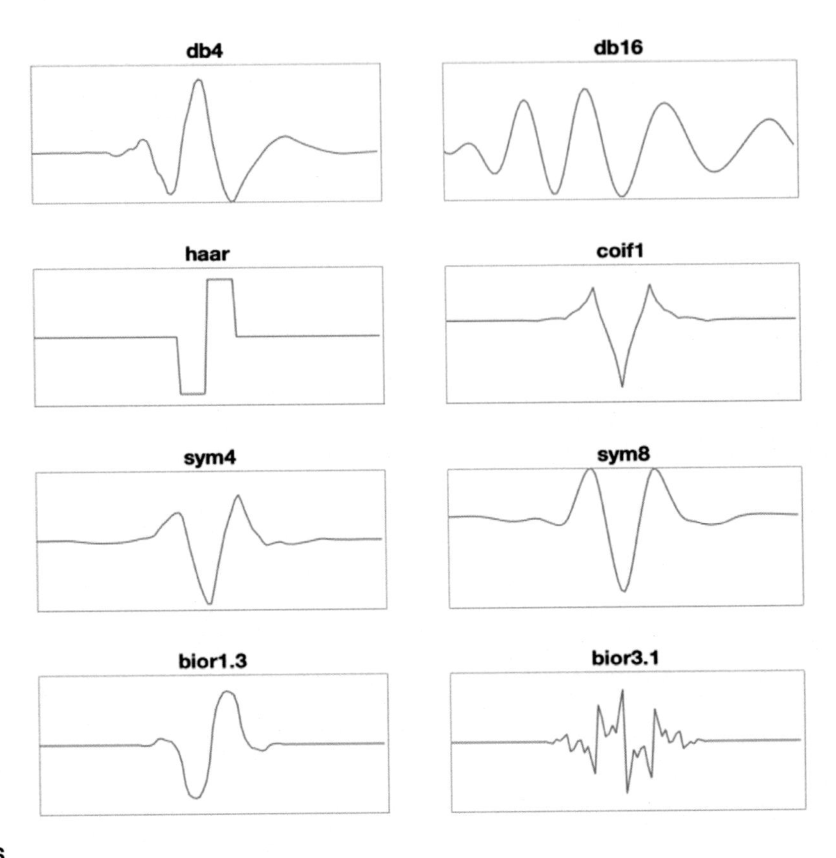

FIGURE 8.6

Graphical representation for various types of wavelets.

and the positions are available only for discrete instants of time. The various types of wavelets are graphically depicted in Fig. 8.6.

The wavelet coefficients serve as the feature set for training the CNN model to detect the abnormal conditions from the baby's cry signal (Dwivedi, Dey, Chakraborty, & Tiwari, 2021).

8.7 Results and discussion

8.7.1 Baby's cry signal collection

The baby's cry signal was recorded using the smartphone. Various models and brands of smartphones were used to record the baby's cry signal. Pediatric Hospitals were approached to collect the cry signal of the babies. Cases where babies were detected with colic pain, cold, earache, diaper rashes, hunger, and excretion were considered and recorded. Correspondingly, the diagnosis was made by a pediatrician and his opinion was recorded. For the sample collection in real-time, various pediatric hospitals were approached and nearly 1,40,000 samples were collected. Nearly, 70,000 samples were used for training, 40,000 samples were used for testing and the remaining 30,000 samples were used for validation after the smartphone app is developed (Bachu, Kopparthi, Adapa, & Barkana, 2010).

The samples of the baby's cry signal for this investigation include the recorded version of the baby's cry for a number of hours from various child care hospitals in and around the city. The babies that were considered for investigation were nearly in the range of 0–6 months of birth. The recording of the audio signal corresponding to the baby's cry was done meticulously 24×7 in a household atmosphere. These audio signals included a variety of types of cry sounds corresponding to colic pain, cold, earache, diaper rashes, hunger, and excretion. The sampling frequency for recording the audio signal of the baby's cry was fixed to be 44.1 kHz.

8.7.2 Methodology—signal processing technique for analysis of baby's cry signal

The various types of baby's cry signals are pre-processed for the removal of noise. Then WT is used to obtain the details and approximation coefficients for the various baby's cry signals. The project's research highlight is the development of a smartphone app that uses signal processing algorithms like WT for feature extraction, where these wavelet coefficients are found to show variation with respect to problems like earache, colic pain, cold, diaper rashes, fever, respiratory problem or hunger from the baby's cry signal. The various problems like earache, colic pain, cold, diaper rashes, fever, respiratory problems, or hunger associated with the infants are invariably classified or clustered using intelligent algorithms like CNN. These signals will be processed by two algorithms; one is the WT for feature extraction and the other one is the CNN, which is attempted, to evaluate the detection efficacy during diagnosis. The features (Wavelet coefficients) are extracted from the cry signals of the infants, whose signals will be recorded using the voice recorder in the smartphone.

The motivation of the proposed method is to develop a noninvasive, contactless, online, continuous, and cost-effective smartphone app for detecting the conditions like cold, colic pain, diaper rashes, fever, respiratory problem, and hunger from the baby's cry signal analysis. This kind of smartphone app relies on intelligent signal processing algorithms and can examine thousands and thousands of infants (samples) in a very short duration of time with high accuracy, thereby resolving time complexity, amidst the pandemic, COVID-19. This indigenous system will offer an effective detection system and is user-friendly, so young mothers possessing a smartphone will be capable of self-oriented diagnosis without the help of a clinician to offer first aid measures to their babies. These values of details and approximation coefficients are used as inputs to train the CNN classifier. The illustration for the block diagram is depicted in Fig. 8.7. The signals for the baby's cry corresponding to the conditions like cold, colic pain, diaper rashes, fever, respiratory problem, hunger, and excretion are recorded and tabulated below. Table 8.1 denotes the audio signal for various abnormalities associated with a baby's cry. The flowchart for training, testing, and validation using CNN is depicted in Figs. 8.8, 8.9, 8.10 respectively.

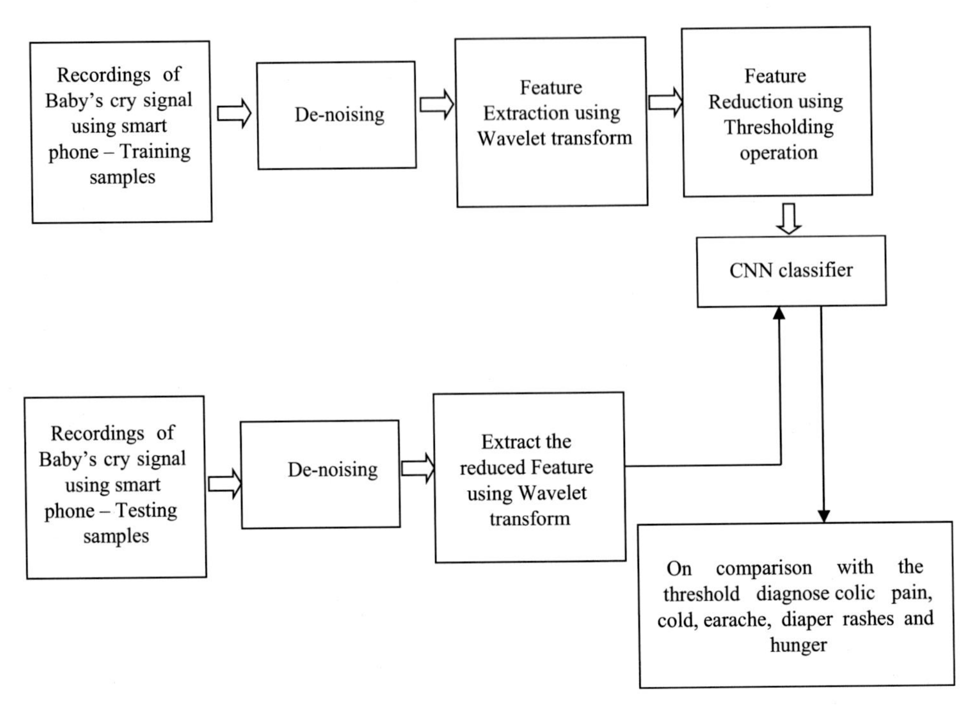

FIGURE 8.7

Block diagram for baby's cry signal analysis.

8.7.3 Principle of operation

The principle of operation of the customized smartphone app for measurement of the baby's cry corresponding to the conditions like cold, colic pain, diaper rashes, fever, respiratory problem, and hunger is discussed here. The proposed algorithm will have high accuracy with excellent screening capacities. Once the smartphone app is developed, then the calibration will be done by asking the young mothers to install the app onto their smartphones. After that, the patient needs to open the App, which will in turn enable the camera in the smartphone to be switched ON. The audio signal of the baby's cry is captured and then processed by the app in the smartphone which uses intelligent signal processing algorithms.

The signal processing algorithms which are built now will select the RoI by a tap on the mobile screen. Audio signals of the baby's cry that are captured using the voice recorder are the actual inputs for this app. Any noise due to interference will be automatically rejected using the WT. For this purpose, only, CNN will be used as a classifier which has the capacity to generalize and robustly detect the colic pain, cold, earache, diaper rashes, hunger, and excretion thereby preserving the accuracy level close to the standard level. When the baby's cry signal is captured by the young mother using the voice recorder in their smartphones, it will identify the RoI, present in the baby's cry signal, and details and approximations obtained from the WT that lie outside the threshold will be barred from detecting the colic pain, cold, earache, diaper rashes, hunger, and excretion. Once

Table 8.1 Sample database for baby's cry signal.

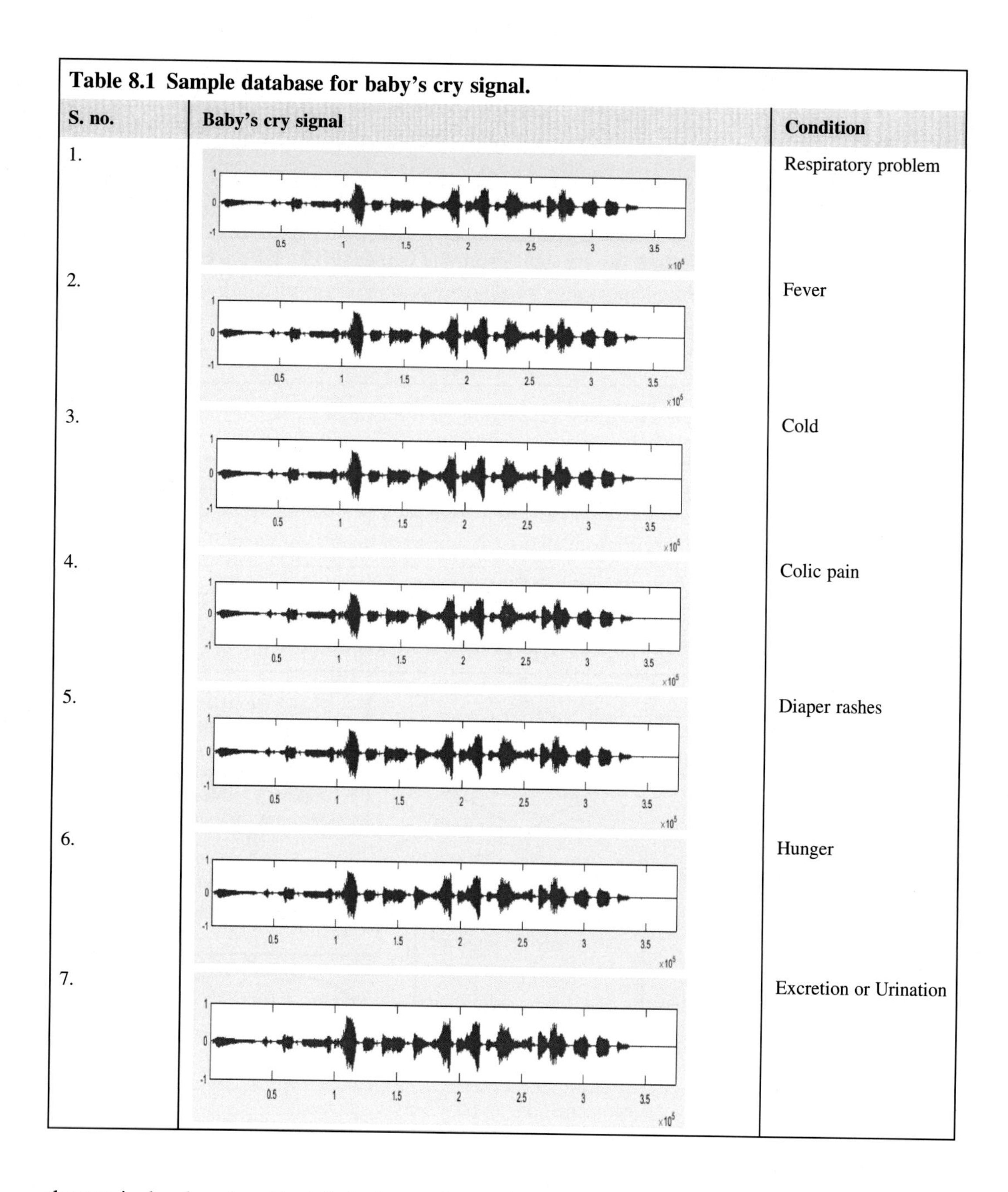

S. no.	Baby's cry signal	Condition
1.		Respiratory problem
2.		Fever
3.		Cold
4.		Colic pain
5.		Diaper rashes
6.		Hunger
7.		Excretion or Urination

the app is developed and installed, the results can be compared for measurements of various baby's cry signal conditions like colic pain, cold, earache, diaper rashes, hunger, and excretion by using the CNN classifier which uses convolution operation as the foundation that exists in the field of

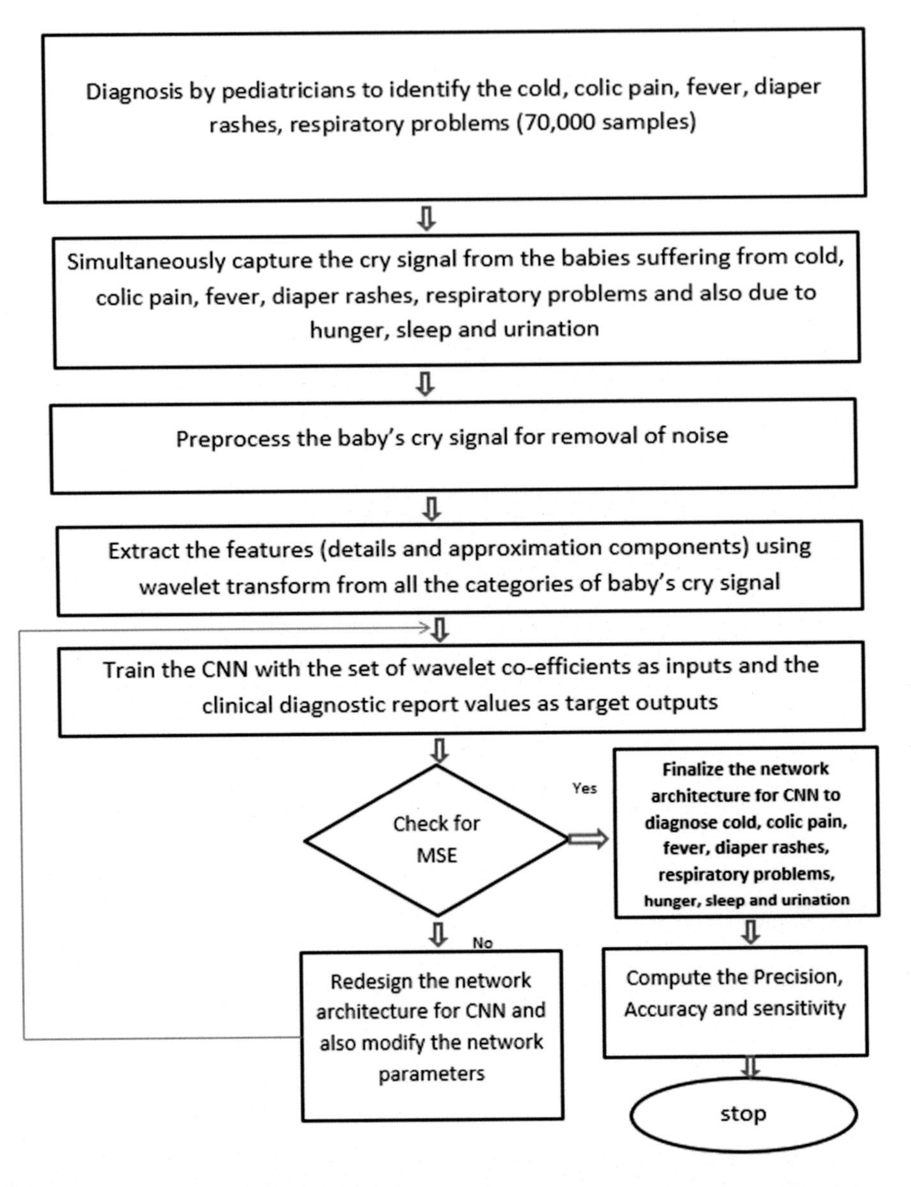

FIGURE 8.8

Flow chart for training the convolutional neural network for baby's cry signal Analysis.

signal processing. These measurements pertaining to the conditions of the baby's cry like colic pain, cold, earache, diaper rashes, hunger, and excretion can be transferred to a cloud-based server for remote monitoring rather than on-board examinations.

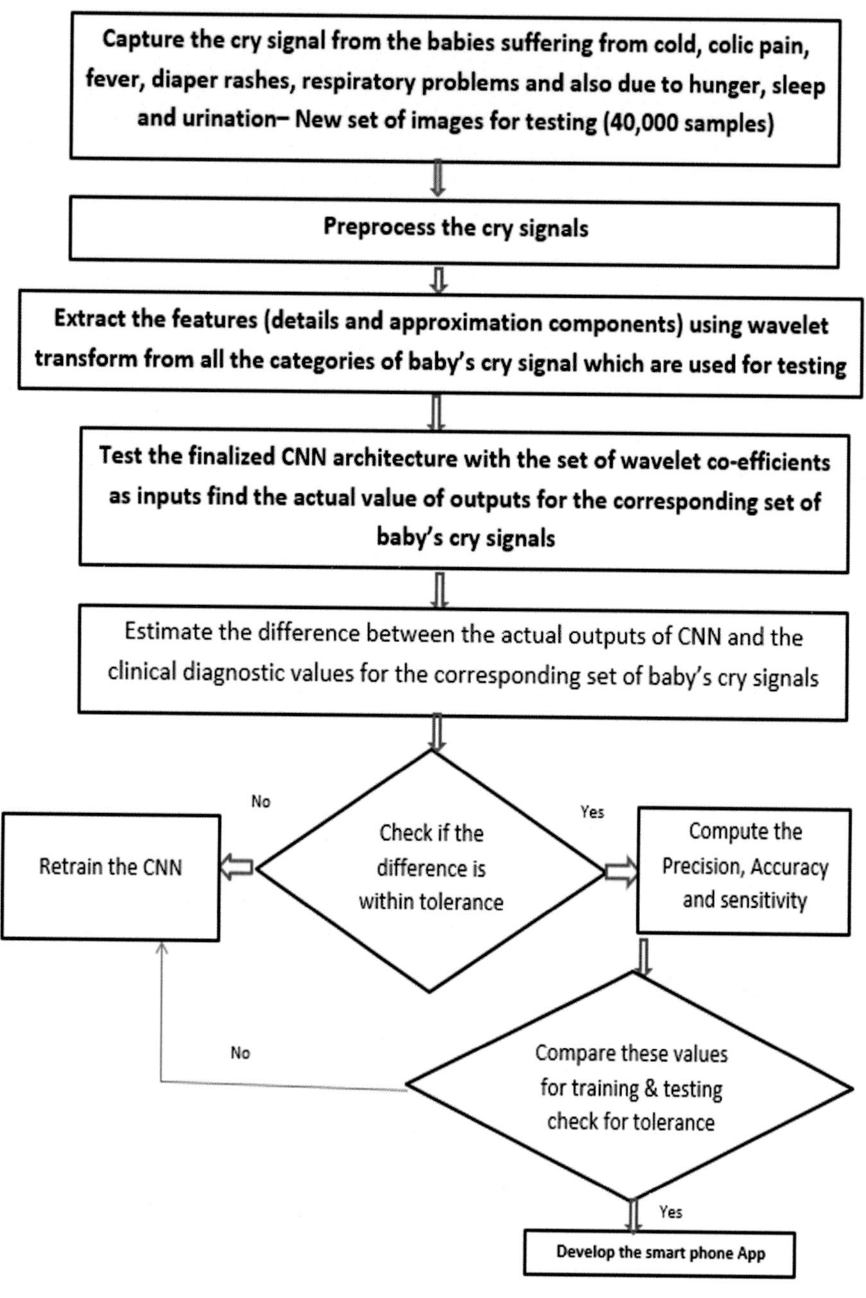

FIGURE 8.9

Flow chart for testing the convolutional neural network for baby's cry signal Analysis.

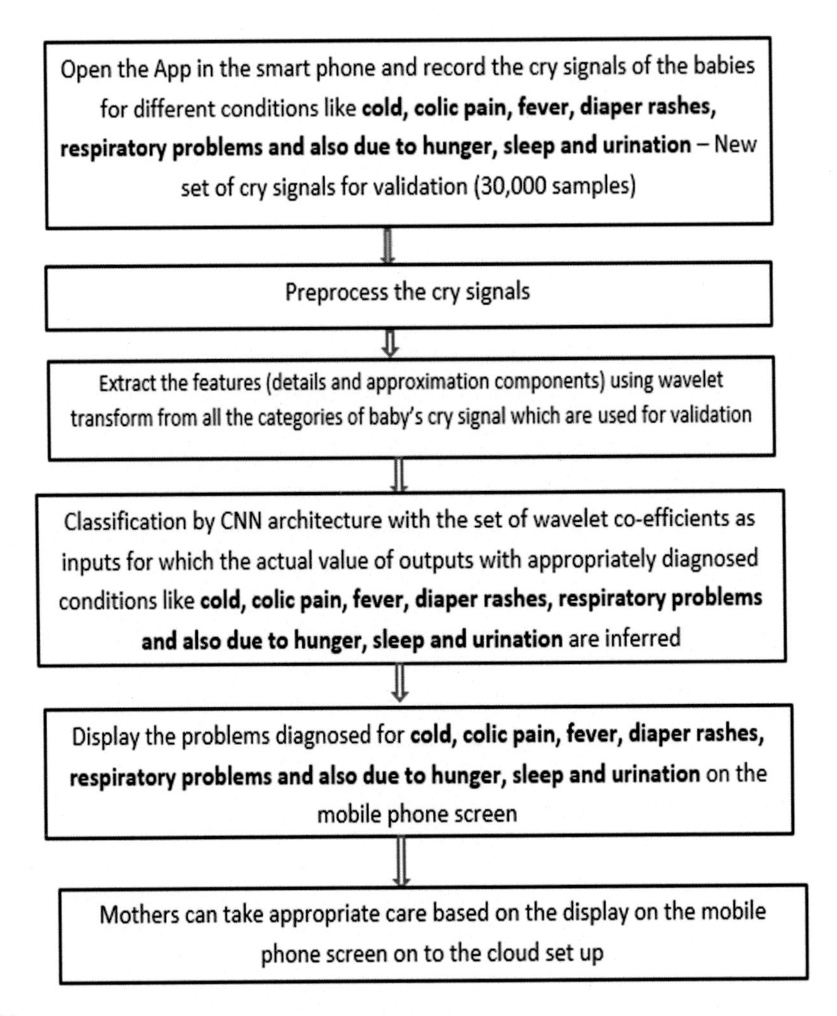

FIGURE 8.10

Flow chart for validation of convolutional neural network for baby's cry signal Analysis.

8.7.4 Preprocessing

The noise removal is also done using the wavelet toolbox. De-noising helps to improve the quality of the signal so that appropriate and exact values of the features alone can be extracted. A high pass filter (HPF) allows the signal portion corresponding to high-frequency values alone to pass through it. The transfer function for the HPF in a discrete domain is given by $1-0.99$ z.

8.7.5 Feature extraction using wavelet transform

An extensive variety of features are extorted from the infant cry signal using a one-dimensional discrete WT as illustrated in Fig. 8.11. The preferred feature set serves as the input and

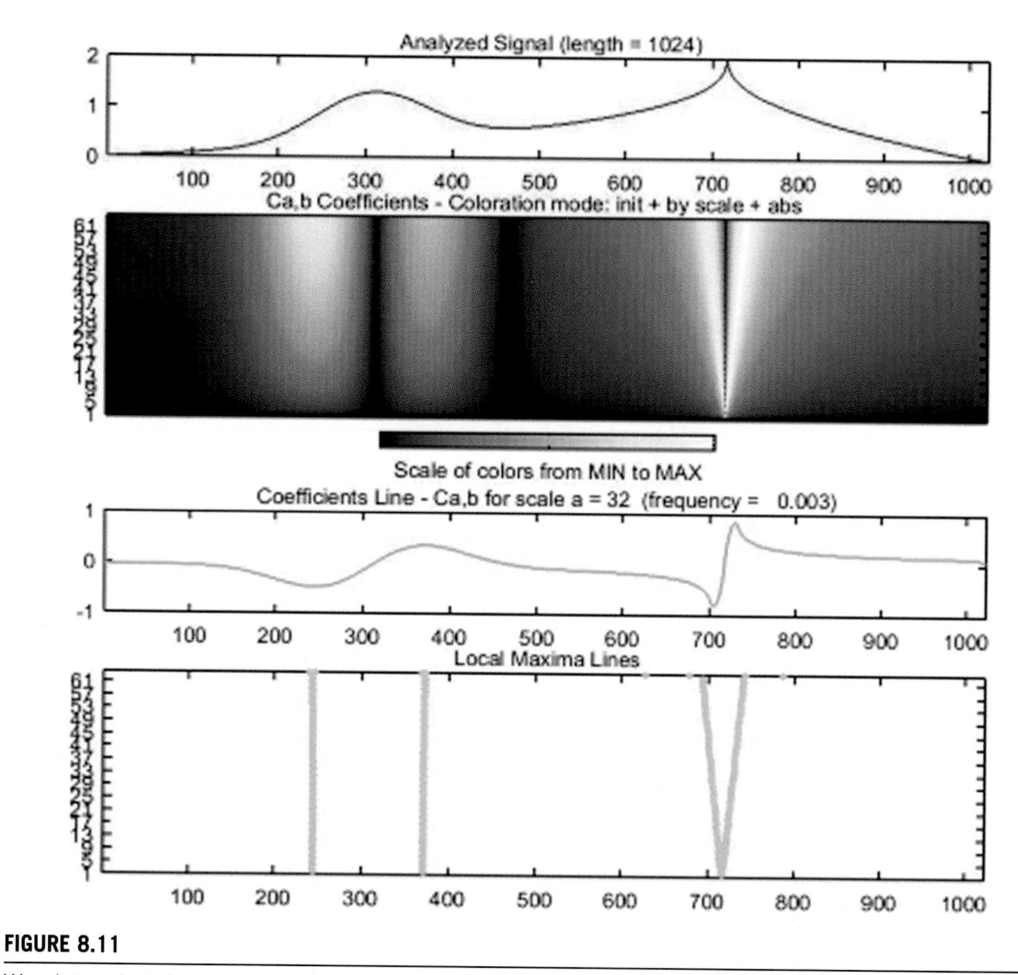

FIGURE 8.11

Wavelet analysis for baby's cry signal with a respiratory problem.

includes the details and approximation components to train and test the CNN as shown in Fig. 8.12.

8.7.6 Identification using convolutional neural network

A supervisory infant cry signal segmentation scheme, to trace and capture the baby's cry signal even in a noisy environment can be facilitated by implementing CNN to detect the abnormality in the training phase by using 70% of the collected signals itself so that an appropriate model for cry signal analysis can be launched. During the testing of CNN, only 20% of the collected sample signals are used for the detection of abnormality. The remaining 10% can be used for validation after the development of the mobile phone app.

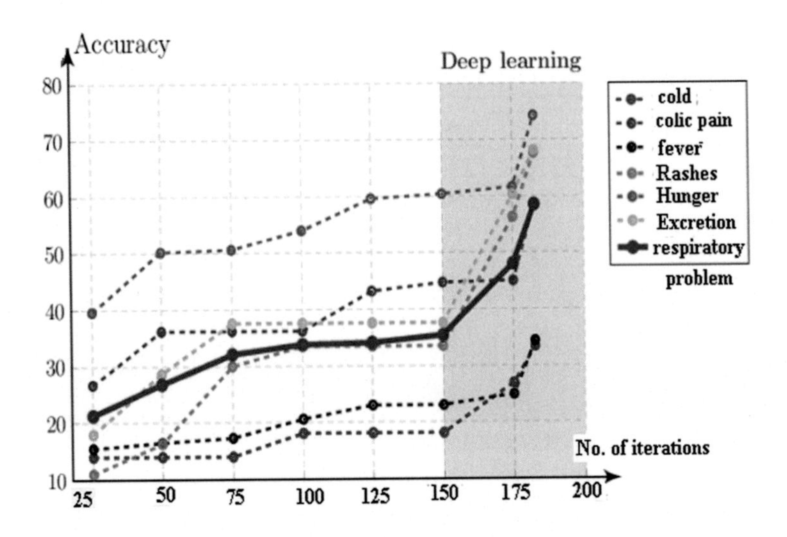

FIGURE 8.12

Determination of Various conditions from baby's cry sound using convolutional neural network.

Convolution neural networks include a variety of applications in the domain of machine vision, Big data analytics, and a lot of other classification applications where an enormous quantity of data is processed and classified. Similar to the conventional ANN, the architecture has a number of interconnected layers of processing elements that are related by random numbers called weights. Since convolution operation is performed in the layers of CNN, it performs the function of the filter which facilitates the noise removal. CNNs learn the filters during the training process, which can be thought of as a way to generate important features out of the data. The CNN requires a priori knowledge about the database for obtaining accurate classification.

8.7.7 Development of mobile app

Once the simulation using MATLAB is over, the next stage is the development of a smartphone App.

Mobile application development architecture

The following block diagram explains the high-level architecture of the mobile application (to be developed). This architecture will have –

1. Mobile client—which will act as a user interface, and can be installed on their mobile phones?
2. Server—which will be deployed on AWS as an API and this will be responsible for the data processing and notifications.

The following are the various components present on the mobile application:

Mobile application—is the container that enables the user to authenticate, and capture a baby's cry sounds. This application will consume the various APIs developed, to process the captured images and detail the deficiencies based on the algorithm. Server APIs are the connectors that bridge the mobile application and the data processing engine hosted on the application server. These API utilize their authentication mechanisms for secured data transport.

A data processing engine is developed with a pre-approved algorithm to process the captured signals, match them with the pre-built models, and provide a response back with the details (Chinmay, Gupta, & Ghosh, 2014). This module also has its data ingestion module for the administrator to upload more samples, or tweak sample results. The error handling module provides the base framework to capture erroneous data or any server related downtime. This module will be responsible for providing user-readable messages. Application Servers are the containers to host the server APIs.

The main motive behind developing accurate technology is to diagnose the reason for a baby's cry was proposed initially by the nurses and the infant's mother depending on the circumstances. Health conditions were recognized by the physician who scrutinized the newborn for which various laboratory testing has been carried out to identify the exact reasons in the maternity homes after the birth of the newborn.

8.7.8 Extraction of wavelet coefficients

The proposed cry signal based infant diagnostic system (CSIDS) was validated with nearly 10% of the total audio signals collected when the infants were crying due to earache, colic pain, cold, diaper rashes, fever, urination, respiratory problems, or due to hunger. The outcome was predictable from the segment of the audio signal of the baby's cry captured using the mobile phones. Then these cry signals from the babies were pre-processed for noise removal and split up manually in consultation with the experts. The main benefit of this method is that the proposed CNN is robust enough that it is capable of restoring even the lost portion of the cry signal.

The training of CNN is done using gradient descent rule with optimal values of learning rate and momentum. The change in weights after each iteration is calculated and each time, the new weights are found. The training includes the filter interpretation which is derived at the first layer of the CNN. The variation in the signal amplitude is tracked and captured during this phase. Once the network is trained and the CNN architecture is finalized, the same network topology can be used for testing and validation.

The measures like true positive (TP), false positive (FP), true negative (TN), and false-negative (FN) were found to suitably address the need. TP denotes the correct number of baby's cry signals detected; FP denotes all cry signals that are erroneously detected. TN denotes the number of baby's cry signals rejected correctly; FN denotes all the cry signals that are not all detected. The performance evaluation is done by calculating the metrics (Table 8.2) like true positive rate (TPR), false positive rate (FPR), false negative rate (FNR), and accuracy. It is inferred from Table 8.2 that, by the proposed method (WT-CNN) for analysis of baby's cry signal analysis, the TPR value is 98.23% and 85.56% during training and Validation. The overall accuracy is calculated to be nearly 96.79% from Fig. 8.12.

Once the smartphone app is developed, an entirely new set of audio signals corresponding to the baby's cry will be used for the detection of earache, colic pain, cold, diaper rashes, fever, urination, respiratory problems, or due to hunger using the proposed smartphone app which is installed in the smartphone device (Apple make) will be used as the masterpiece for calibration. Calibration of the developed smartphone app can be done by cross verifying with the results which are

Table 8.2 Performance metrics.

	Training the CNN			Testing the CNN			Validation of CNN			
%	TPR	FPR	FNR	TPR	FPR	FNR	TPR	FPR	FNR	Accuracy
WT-CNN	98.23	8.12	3.97	84.56	19.65	9.44	97.36	3.13	1.74	96.79
K-means	88.12	24.7	12.69	91.25	29.72	10.65	82.85	29.6	17.15	85.64
FFT-GMM [Abou-Abbas]	93.03	18.52	6.97	84.56	29.65	15.44	95.63	10.36	3.37	91.01
EMD-HMM [Abou-Abbas]	88.32	23.6	11.68	90.35	27.74	9.65	92.85	19.6	7.15	88.94
FFT-GMM (post-processing) [Abou-Abbas]	95.6	10.32	4.4	89.52	20.46	10.48	97.26	3.23	2.74	94.29
EMD + HMM (Post-processing) [Abou-Abbas]	93.7	15.08	6.3	92.47	18.21	7.53	95.07	7.84	4.93	92.16

CNN, convolutional neural network; *TPR*, true positive rate; *FPR*, false positive rate; *FNR*, false negative rate; *WT*, wavelet transform; *GMM*, Gaussian mixture model; *HMM*, hidden Markov model.

obtained from the Child care hospitals recorded for diagnostics. A comparative analysis will be done as indicated for various models and make of the mobile phones. If the deviation is within tolerance on consultation with the recognized, medical council authorized physician, the smartphone app will be hosted on the server for the free usage of the young mothers who need assistance in neonatal care frequently. During the testing phase, randomly few brands of the smartphone will be used to detect the earache, colic pain, cold, diaper rashes, fever, urination, respiratory problems, or due to hunger by capturing the baby's cry signal and then identifying the appropriate reasons using the signal processing techniques and will be compared with the diagnosis done by a child specialist.

The proposed technology to be developed is an app for smartphones that can cater to as many numbers of users as possible and will be able to download the app from internet sources free of cost. They only have to install the app on their smartphone, open the app to capture the baby's cry signals, and simply use it. The major advantage of the proposed technology for the detection of earache, colic pain, cold, diaper rashes, fever, urination, respiratory problems from the audio signals of the baby's cry recorded using a voice recorder. A smartphone app once developed, does not require a separate device instead, the app which is available in the internet sources free of cost can be downloaded on a smartphone with a voice recorder and Android technology.

In the present scenario, the internet is being used to connect various medical-related devices to increase the diagnosis efficiency, reduced cost, and achieve better results in the domain of healthcare. This proposed technology is a wireless technology, with high computing levels to integrate with the internet and develop an IoMT technology. This IoMT technology is capable to gather, transmit and track the medical data related to newborn babies to evolve into state-of-the-art technology.

8.8 Conclusion and future scope

Segmentation of an infant's cry signal involves recording the cry signal in the hospitals and segmenting the homogenous parts in the signal. CSIDS is used to detect the RoI from the cry signal of babies. The performance can be considerably increased for the proposed CSIDS, even if the signal is corrupted with noise. The key objective behind this work is to identify the useful component and the noisy component from the cry signal. In the future, this method of cry signal analysis can be used for developing the database that facilitates the following applications like differentiating the diversified features present in the audio signal of the baby's cry due to earache, colic pain, cold, diaper rashes, or hunger and to categorize the infants who are healthy and unable to hear and suffering from diseases like asphyxia and meningitis.

The objective of this CSIDS is achieved by developing a basement for initiating research work pertaining to psychosomatic activity and coregulatory patterns existing between the infants and the mothers. Thus, the significance of achieving a very high recognition rate is evident. However, a less value of FPR is conceivably vital, to prevent the data from being corrupted by noise, which may stop significant conclusions. Thus it is very clear from this smartphone-based application, that the various intelligent algorithms, with their acquired intelligence from the data patterns extracted from the baby's cry signal, were helpful to develop a technology that can facilitate the young mothers in their tenure of child care without mental stress. Thus the proposed IoMT strategy is not only helping to improve the experience of young mother's by eliminating the need for in-person medical visits, but it's also helping to reduce costs.

As a part of the future scope of this work, the other reasons for the baby's cry can also be analyzed and estimated by using this smartphone App, where the baby's cry signals are captured and analyzed. The COVID-19 symptoms can also be obtained using this smartphone app which captures the audio signal of the baby's cry signal along with the respiratory and inspiratory breathing sounds using signal processing algorithms and uses a smartphone app with inbuilt intelligent signal processing algorithms, so that it also possible to diagnose COVID-19 by maintaining the social distancing and avoiding the swab test. This smartphone-based app can also be extended to identify the presence of Corona Virus by measuring the physiological parameters from the coughing sounds using an improvised version of the smartphone-based App.

References

Abou-Abbas, L., Alaie, H., & Tadj, C. (2015a). Segmentation of voiced newborns' cry sounds using wavelet packet based features. In Proceedings of the *IEEE twenty-eighth Canadian conference on electrical and computer engineering (CCECE)*, Halifax, Canada, pp. 796–800.

Abou-Abbas, L., FersaieAlaie, H., & Tadj, C. (2015b). Automatic detection of the expiratory and inspiratory phases in newborn cry signals. *Biomedical Signal Processing and Control*, *19*, 35–43. Available from https://doi.org/10.1016/j.bspc.2015.03.007.

Abou-Abbas, L., Montazeri, L., Gargour, C., & Tadj, C. (2015c). On the use of EMD for automatic newborn cry segmentation in 2015 international conference on advances in biomedical engineering (ICABME), Beirut, Lebanon, pp. 262–265.

Abou-Abbas, L., Tadj, C., Gargour, C., & Montazeri, L. (2017). Expiratory and inspiratory cries detection using different signals' decomposition techniques. *Journal of Voice: Official Journal of the Voice Foundation*, *31*, 259, e13-259.e28. Available from https://doi.org/10.1016/j.jvoice.2016.05.015.

Aucouturier, J. J., Nonaka, Y., Katahira, K., & Okanoya, K. (2011). Segmentation of expiratory and inspiratory sounds in baby cry audio recordings using hidden Markov models. *The Journal of the Acoustical Society of America*, *130*(5), 2969–2977. Available from https://doi.org/10.1121/1.3641377.

Bachu, R., Kopparthi, S., Adapa, B., & Barkana, B. (2010). *Voiced/unvoiced decision for speech signals based on zero-crossing rate and energy. Advanced techniques. computing sciences and software engineering* (pp. 279–282). Dordrecht: Springer. Available from https://doi.org/10.1007/978-90-481-3660-5.

Barr, R. G. (2006). Cryingbehavior and its importance for psychosocial development in children. In *Encyclopedia on early childhood development*, Centre of Excellence for Early Childhood Development, Montreal, Quebec, pp. 1–10.

Chinmay, C., Gupta, B., & Ghosh, S. K. (2014). Mobile metadata assisted community database of chronic wound. *Elsevier International Journal of Wound Medicine*, *6*(34–42), 2014. Available from https://doi.org/10.1016/j.wndm.2014.09.002, ISSN: 2213–9095.

Dwivedi, R., Dey, S., Chakraborty, C., & Tiwari, S. (2021). Grape disease detection network based on multi-task learning and attention features. *IEEE Sensors Journal*, 1–8.

Farsaie Alaie, H., Abou-Abbas, L., & Tadj, C. (2015). Cry-Based Infant Pathology Classification Using GMMs. *Speech Communication*, *77*, 2.

Golub, H. (1979). Aphysioacoustic model of the infant cry and its use for medical diagnosis and prognosis. *The Journal of the Acoustical Society of America*, *65*, S25–S26. Available from https://doi.org/10.1121/1.2017179.

Hariharan, M., Sindhu, R., & Yaacob, S. (2012). Normal and hypoacoustic infant cry signal classification using time–frequency analysis and general regression neural network. *Comput. Methods Programs Biomed.*, *108*, 559–569. Available from https://doi.org/10.1016/j.cmpb.2011.07.010.

Kim, M., Kim, Y., Hong, S., & Kim, H. (2013). ROBUST detection of infant crying in adverse environments using weighted segmental two-dimensional linear frequency cepstral coefficients. In *Proceedings of the IEEE international conference on multimedia and expo workshops (ICMEW)*, pp. 1–4.

Kuo, K. (2010). Feature extraction and recognition of infant cries. In *Proceedings of the IEEE international conference on electro/information technology*, pp. 1–5.

Ntalampiras, S. (2013). A novel holistic modeling approach for generalized sound recognition. *IEEE Signal Processing Letters*, *20*(2), 185–188.

Orozco-García, J., & Reyes-García, C. A. (2003). A Study on the recognition of patterns of infant cry for the identification of deafness in just born babies with neural networks. *Lecture Notes in Computer Science book series (LNCS)*, *2905*, 3.

Petroni, S., Bono, G., Marconi, M., & Stellingwerf, R. F. (2003). Classical cepheid pulsation models. *IX. New Input Physics, The Astrophysical Journal*, *599*, 522–536.

Petroni, M., Malowany, A., Johnston, C., and Stevens, B. (1995). Classification of infant cry vocalizations using artificial neural networks (ANNs). In *Proceedings of the international conference on acoustics, speech, and signal processing*, pp. 3475–3478.

Reyes-Galaviz, O., Cano-Ortiz, S., and Reyes-García, C. (2008). Evolutionary-neural system to classify infant cry units for pathologies identification in recently born babies. In *Proceedings of the seventh Mexican international conference on artificial intelligence*, pp. 330–335.

Ruiz, M., Altamirano, L., Reyes, C., and Herrera, O. (2010). Automatic identification of qualitatives characteristics in infant cry. In *Proceedings of the IEEE spoken language technology workshop*, pp. 442–447.

Sahak, R., Lee, Y., Mansor, W., Yassin, A., and Zabidi, A. (2010). Optimized support vector machine for classifying infant cries with asphyxia using orthogonal least square. In *Proceedings of the international conference on computer applications and industrial electronics*, pp. 692–696.

Soltis, J. (2004). The signal functions of early infant crying. *The Behavioral and Brain Sciences*, *27*(4), 443–458. Available from https://doi.org/10.1017/S0140525X0400010X.

Sravanth, K. R., Kodali, M., Bharat, G., Anudeep, P., Kalyana, K. S., & Chinmay, C. (2020). A survey on recent trends in brain computer interface classification and applications. *Journal of Critical Reviews*, *7* (11), 650–658.

Várallyay, G. (2006). Future prospects of the application of the infant cry in the medicine. *Electrical Engineering*, *50*(1–2), 47–62.

Várallyay, É., Lichner, Z., Sáfrány, J., Havelda, Z., Salamon, P., Bisztray, G., & Burgyán, J. (2010). Development of a virus induced gene silencing vector from a legumes infecting tobamovirus. *Acta Biologica Hungarica*, *61*(4), 457–469.

Yamamoto, S., Bayat, V., Bellen, H. J., & Tan, C. (2013). Protein phosphatase 1ß limits ring canal constriction during drosophila germline cyst formation. *PLoS ONE*, *8*(7), e70502.

Yamamoto, S., Yoshitomi, Y., Tabuse, M., Kushida, K., & Asada, T. (2013). Recognition of a baby's emotional cry towards robotics baby caregiver. *International Journal of Advanced Robotic Systems*, *10*(2), 86–92. Available from https://doi.org/10.5772/55406.

Yogesh, S., & Chinmay, C. (2020). Augmented reality and virtual reality transforming spinal imaging landscape: A feasibility study. *IEEE Computer Graphics and Applications*, 1–18.

Zabidi, M. A., Arnold, C. D., Schernhuber, K., Pagani, M., Rath, M., Frank, O., & Stark, A. (2015). Enhancer-core-promoter specificity separates developmental and housekeeping gene regulation. *Nature*, *518*(7540), 556–559.

Zabidi, A., Khuan, L., Mansor, W., Yassin, I., & Sahak, R. (2010). Classification of infant cries with asphyxia using multilayer perceptron neural network. In *Proceedings of the second international conference on computer engineering and applications*, Vol. 1, pp. 204–208.

CHAPTER

9 Internet of things based effective wearable healthcare monitoring system for remote areas

G. Boopathi Raja[1] and Chinmay Chakraborty[2]

[1]Department of Electronics and Communication Engineering, Velalar College of Engineering and Technology, Erode, Tamil Nadu, India [2]Department of Electronics and Communication Engineering, Birla Institute of Technology, Mesra, Jharkhand, India

9.1 Introduction

Remote health management services are becoming more popular as the aged population and health care prices increase. The modern innovations in physiological sensing devices and the recent developments in power-efficient wireless communication networks may provide a lot of health tracking solutions for patients who are suffering from illness in remote areas, (Ahmed, Bjorkman, Causevic, Fotouhi, & Linden, 2015).

Technological developments in physiological sensing systems, as well as IoT-based wireless communication, could allow major changes in how health tracking and remote healthcare are conducted in the future. But recently several enabling technologies were preferred in providing healthcare services. Since neither the patients nor the community will accept IoT technologies that do not adhere to current healthcare best practices.

IoT—based health monitoring systems can provide new opportunities for patients who are currently inaccessible, especially for those who are not necessary to monitor continuously in the hospital. IoT can facilitate surveillance of such patients by offering low-cost options for in-home care, allowing for early identification of signs of declining health and prompter responses and care. For at-home supervised patients to feel comfortable and protected while living at home, IoT systems must ensure protection and security on a more technological basis.

Fig. 9.1 represents the example of a smart wearable gadget used to measure physiological parameters. The smartwatch was used to measure pulse rate, body temperature, blood pressure, SpO_2, Calorie, etc. This type of watch is a wearable computer that performs multitasking such as WiFi/Bluetooth connectivity, mobile app support, FM radio, Bluetooth headset, portable media players, watch phones, etc. It may include peripheral hardware devices such as thermometers, accelerometers, barometers, GPS receivers, speakers, MicroSD cards, altimeters, heart rate monitors, and digital cameras. Along with hardware peripherals, it needs software resources to perform these tasks effectively. Some of them are calculators, different sensors, wireless headsets, displays, personal organizers, schedulers, digital maps, etc.

Implementation of Smart Healthcare Systems using AI, IoT, and Blockchain. DOI: https://doi.org/10.1016/B978-0-323-91916-6.00004-7

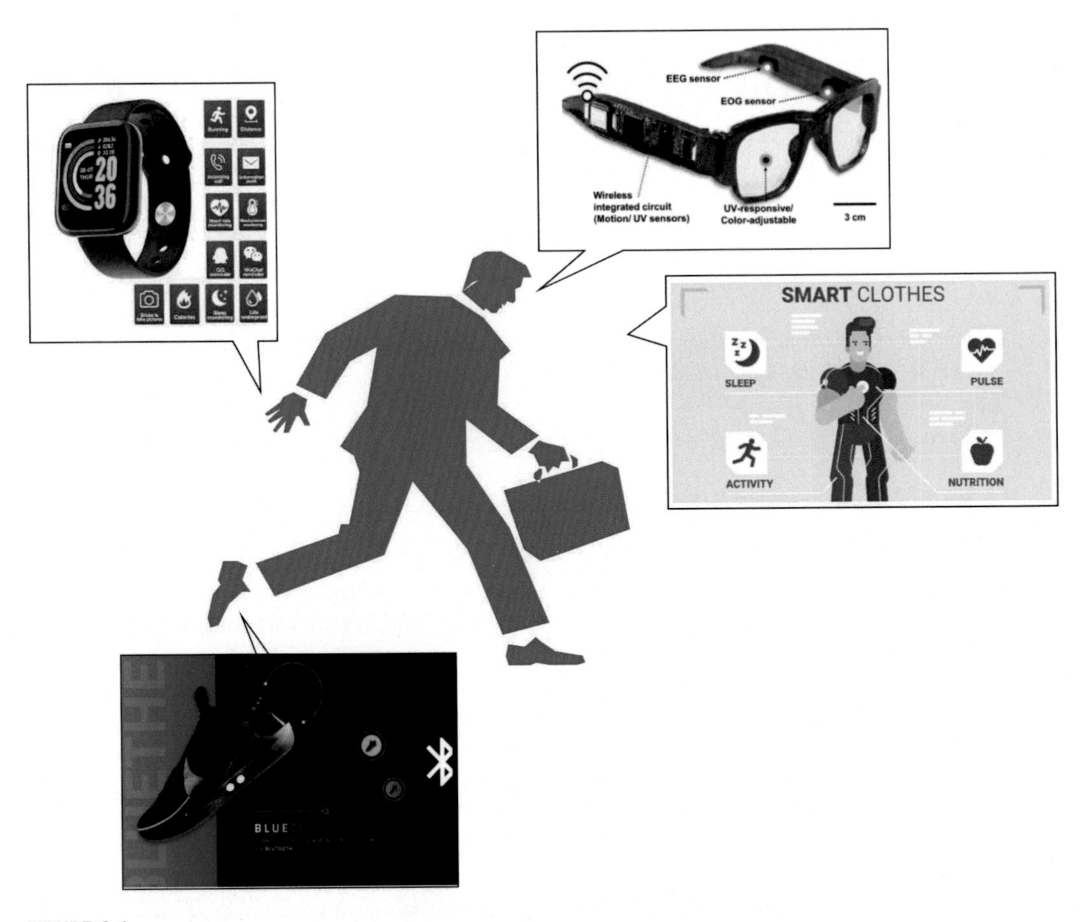

FIGURE 9.1

Smart wearable gadget to measure physiological parameters.

Smart glasses or goggles are wearable glasses that can alter their optical properties. It may support wireless technologies such as WiFi, GPS, and Bluetooth. It may collect data from internal or external sensors. Smart clothes are otherwise called electronic clothes or E-clothes. Sensors and other electronic components are embedded into the fabrics to measure the physiological parameters of the human such as body temperature, heartbeat, etc. A smart shoe is an electronic shoe used to measure athletic performance monitor fitness and analyze the various health metrics.

There are several innovations and developments in the wearable healthcare monitoring system to track the status of physiological parameters continuously. Most of the existing system provides better performance but it has certain limitations such as battery lifetime, size, communication, integration with other platforms, knowledge about the handling of equipment, etc. However, these limitations are carefully considered while designing the proposed system. The major contribution of the proposed framework was described below. The battery life was increased as the entire framework was split into several sub-modules. Each sub-modules are powered separately from the exclusive power source. The entire framework was divided into several modules, hence the size of the framework is not a problem and it is easy to fix various portions of the body. Each sub-module can communicate with the IoT module directly. Suppose if one module failed in transmission does not affect the transmission of other parameters. The proposed framework can be used for tracking the parameters of persons with disabilities, aged communities, drivers, and pandemic patients and even to track the changes in physiological parameters of sportspersons such as athletes, wrestlers, boxers, etc.

The overview of this chapter is described below. Section 9.2 elaborates on the various parameters such as Blood Pressure, body temperature, Pulse Rate, etc. required for healthcare monitoring. Section 9.3 describes the comparison of physiological parameters under in-vitro and in-vivo measurements. Section 9.4 discusses several works related to the measurement of physiological parameters in remote areas. The hardware requirements and the necessity of wearable sensors are discussed in Sections 9.5 and 9.6. The proposed methodology was introduced in Section 9.7 with a detailed description and workflow diagram. Section 9.8 describes the results and discussion of the proposed healthcare monitoring system. The chapter ended with the proper conclusion and future scope (Section 9.9).

9.2 Parameters of healthcare monitoring

Remote health management services are becoming more popular for the aged population and healthcare. Several physiological parameters can be tracked using remote health control systems. The most common healthcare parameters are pulse rate, blood pressure, the temperature of the human body, respiratory rate, etc. These common parameters are essential before providing treatment to any patients and hence these are used in health tracking systems (Ahmed et al., 2015; Ahmed, Banaee, Loutfi, & Rafael-Palou, 2014). But apart from these, some other parameters have been frequently required, such as blood glucose level, pulse rate, body mass index, activity, oxygen saturation level, body weight, and medication compliance (Yang et al., 2014).

Recently, remote-based electrocardiography (ECG) and electromyography (EMG) monitoring have been successfully implemented in a healthcare monitoring system (Xavier & Dahikar, 2013). However, remote-based electroencephalogram (EEG) monitoring was limited due to the complexity of the design, (Tomasic, Avbelj, & Trobec, 2014). Some health metrics, such as blood glucose level, body weight, pulse rate, respiratory rate, heart rate, Blood Pressure, and body temperature are calculated frequently, while electrocardiogram, electromyogram, and EEG are constantly tracked over some time.

9.2.1 Physiological parameters

9.2.1.1 Body temperature

The nominal temperature of a human body varies based on age, behavior, and time of day. The average human body temperature is 98.6 degrees (37°C). According to some sources, the temperature range of the "normal" human body is between 97°F and 99°F (i.e., 36.1°C–37.2°C). A fever of more than 100.4°F (i.e., around 38°C) is usually the result of an infection or disease, (Greger & Bleich, 1996).

The average body temperature of a healthy adult varies from 97°F to 99°F. The temperature range for babies and children is slightly higher: 97.9°F to 100.4°F.

9.2.1.2 Pulse rate

The functionality of the heart is analyzed by an important parameter, that is, heart rate. It is also known as pulse rate. Pulse rate is simply calculated by counting the number of times the heartbeats per minute. When the heart supplies blood into the vessels, it will contract and expand simultaneously based on the flow of blood. This activity is recorded as the pulse; this will reveal the following details in addition to the pulse rate:

1. Pulse strength
2. Rhythm of heart

The average pulse beat rate of any healthy person varies around 60 to 120 beats per minute. The heart rate will rise and fall as a result of exercise, cancer, an injury, or anxiety. Females have a higher heart rate than males at a heart rate of 12 and up. Athletes who engage in a lot of physical conditioning, such as athletes, may get heart rates as high as 40 beats per minute without experiencing any symptoms.

9.2.1.3 Blood pressure

It is defined as the force through which blood is pushed through the blood vessels probably arteries. Arteries transport blood from the ventricle chamber of the heart to other areas of the body. There are two quantities are used to calculate blood pressure. They are:

1. Systolic blood pressure
2. Diastolic blood pressure

The first term, systolic BP, is a measurement of the blood pressure developed in the arteries when the heartbeats. The diastolic BP measurement determines the blood pressure in the arteries during heartbeats. For a healthy adult, the nominal pressure value must be in the range of 120/80 mm Hg. Table 9.1 describes the nominal values of systolic and diastolic pressure for a person under normal and abnormal cases, (Greger & Bleich, 1996).

Hypertension is another word used often for indicating higher blood pressure. It is a condition where the blood pressure is abnormally high. The blood pressure fluctuates during the day depending on the activity to be performed. Hypertension or elevated blood pressure is usually recorded in the situation where the blood pressure readings are regularly higher than average.

Table 9.1 Nominal range of systolic and diastolic pressure.

	Normal	Prehypertension (abnormal probably at risk)	Hypertension (extreme high BP)
Systolic	≤ 120 mm Hg	In the range of 120–139 mm Hg	140 mm Hg and above
Diastolic	≤ 80 mm Hg	Around 80–89 mm Hg	90 mm Hg and above

Hypertension may have a lot of critical health consequences. It has the potential to damage vital organs like the brain, kidneys, heart, liver, and eyes.

9.2.1.4 Electrocardiogram

Table 9.2 indicates the average durations of ECG waves, (Greger & Bleich, 1996). The frequency has a big impact on the QT period. The amplitude of any dielectric signal is usually measured only in the millivolt range. Since these signals are weak in amplitude and have a lower frequency. ECG signal also falls under this category. It may differ among the different lead systems used in ECG measurement. The chest wall leads and three orthogonal leads on the ECG have normal limits in the amplitude range. Adults under 65 years old have a diastolic pressure of less than 100 mm of mercury and adults over 65 years old have a diastolic pressure of around 105 mm of mercury.

9.2.1.4.1 The heart's pressures and volumes

Table 9.3 describes the various parameters needed to analyze the performance of the heart such as stroke volume, systolic and diastolic pressure, etc. Greger and Bleich (1996). The average heart rate of a healthy adult is around in the range of 60 to 120 beats per minute.

Table 9.2 Time duration of each segment in electrocardiography waveform.

	Men	Women
Duration of P wave	0.102 s	0.106 s
PR interval	0.153 s	0.154 s
PR segment	0.051 s	0.048 s
Duration of QRS	0.093 s	0.084 s
QT interval	0.367 s	0.372 s

Table 9.3 Volume and pressure in the heart of a healthy human.

Parameter	The normal range for a healthy person
VOLUME (mL/m²)	
Stroke volume	40–70
Left ventricular end-diastolic volume	70–100
Left ventricular end-systolic volume	25–35
PRESSURE (mmHg)	
Left atrial mean pressure	Less than or equal to 12
Right atrial mean pressure	Less than or equal to 6
Left ventricular peak systolic pressure	100–150
Right ventricular peak systolic pressure	15–30
Left ventricular end-systolic pressure	Less than or equal to 12
Right ventricular end-systolic pressure	Less than or equal to 6
Pulmonary artery peak systolic pressure	15–30
Pulmonary artery diastolic pressure	4–12

9.2.1.4.2 Heart attack and heart-related issues

Hypertension inhibits heart failure by making the blood vessels less flexible, restricting blood, and supplying oxygen to the heart. Also, excessive flow to the heart causes the following problems:

1. Angina is a form of chest pain.
2. A heart attack occurs due to the interruption of the normal blood supply to the heart, and the heart muscle starts to fail which causes a shortage of oxygen. The excess blood supply of the heart is disrupted, hence the more damage this will takes.
3. Heart failure happens when the heart is unable to sufficiently pump oxygen and blood to other organs.

9.2.1.5 Electroencephalogram

The various EEG patterns are categorized depending on their frequency. Table 9.4 describes the amplitude and corresponding frequency range of the EEG pattern.

9.2.1.5.1 Nominal range of brain function tests

Cerebral blood flow. The average CBF per 100 g is 50 mL/min. The procedure has an impact on regional blood supply. The Gray region also has a greater density than the white region. The normal rate of extraction of oxygen is 40%, which is hindered by activity. The blood supply to the brain for aged people and also oxygen consumption may decrease by 0.5% each year.

At rest, the normal intake of oxygen in the brain was around 25% of the entire absorption of oxygen. It is approximately 4.2 mL/min per 100 g. Almost CBF is fully auto-regulated in the 60–160 mm mercury range.

Stroke and brain related issues. Hypertension may cause a stroke by bursting or blocking the arteries. These blood vessels are responsible for supplying nutrition to the brain in the form of oxygen and blood. A lack of oxygen leads to the death of brain cells during a stroke. Strokes may also cause severe impairments of voice, movement, and other essential functions. High blood pressure is associated with decreased intellectual capacity and depression, lifelong, especially in middle age.

9.2.1.6 Kidney disease

Patients having diabetes or hypertension, and/or both are more likely than someone without to acquire the chronic renal disease.

Table 9.4 Amplitude and the frequency value of electroencephalogram pattern.

Pattern	Amplitude	Frequency (Hz)	State
Delta	High	0.5–4	Deep sleep
Theta	Low	4–7	Rapid eye movement, falling asleep
Alpha	Low	8–13	Awake but relaxed
Beta	Low	14–30	Intense mental activity
Gamma	Low	30–40	Actively focused or intensely focus

9.3 Types of physiological parameters measurement

The physiological parameters of any human can be measured in two ways. They are in vitro measurements and in vivo measurements. Table 9.5 describes the comparison of in vitro measurements with in vivo measurements.

Bioelectric signals are measured either externally or internally. The internal measurements of these biosignals are more accurate than external measurements. However, in this framework, external measurement is used with the help of sensors.

9.4 Related work

Chakraborty has proposed a smartphone-based mobile telemedicine device for chronic wound (CW) tracking, (Chakraborty, Gupta, Ghosh, & Management Association, I., 2020). The method proved to be fast and dependable in terms of bringing health care to people's doorsteps. The tele-wound technology network (TWTN) architecture was proposed for telemedicine networks that use smartphones for remote wound control. This system is beneficial to both rural and urban communities, as it allows for effective wound tracking and advanced diagnostic results. The purpose of this work is to develop and create a TWTN device model. This model has the ability to receive, process, and monitor CW-related problems using a low-cost smartphone to improve the efficiency of the system.

Boopathi Raja proposed an effective biometric-based emergency assist framework using IoT. The main objective is to store and manage the medical records of patients effectively, (Boopathi Raja, 2021). This designed framework shares the medical history of the patients with the Doctors to save the patient from risk through a biometric-based fingerprint methodology. The privacy and confidentiality of any patient are maintained by using specific methodologies. Thus the transfer of

Table 9.5 Comparison of in vitro measurement and in vivo measurement.

S. no.	In vitro measurements	In vivo measurements
1.	In vitro measurements refers to a phenomenon in which a given procedure is performed in a controlled environment outside of a living organism.	The term "in vivo calculation" refers to a phenomenon in which experiments are carried out on a living organism as a whole.
2.	Performed under controlled laboratory conditions	Performed under physiological conditions
3.	Dead organisms or isolated cellular components are used	A whole living organism is used
4.	Less expensive	Expensive
5.	Less time-consuming	More time-consuming
6.	Less precise	More precise
7.	Tests in Petri dishes and test tubes, for example, are examples of cell culture experiments.	Experiments on medications using model animals such as rats, rabbits, and primates.

essential health information to a doctor was allowed. This information includes basic health details such as a blood group, blood pressure (BP), glucose level, etc. By providing recorded patients' medical information, the designed biometric-based framework completely replaces the existing electronic-based and usual paper-based records systems.

According to Yen-Cheng Chen, medical images are also part of electronic medical records that must be protected from unauthorized users, (Chen et al., 2008). Based on the research, only authorized parties to have access to the hospital Image Archiving and Communication System or radiology information system, which incorporates fingerprint authentication, digital signature, DICOM object, and digital envelope techniques.

Abidoye et al. recommend that a wireless body area network (WBAN) be created by designing low-cost, intelligent, small, and lightweight medical sensor nodes, (Abidoye, Azeez, Adesina, Agbele, & Nyongesa, 2011). This device can be strategically placed on the human body to monitor various physiological vital signs over time while providing real-time feedback to the patient and medical staff. WBANs have the potential to change the way health monitoring is done.

Boopathi Raja suggested a system for interpreting hand signals and operating equipment using sign language translation techniques, (Boopathi Raja, 2021). Wearable sensors are preferred for implementing this system. The objective of this framework is to remove the barrier to communication. Speech impaired people may encounter these issues while talking with others. This may be accomplished by using an interactive computer that can decode sign language. This interactive-based system along with wearable sensors is used in the proposed design.

Amit et al. suggested a reinforcement learning-based multimedia data segregation algorithm and Computing QoS in Medical Information System Using a Fuzzy algorithm to increase the quality of service over a heterogeneous network, (Amit, Chinmay, & Wilson, 2021). These algorithms perform in 3 stages: classification of healthcare data, collection of optimum data transmission gateways, and optimizing transmission efficiency by taking into account parameters like latency, end-to-end latency, and jitter. These algorithms are used to identify healthcare data and pass categorized high-risk data to end-users using the best portal available.

The idea of accessing the patient information is changing all the time by the experts and also input provided to users. Gupta et al. built a device that monitors patients in real-time and makes recommendations, (Gupta, Chinmay, & Gupta, 2019). Even though data protection is a concern, reliable transfer of EEG data is now possible thanks to the use of watermarking technologies, which provide patients with a stable and superior level of service. In this study, DWT-SVD-based watermarking on time-frequency domain EEG data was successfully implemented, with good watermarking output in terms of SNR and PSNR.

IoT-based hardware model was proposed by Lalit Garg et al. to gather details on object movements and touch, (Lalit, Emeka, Nasser, Chinmay, & Garg, 2020). This model means that this is carried out anonymously before the holders have tested positive for an infectious condition such as COVID-19. For the youth component hardware, this device uses a passive RFID transceiver. Without the use of a cell phone, animals and individuals may wear passive RFID tags. It is best used to enter service when collecting points to ensure usage. This is the first approach to propose IoT and, more precisely, RFID for anonymous RFID touch tracing of infection transmission. This prototype has recommended the use of blockchain for data storage to protect data integrity by distributing ownership and power, (Boopathi Raja, 2021). The RFID scanner, which can be used in a building or a control unit such as a car, takes the readings. Based on the existing works, the various health parameters of the patients can be measured effectively by implementing a multisensor-based approach.

9.5 Hardware and software requirements

9.5.1 Hardware requirements

The modules required to construct the remote-based healthcare monitoring system are listed as follows:

1. Sensing module
2. Control module
3. Communication module
4. Remote server module and
5. User interface module

A sensing module is made up of a series of battery-powered sensors for various health parameters. Also, it includes a dedicated processor or general microcontroller for processing data sensed from the environment and uses an antenna for communication. The majority of healthcare monitoring system uses commercially available sensors.

The main functionality of the control module used in this healthcare monitoring system is providing coordination and global controllability of the system. This is done by the development of the control unit in two ways:

1. Based on Hardware, and
2. Based on software.

The main hardware components are the CPU, communication module, memory element, and related sensor(s) (Parra, Hossain, Uribarren, & Jacob, 2014), while the associated mobile application was specifically dedicated so that it collects various measurements from sensors on a host device, such as an Android operating system (Ahmed et al., 2014, 2015). In the cloud, a remote server is normally composed of a Portal and storage.

In terms of data transmission, the Portal allows transmitting data from one medium to another medium through a wireless medium. This portal prioritizes security, privacy, and safety matters, as well as user and request management. Both user-related records, as well as health metrics, are kept in the storage. It also contains import and export features, as well as data encryption.

Except for a mobile TV-based implementation (Chakraborty et al., 2020), mostly the existing user interfaces are designed for smart devices such as mobile phones or tablets (Ahmed et al., 2014; Yang et al., 2014), and (Tomasic et al., 2014), or laptop- or desktop-based platforms, (Xavier & Dahikar, 2013). Data exchange is looked at from two perspectives: local and worldwide. Local connectivity happens between sensor and coordinator devices and is usually achieved using Bluetooth or IEEE 802.15.4. The organizer, user interface, and remote-based server are all connected by a global networking infrastructure that uses either HTTPS web servers or cellular networks. The researchers suggested a generalized system-level structure of health surveillance systems, in which they sought to incorporate a variety of different techniques, (Ahmed, Bjorkman, & Linden, 2012; Islam, Rahaman, & Islam, 2020).

9.5.2 Software requirements

The application, which is developed with a graphical user interface, may interact with various physiological parameters from each patient sequentially with a predetermined time interval for each patient. Any doctor or healthcare professional may log on to the computer interfaced with the cloud and review the history of vital parameters of any patient connected to the network.

In the case of emergencies demanding the immediate attention of physicians or nurses for any of the patients, the custom software will direct the Control unit to allow the GSM modem to send an SMS with the patient ID. A voice call is also made to the hospital's physicians and personnel. The status of the patient's physiological state is also included in the SMS. The doctor can quickly identify and treat the patient with the aid of the patient ID.

9.6 Need for wearable sensors

9.6.1 Sensors for health monitoring

The handling of wearable equipment, as well as the sensory setup, was utilized in healthcare sectors. This may be expanded significantly as a result of the modern innovations and unpredictable developments in sensors, inexpensive integrated chips, and the advancement of communication networks, whether for well-being, tracking of a sport-person, or medical recovery. This chapter examines the most up-to-date sensors and devices for healing and monitoring the status of health in this situation, (Nasiri & Khosravani, 2020; Wen et al., 2020). Despite the growing expectations of data handling techniques, the objective was to research sensor deployment and biomedical applications. Even though all of the subjects overlap, the analysis was divided into the following 3 categories:

1. Sensors in the medical field, ongoing health monitoring, and residential medical support;
2. Physical rehabilitation devices and sensors; and
3. Assistive systems.

The design of comfortable products has been significantly influenced by recent developments in electronics and communication technologies. This is due to factors such as readily available technology, power systems that give computers more control, and miniaturization of electronic circuits, both of which allow continuous surveillance of individuals.

Bioelectrical signs and gestures may be monitored and analyzed to help with diagnosis, avoidance, and examination of a variety of problems. In the healthcare sector, remote surveillance and ambulatory monitoring are becoming increasingly relevant. These processes go through preanalysis and function extraction steps, with orthopedic, psychological, cardiovascular, and pulmonary issues. It is possible to derive certain characteristics from those to make for standardization and diagnostic assistance. Furthermore, remote tracking offers improved data volume and may aid in disease detection, athletic success, and/or patient recovery.

Wearable sensors are essential in this situation for predicting health problems. Body temperature, blood pressure, pH value, PO_2, PCO_2, etc., is coming under nonbioelectric signals, whereas ECGs, electrooculogram (EOG), EMGs, and EEGs are considered as bioelectric signals. It is all applicable to analyze the status of health, personalization of treatment plans, and health management through mobile health (m-health) applications.

The exponential development in the internet of things (IoT) along with recent innovations in these technologies has resulted in the idea of electronic healthcare (e-Health). It is the collection and transmission of data through mobile sensors that are normally installed or placed closer to the patient's body. The status of various parameters of that patient can be viewed remotely. Health 4.0, similar to Industry 4.0, aims to promote radical virtualization of the patient, and medical equipment, and provide medical services through personalized online health. This is done based on the concept of teleconferencing and cloud computing. This idea will pave the way for Enterprise 4.0 concepts like cyber-physical systems (CPS), internet of everything, wireless networking, and artificial intelligence to be used to solve problems. The convergence of a virtual universe with physical processes is possible with CPS to depend on embedded device tracking and physical management. Furthermore, CPS facilitates the interaction between these two areas: cyber and real, to support IoT which is the foundation of Health 4.0.

Simultaneously, various numerical intelligence methods play a critical role in delivering advanced and informative interpretations of obtaining bio-electric and biometric signals, including signal pre-processing, collecting characteristics, classifying or clustering, and statistical analysis.

This analysis, on the other hand, seeks to evaluate, measure, and illustrate the most commonly used sensors, as well as the pattern in their use in the different applications. The importance of this analysis was to understand the role of sensors used in the patient tracking systems and improved physical treatment to each patient on their health status, as well as assistive systems that recorded experimental results, from this viewpoint.

However, the objective of this study is to examine, measure, and highlight the most commonly used sensors, as well as the pattern in their use in the various applications. From this direction, researchers were interested in sensors and technologies (hardware) that are widely used in patient tracking equipment and adapted to physical therapy, as well as assistive systems that recorded experimental outcomes.

9.6.2 Sensors needed for health monitoring and rehabilitation

To accurately present and correlate the research, this study is divided into three main topics:

1. It discusses the presently available and applicable sensors and devices used for residential health monitoring, personal health care, and continuous health tracking applications for patients, Medical Assistance in-home, and periodic Health Monitoring for drivers, workers, and/or sportperson. Also, researchers include wearable equipment, devices for tracking patients at home, and athletes doing workouts as part of the study.
2. Sensors and devices for physical therapy, with an emphasis on health care applications in recovery and;
3. Assistive technologies include the most recent research on devices that aid in the contact and/or relocation of persons with movement or speech issues such as movement, voice, sign, etc.

It is important to remember that there might be some overlap in all three sub-areas from a broader perspective, but we made sure to keep an organizational structure in mind. Topic 2 was created to group equipment and techniques that help in recovery practices and workouts, avoiding the mention of constant home monitoring techniques. For continuous home or hospital surveillance and medical assistance, certain monitoring cases, including crashes were coordinated and reported in group I. Similarly; the sensors described in the descriptions of assistive devices were not mentioned in the descriptions of sensors used in healthcare, preventing repetition.

9.6.3 Sensors in continuous health monitoring and medical assistance in home

The multi-sensor interpretation and data fusion concepts were used in E-Health networks, allowing for interactions among several experts from biomedical parameter evaluations. It has a direct effect on the health sector, especially in domestic circumstances that require multiple sensors and wireless communication support. Based on wearable and mobile devices, it is easy to obtain continuous and parallel medical support. But the wearable devices have been essential for tracking parameters of the patients for diagnosing the category of disease and athletes during training sessions which are known as m-health devices. They encourage their doctors to make a detailed study on their report and diagnose at a remote distance, in this case, aided by communication capabilities.

The health field is constantly changing based on innovations in both healthcare services and medical equipment for assistance and diagnosis to be developed and improved. This transformation aims to help doctors, technicians, and other professionals with treatment procedures, continuous monitoring, and patient care, as well as finding new ways to reduce the device's cost. Microelectronics, networking, sensors, and data processing advancements have allowed the emergence of modern healthcare systems and user assistance devices that are substantial and far-reaching. Wearable sensors belong to these innovations, and now they were considered one of the essential and most irreplaceable parts of the medical field. It has a wide variety of tracking features such as pulse rate, body temperature, blood oxygen saturation, etc. Along with this equipment, Smartphone apps can be used to monitor the parameters of the user in physical activities like running, sleeping, biking, and cycling, as well as the execution of workouts during disease care. Owing to the ease of buying and subsequent cost reductions, more consumers are getting access to these gadgets and software.

9.6.4 Sensors in physical rehabilitation

The researchers described a system for remotely measuring the efficiency of recovery workouts by using several sensors, including IMU and sEMG sensors to detect people who were doing physical activities. The technique was tested on 17 physiotherapy participants, with an overall precision of 96% in identifying problems, including acceleration, movement, angular velocity, and standardizing detail in the exercises, according to the researchers. The effects of device reliability play experience and training impact were presented using accelerometers and sEMG sensors in a wearable application.

The use of force sensors to achieve pressure foot measurement has been addressed in several articles. The framework can support patients with poor physical capacity, gait, and balance, such as the aged, people who have had a stroke, people who have multiple sclerosis, and people who have

Parkinson's disease. Pressure sensors installed minimally under the heel and foot were found to be a lightweight and inconspicuous alternative to F&M sensing for predicting anesthesia gait and focusing on quality progress in the report.

Gait, knee tilt, and foot plantar pressure were all measured using inertial sensors and force sensors. An instrumented jacket with six force sensors clustered according to movement phases at various pressure points on the foot was used to obtain foot plantar pressure signals. On the other hand, four inertial sensors were used to determine the knee angle, which is placed as one above and one below the knee on each leg. It calculates the angle between them by obtaining the angle of limbs from each sensor and combining them. The software captures the data, displays it to users, and analyses normal and irregular gait, allowing a substantial variation in the minimum angle to be distinguished. The movement of the upper and lower limbs may be hampered as a result of a negative situation.

9.6.5 Assistive systems

This section aims to bring together the most current published literature on sensors and technology for health care and monitoring. It focuses on assistive technologies and assistive devices for people with motor disorders, either by pattern recognition model or framework that reads emotions and gestures.

Several assistive device choices centered on a subject with reduced mobility. A wheelchair depicts the possibilities of detecting locomotion and barriers, as well as the possibilities of control by voice order. A series of sensors on wearable devices may be connected to the consumer in several ways at the same time. Sensors for vital sign monitoring, limb or head movement assessment, or EMG prosthesis control, as well as sensors placed on the head, such as virtual reality glasses or other functional elements in facial sEMG, EOG, or EEG signal acquisition.

9.7 Proposed system

The proposed IoT-based smart healthcare monitoring system consists of multiple nodes that are placed in the various parts of the human body. Each node in this system is placed on a specific portion of the body based on requirements. Fig. 9.2 describes the methodology used in the proposed framework. It is not necessary to use the same type of sensors on each sensor node.

These sensors are used to measure basic physiological parameters such as body temperature, blood pressure, pulse rate, etc. Fig. 9.3 shows the block diagram of the transmitter node placed on the arm. This transmitter node A consists of four different sensors such as a Room temperature sensor, Body temperature sensor, Heartbeat sensor, and blood pressure sensor. The microcontroller used in the transmitter module is ARM microcontroller LPC2148.

The microcontroller can collect the information from each sensor periodically and update the information to the cloud through the IoT module. The sensed information will be displayed immediately in the LCD module which is attached to each transmitter node. It also transmits the collected information to the Master node through a wireless medium, which may be Bluetooth or Zigbee based on requirements. This master node act as a receiver section in this framework.

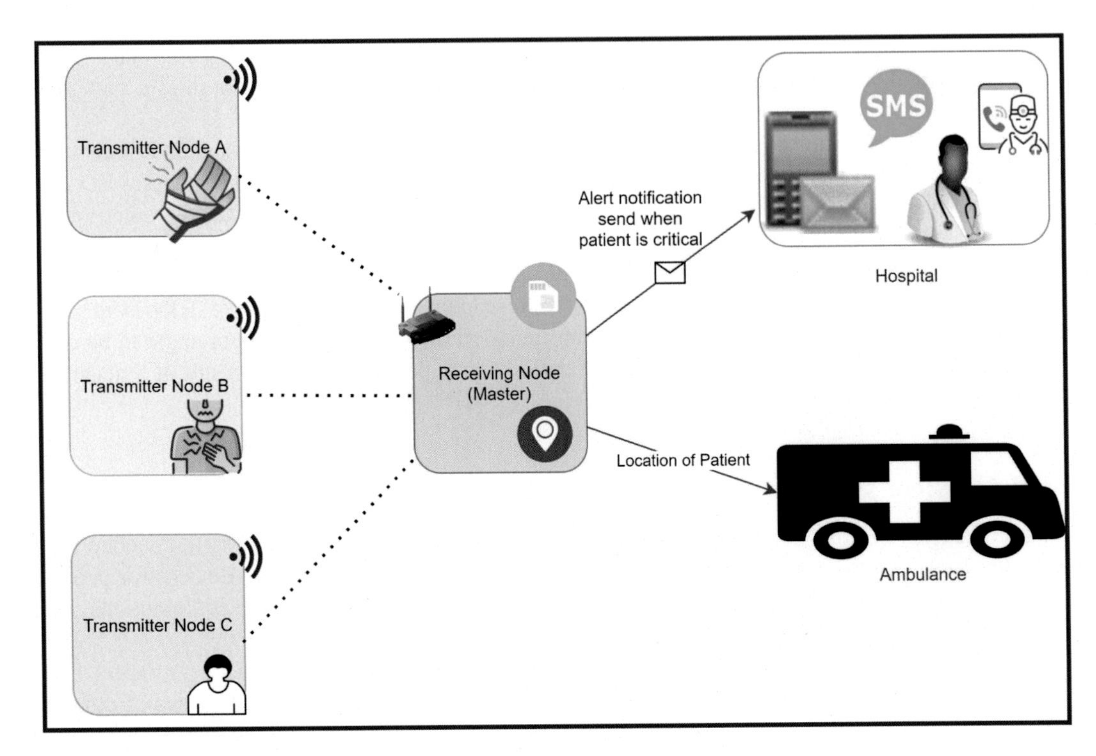

FIGURE 9.2

Methodology used in the proposed framework.

Fig. 9.4 shows the block diagram of transmitter node B placed on the Chest. Tracking the status of the heart is essential for many situations, including aged communities, heart patients, etc. For that, it is necessary to record ECG signals. One simple way to perform this is to use an ECG drive circuit.

Transmitter node B consists of ARM microcontroller LPC2148 along with a heartbeat driver circuit, ECG driver circuit, and temperature sensor. An ECG sensor collects the amplitude of the ECG signal and sends these values continuously to the ARM microcontroller. In parallel, the heartbeat sensor counts and updates the total number of beats that occurred in each minute. These values are periodically updated in Cloud by using the IoT module. Also, the node transmits this information to the master node.

Fig. 9.5 shows the block diagram of transmitter node C attached to the head in the form of the headband. It consists of an ARM microcontroller LPC2148, EEG driver circuit, and temperature sensor. This setup is placed in the head by embedding this transmitter node in a wearable headband.

The EEG sensor used in this framework provides the amplitude value of the EEG signal sensed at that time. Based on the amplitude level, it is easy to identify the band of EEG waveform. This information is also essential in the treatment. However, this type of measurement is limited in hospitals. Since it may cause a delay in transmission and does not provide a complete waveform at a time.

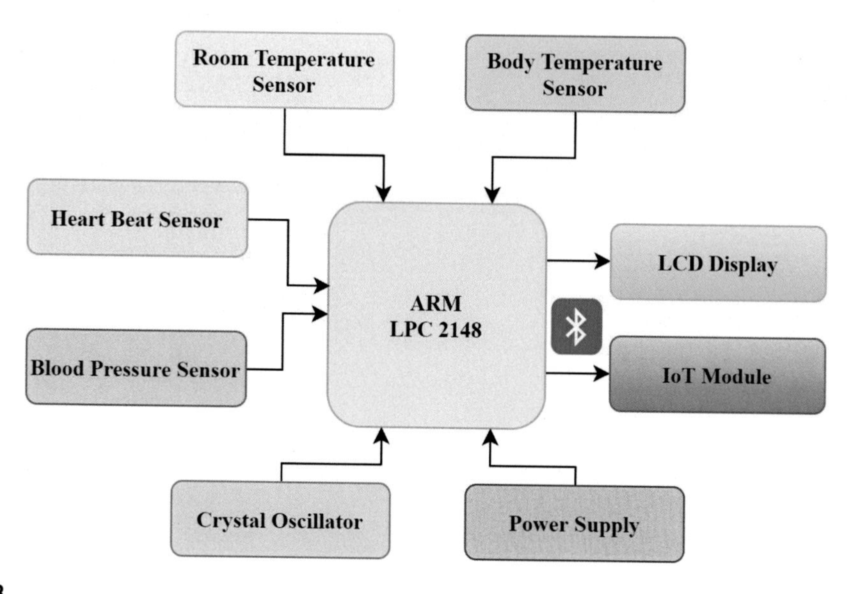

FIGURE 9.3

Transmitter node A placed near to chest.

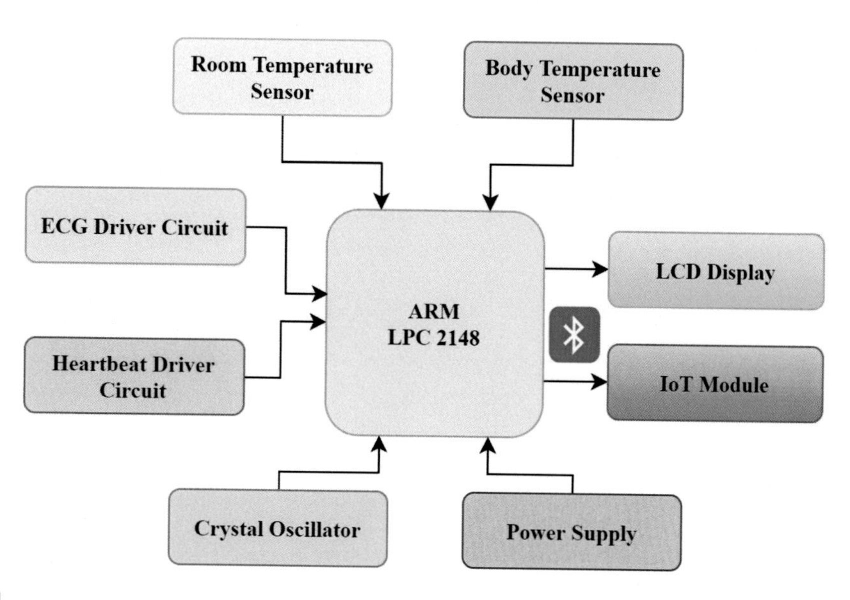

FIGURE 9.4

Transmitter node B placed near to chest.

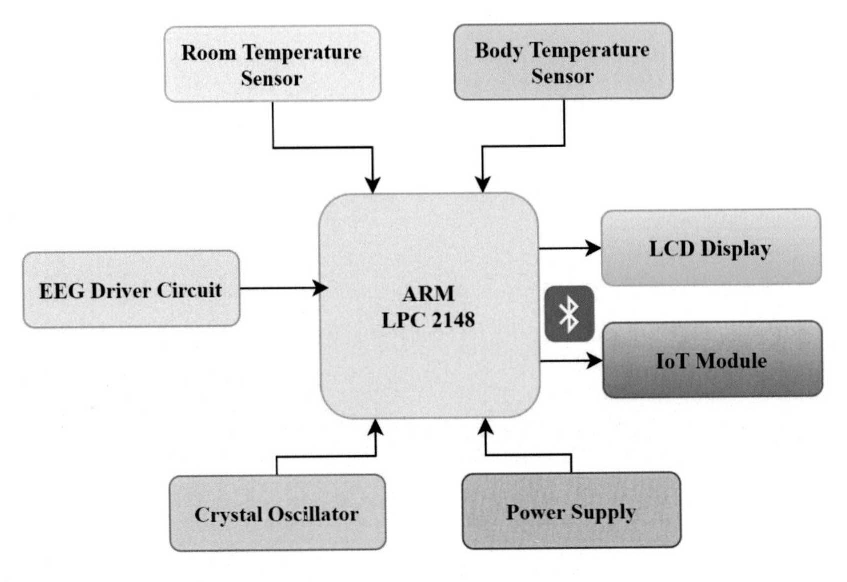

FIGURE 9.5

Transmitter node C attached with the head.

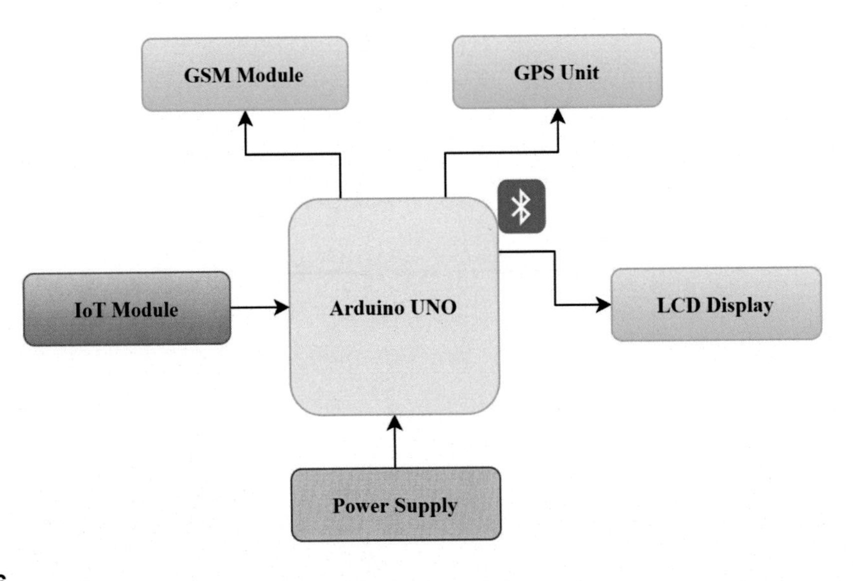

FIGURE 9.6

Receiver node.

Fig. 9.6 shows the block diagram of the receiver node used in IoT based effective health-care monitoring system. It consists of Arduino UNO, IoT module, GSM module, and GPS unit.

In this framework, three transmitter nodes A, B, C, and one receiver node were used. These transmitter nodes are placed in different regions of the human body. Transmitter node A is placed in one arm in the form of a hand glove. Transmitter node B is placed in the chest by attaching to the wearable belt. Transmitter node C is embedded with the headband by continuously monitoring EEG signals. In this framework, three transmitter nodes A, B, C, and one receiver node were used. These transmitter nodes are placed in different regions of the human body. Transmitter node A is placed in one arm in the form of a hand glove. Transmitter node B is placed in the chest by attaching to the wearable belt. Transmitter node C is embedded with the headband by continuously monitoring EEG signals.

The purpose of the receiver node is to collect the information and checks the received data along with the threshold value. If the sensed data falls below the normal level, the master node (or receiver section) is idle. Suppose if any parameter is found to be abnormal, the status is forwarded as an SMS to the doctor with the help of the GSM module. Also, the location is shared by Ambulance drivers or rescue workers with the help of a GPS module.

9.7.1 Working principle

The proposed framework performs different tasks and finally provides medical services with a set of predefined procedures. Fig. 9.7 shows the workflow of the proposed framework. The steps involved in the procedure are listed as follows:

1. Collecting physiological parameters with the help of sensors.
2. Performing a suitable data encryption scheme on obtained physiological data.
3. Transferring secured medical data to the communication service providers.
4. Transferring and storing data in the cloud in the form of distributed data storage.
5. Perform data decryption on secured data based on the need.
6. Data processing.
7. Diagnosing whether any abnormalities found in the processed data.

If all the parameters received from the transmitter node are found to be normal, then there is no action to be taken further and the same procedure to be followed. If the parameter is abnormal, then it follows the following steps:

1. Update the status of the patient to the doctor.
2. Based on the parameters provided to the doctor, he/she decides whether the patient is critical.
3. If the patient is not critical, the extra care to be taken to recover soon from that problem by providing proper consultation to that patient.
4. If the patient is critical, it is necessary to send warning messages to the caretaker of the patient, and also notification to the doctor.
5. The status of the patient is forwarded to the doctors through SMS whenever an emergency arises and/or he/she can able to monitor the status of the patient continuously through the portal installed on the computer or smartphone.
6. The location of the patient was also shared with the Ambulance driver through a GPS module.

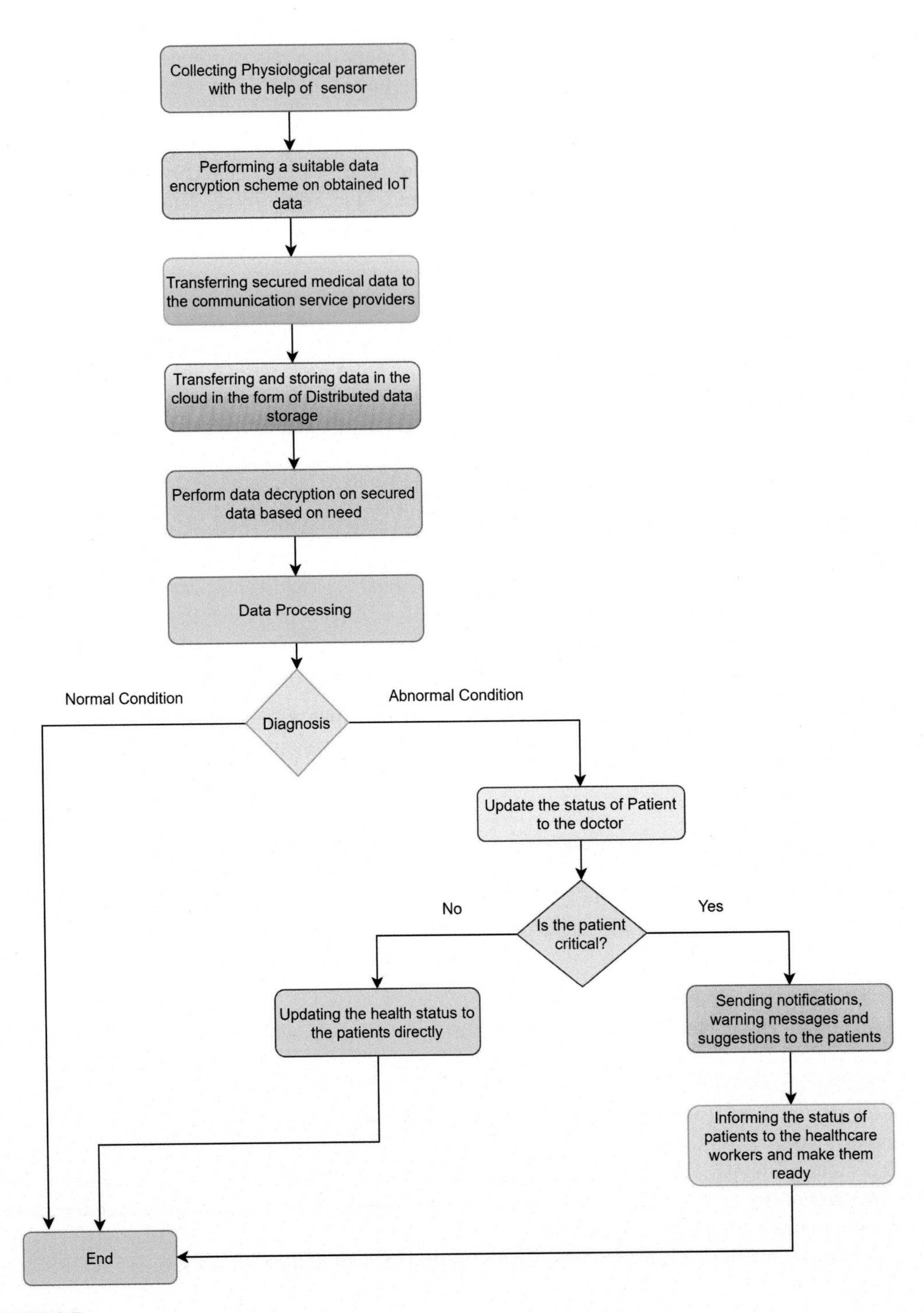

FIGURE 9.7

Workflow of the proposed method.

9.7.2 Hardware components

The proposed framework makes use of hardware components of some kind. The following are the modules that were used to build the system.

9.7.2.1 *ARM microcontroller—LPC2148 microcontroller*

For embedded system manufacturers, an ARM processor is one of the best choices available. The ARM architecture has grown in popularity in recent years, and it is now available from a variety of IC suppliers. ARM processors are used in mobile phones, car braking systems, and other applications. Engineers, designers, and developers work for ARM's worldwide community partners, which have established semiconductor and product design companies.

An ARM is an Advanced RISC Machine. It is a microcontroller with a 32-bit processor architecture that has been extended by ARM holdings. Because of their attractive features, ARM processors are preferred for numerous applications, including microcontrollers and processors. Many companies have licensed the architecture of an ARM processor to design ARM processor-based designs and products. This enables businesses to produce products based on the ARM architecture. Similarly, all major semiconductor firms, such as Samsung, Atmel, and TI, will produce ARM-based SOCs.

In embedded systems, the ARM7 processor is widely used. It also bridges the gap between the original and new-Cortex sequences. This is excellent for transferring resources on the internet.

Philips (NXP Semiconductor) manufactured the LPC2148 microcontroller, which has many built-in features and peripherals. LPC2148 is an ARM7-based microcontroller and is available in both16-bit and 32-bit architecture. Because of these factors, it would be a more dependable and effective choice for an application developer.

9.7.2.1.1 Features of LPC2148

The following are the major characteristics of LPC2148.

1. The LPC2148 is a microcontroller based on the ARM7 family.
2. On-chip flash memory is 32–512 kB and On-chip SRAM is 8–40 kB, and the accelerator can operate at 60 MHz.
3. It requires around 400 ms to erase all of the data on a full chip and 1 ms to program 256 bytes.
4. On-chip real monitor software is possible.
5. Real-time debugging is available with the high-speed tracing of instruction execution.
6. It has a USB 2.0 full-speed device controller and 2 kB of endpoint RAM.
7. This microcontroller has an on-chip RAM of 8 kB and is USB compatible with DMA.
8. 10-bit ADCs are available to provide 6 or 14 analog input/output channels with a conversion time of 2.44 s per channel.
9. Only a 10-bit DAC has a switchable analog output.
10. External event counter with two 32-bit timers, PWM unit, and watchdog.
11. 32 kHz clock input and low-power RTC (real-time clock).
12. Quick general-purpose input/output pins that can withstand 5 volts in a small LQFP64 package.

13. Pins 21 and 22 on the outside interrupt.
14. The maximum Clock obtained from the programmable-on-chip PLL is 60 MHz, and the resolving time is 100 s.
15. The chip's built-in oscillator will be powered by a 1 to 25 MHz external crystal.
16. Idle and power down is the two most common power-saving modes.
17. Individual peripheral function enable/disable and peripheral CLK scaling is available for additional power optimization.

9.7.2.2 ESP32 processor

One of the most common IoT learning platforms is ESP32. The ESP32 GPIO pins can be used to connect device sensors and actuators. ESP32 provides the internet of things (IoT) to come together to form a game-changing technology for healthcare development.

The ESP32 is extremely well-designed, with integrated antenna switches, low-noise amplifiers, control amplification, and filters, as well as power management modules. This module may operate as a stand-alone scheme or as a slave to a host microcontroller unit, which cuts down on the time spent dealing with the main application processor.

The SPI/SDIO and I2C/UART interfaces on the EPS32 allow it to connect with other Wi-Fi and Bluetooth devices.

9.7.2.3 Heartbeat sensor

The plethysmography principle was used to design the heartbeat sensor. It measures the change in blood supply in any organ that requires light to pass at a certain intensity.

The synchronization of the signals is much more important in devices that measure the heart rhythm. The rate of heartbeats affects blood volume, distribution, and light absorption by the blood, and signal pulses are identical to heartbeat pulses.

For photoplethysmography, the reflective optical sensor TCRT1000 was preferred to construct a heartbeat sensor. The use of TCRT100 simplifies the sensor portion during the implementation process of the heartbeat sensor. Since, both the infrared LED and the photodetector are grouped side by side in a leaded package, blocking ambient light from affecting sensor performance. The output of this sensor is a digital pulse that is proportional to the heartbeat. For further processing and retrieval of the heart rate in beats per minute (BPM), the output pulse may be directly fed into either an ADC block or by using a digital input pin of a microcontroller.

9.7.2.4 Heart rate monitor kit with AD8232 electrocardiography sensor module

The ECG sensor module AD8232 is a low-cost board for measuring the electrical activity of the heart. This electrical activity will be recorded as an ECG (electrocardiogram) and read as an analog number.

Because the obtained ECG signal was corrupted by a lot of noises, the single lead heart rate monitor AD8232 behaves as an operational amplifier to make it easier to get a clear signal from the PR and QT intervals. Fig. 9.8 shows the picture of the AD8232 ECG Sensor Module obtained from SparkFun electronics.

The AD8232 Cardiac ECG Monitoring Sensor Module is designed to extract, amplify, and process small biopotential signals in the presence of noisy environments like motion or the location of distant electrodes.

FIGURE 9.8

AD8232 electrocardiography sensor module.

9.7.2.5 Body temperature sensor (LM35)

The LM35 series of temperature circuits has a relative output voltage to the temperature in degrees Celsius. In comparison to Kelvin's linear temperature sensors, the LM35 has the advantage of realistic centigrade scaling, which prevents the consumer from removing the high constant voltage from the display. Fig. 9.9 shows the image of the body temperature sensor LM35 obtained from ElectronicWings.

9.7.2.6 Room temperature sensor (DHT11)

The DHT11 is a temperature and humidity sensor that is widely used. The serial processing of temperature and humidity values was done by an 8-bit microcontroller present in that sensor. This sensor is more convenient for all types of microcontrollers since it was already calibrated by the manufacturer.

The most popular sensor used to measure humidity and temperature is DHT11. It is a digital humidity and temperature sensor with a single wire that provides values of humidity and temperature serially. These sensors can measure relative humidity as a percentage mostly in the range of 20% to 90% RH and also temperature as a degree Celsius up to 50°C. Fig. 9.10 shows the image of the room temperature sensor DHT11, a product taken from Adafruit industries.

It consists of four pins. Out of these, two pins are allotted for power supply and ground, one is not used, and the fourth is used for receiving data. For communicating, only the data pin is used. Different TON and TOFF pulses are decoded as logic 1 or logic 0 or start pulse or end of the frame, respectively.

9.7.2.7 CO sensor (MQ-9)

Gas sensor MQ-9 is considered an excellent sensor for detecting Liquified Petroleum Gas (LP), Carbon monoxide (CO), and methane gas (CH4). Measurements are easy due to their high

FIGURE 9.9

Body temperature sensor—LM35.

FIGURE 9.10

DHT11.

sensitivity and fast reaction time, (Bag & Lee, 2021). The sensitivity of the sensor can be changed using the potentiometer.

9.7.2.8 CO_2 sensor (MQ-135)

The MQ-135 gas sensors are used to detect and measure NH_3, Nicotine, Smoke, Benzene, and CO_2 in air quality control systems. The digital pin present in the MQ-135 sensor module allows it to run independently of a microcontroller. This pin is useful for sensing complex gases. The analog pins are used to measure gas concentrations in PPM. It can be used for practically any modern microcontroller since the analog pin is TTL and runs at 5 V.

9.8 Results and discussion

Fig. 9.11 shows the receiver node setup. It consists of Arduino UNO, IOT module, GSM, and GPS module along with LCD. Each submodule is placed in the appropriate parts of the human body. Each of these transmitting submodules continuously tracks the physiological parameters of the body and updates the value periodically to the experts through the cloud and/or SMS.

Fig. 9.12 shows the LCD response of transmitter node B. It measures the physiological parameters such as Pulse rate, SpO_2, Body temperature, and ECG value.

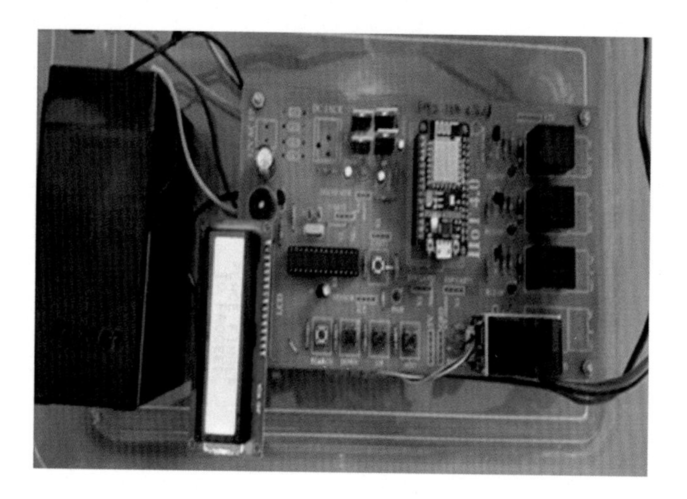

FIGURE 9.11

Receiver module.

FIGURE 9.12

LCD display of transmitter node B.

Fig. 9.13 shows the response obtained from the mobile app. At the receiver end, the status of the patient can be read from the cloud through the electronic gadget with internet connectivity. The sensed information in each node is transferred to the master node. Finally, the master node updated all the collected information into the cloud periodically. Whenever there is a need for the health status of a particular patient, it is easy to utilize these records through the mobile app.

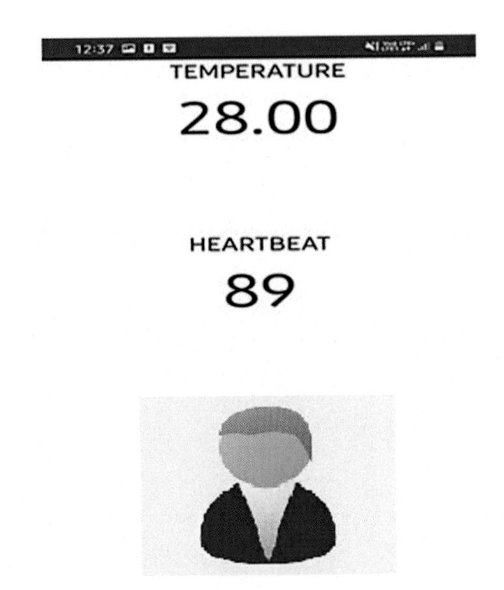

FIGURE 9.13

Output read in mobile app.

Table 9.6 Performance comparison of the proposed framework with existing framework.

Parameter	Proposed system	Biometric-based emergency first aid kit, (Boopathi Raja, 2021)
Temperature	Yes	Yes
Pulse rate	Yes	Yes
Electrocardiography	Yes	No
SpO$_2$	Yes	No
GPS module	Available	Available
Internet of things/ *Wifi module*	Available	Available
Cloud database	Available	Available
Accessibility	Wearable sensors are distributed throughout the body. The information is automatically updated to the cloud	Requires Fingerprint impression each time to read the information

Table 9.6 shows the comparison of features of the proposed system with some of the existing wearable healthcare systems.

9.9 Conclusion and future scope

This work offers an analysis of health monitoring technologies in real life, with a focus on IoT for health. We looked at the most critical facets of health in this report, reflecting on emerging developments and the advancement of IoT-based health management systems. Based on health criteria, nature of the system used, preferred networking, and security concerns, a variety of present health monitoring technologies have been reviewed. The importance of the IoT for interoperability among various devices, networks, and applications has been discussed in this chapter. Based on the results, the growth of IoT research in the healthcare sector is exponentially increasing, but several problems remain unresolved.

Wearable devices provide a comfortable way to track a wide range of physiological functions, opening up plenty of medical possibilities. These technologies are not only simple to use, but they also provide real-time data for specialists to examine. This system is suitable only for patients who are recovered from severe health illnesses. Each submodule is allowed to communicate directly with the cloud and update the status of the parameter whenever there is any variation. The other medical scan reports such are X-ray, MRI, PET, CT, etc., are also automatically updated in the cloud once the person is subject to examination. This may further help the experts to diagnose the problem more accurately for the patients who are located in remote areas.

References

Abidoye, A., Azeez, N., Adesina, A., Agbele, K., & Nyongesa, H. (2011). Using wearable sensors for remote healthcare monitoring system. *Journal of Sensor Technology*, *1*(2), 22–28. Available from https://doi.org/10.4236/jst.2011.12004.

Ahmed, M.U., Banaee, H., Loutfi, A., Rafael-Palou, X. (2014). Intelligent healthcare services to support health monitoring of elderly. In *Healthy IoT*.

M.U. Ahmed, M. Bjorkman, A. Causevic, H. Fotouhi, & M. Linden (2015). An overview on the internet of things for health monitoring systems. In Mandler, B., et al. (Eds.) *IoT 360, Part I, LNICST 169*, pp. 429–436, 2016. Available from https://doi.org/10.1007/978-3-319–47063-444.

Ahmed, M.U., Bjorkman, M., Linden, M. (2012) A generic system-level framework for self-serve health monitoring system through internet of things (IoT). In *pHealth*.

Ahmed, M.U., Espinosa, J.R., Reissner, A., Domingo, 'A., Banaee, H., Loutfi, A., ... Rafael-Palou, X. (2015). Self-serve ICT-based health monitoring to support active ageing. In *HEALTHINF*.

Amit K, Chinmay C, Wilson J, (2021). Reinforcement learning for medical information processing over heterogeneous networks, In Multimedia Tools and Applications, Springer, Available from https://doi.org/10.1007/s11042-021-10840-0.

Bag, A., & Lee, N. -E. (2021). Recent advancements in development of wearable gas sensors. *Advanced Materials Technologies*, *6*, 3. Available from https://doi.org/10.1002/admt.202000883.

Boopathi Raja, G. (2021). Chapter 15*Fingerprint-based smart medical emergency first aid kit using IoT," Electronic devices, circuits, and systems for biomedical applications—Challenges and intelligent approach.* Elsevier. Available from https://doi.org/10.1016/B978-0-323-85172-5.00015-0.

Boopathi Raja, G. (2021). *Appliance control system for physically challenged and elderly persons through hand gesture-based sign language, biomedical signal processing for healthcare applications* (1st ed.). CRC.

Boopathi Raja, G. (2021). Impact of internet of things, artificial intelligence and blockchain technology in industry 4.0. In R. L. Kumar, Y. Wang, T. Poongodi, & A. L. Imoize (Eds.), *Internet of things, artificial intelligence and blockchain technology. Security and Cryptography*. Springer International Publishing. Available from https://doi.org/10.1007/978-3-030-74150-1.

Chakraborty, C., Gupta, B., & Ghosh, S. K. (2020). *Mobile telemedicine systems for remote patient's chronic wound monitoring. Virtual and mobile healthcare: Breakthroughs in research and practice* (pp. 977–1003). IGI Global Management Association, I. (Ed.). Available from https://doi.org/10.4018/978-1-5225-9863-3.ch049.

Chen, Y.-C., Chen, L.-K., Tsai, M.-D., Chiu, H.-C., Chiu, J.-S., & Chong, C.-F. (2008). Fingerprint verification on medical image reporting system. *Computer Methods and Programs in Biomedicine*, *89*(3), 282–288.

Greger, R., & Bleich, M. (1996). Normal values for physiological parameters. In R. Greger, & U. Windhorst (Eds.), *Comprehensive human physiology*. Berlin, Heidelberg: Springer. Available from https://doi.org/10.1007/978-3-642-60946-6_127.

Gupta, A., Chinmay, C., & Gupta, B. (2019). Monitoring of epileptical patients using cloud-enabled health-IoT system. *Traitement du Signal, IIETA*, *36*(5), 425–431. Available from https://doi.org/10.18280/ts.360507.

Islam, M. M., Rahaman, A., & Islam, M. R. (2020). Development of smart healthcare monitoring system in IoT environment. *SN Computer Science*, *1*, 185. Available from https://doi.org/10.1007/s42979-020-00195-y.

Lalit, G., Emeka, C., Nasser, N., Chinmay, C., & Garg, G. (2020). Anonymity preserving IoT-based COVID-19 and other infectious disease contact tracing model. *IEEE Access*, *8*, 159402–159414. Available from https://doi.org/10.1109/ACCESS.2020.3020513, ISSN: 2169–3536.

Nasiri, S., & Khosravani, M. R. (2020). Progress and challenges in fabrication of wearable sensors for health monitoring. *Sensors, and Actuators A: Physical*, 112105. Available from https://doi.org/10.1016/j.sna.2020.112105,.

Parra, J., Hossain, M. A., Uribarren, A., & Jacob, E. (2014). Restful discovery and eventing for service provisioning in assisted living environments. *Sensors*, *14*, 9227–9246.

Tomasic, I., Avbelj, V., & Trobec, R. (2014). Smart wireless sensor for physiological monitoring. *Studies in Health Technology and Informatics*, *211*, 295–301.

Wen, F., He, T., Liu, H., Chen, H.-Y., Zhang, T., & Lee, C. (2020). Advances in chemical sensing technology for enabling the next-generation self-sustainable integrated wearable system in the IoT era. *Nano Energy*, 105155. Available from https://doi.org/10.1016/j.nanoen.2020.105155.

Xavier, B., & Dahikar, P. (2013). A perspective study on patient monitoring systems based on wireless sensor network, its development and future challenges. *International Journal of Computers and Applications*, *65*, 35–38.

Yang, G., Xie, L., Mantysalo, M., Zhou, X., Pang, Z., Da Xu, L., ... Zheng, L. R. (2014). A health-IoT platform based on the integration of intelligent packaging, unobtrusive bio-sensor, and intelligent medicine box. *IEEE Transactions on Industrial Informatics*, *10*, 2180–2191.

CHAPTER

Blockchain for transparent, privacy preserved, and secure health data management 10

Mohsen Hosseini Yekta[1], Ali Shahidinejad[2] and Mostafa Ghobaei-Arani[3]

[1]Department of Engineering, Garmsar Branch, Islamic Azad University, Garmsar, Iran [2]Department of Computing and IT, Global College of Engineering and Technology, Muscat, Oman [3]Department of Computer Engineering, Qom Branch, Islamic Azad University, Qom, Iran

10.1 Introduction

Healthcare data and remote patient monitoring has received significant attention nowadays. Volume, velocity, and variety of data generate and gathered in this era are dramatically high. These data are sensed and captured by IoT devices, wearable sensors, and so on. Then they are sent to the user's local gateway, formatted as electronic medical records (EMR), and then sent to healthcare providers for diagnosis and feedback (Ali et al., 2021). A common approach to assist users' limitations in data storage, data processing, and data sharing is using fog, edge, and cloud infrastructures (Jain, Gupta, Nayyar, & Sharma, 2021). However, cloud infrastructure improves efficiency and decreases costs, but it has its drawbacks:

1. Healthcare data is very sensitive and must be protected from unauthorized access and tampering. Data may leak from cloud servers unintentionally or in a collusion manner.
2. Healthcare treatment is a timeline process, and everyone should be responsible for his/her advice and prescriptions. Users may distrust cloud and fog servers because there are no guaranteed access control policies among all cloud and fog service providers.

Blockchain is used to overcome these challenges and aims to solve the drawbacks of sharing healthcare data like exposing privacy data, leakage, or abuse (Sarkar et al., 2021; Yaqoob, Salah, Jayaraman, & Al-Hammadi, 2021).

Data stored in EMR formats can be divided into three parts, sensitive items (known as IDs), for example, SSN, Zip Code, etc., semisensitive items (known as semi-IDs), for example, age, location, gender, nationality, etc. and nonsensitive items, for example, blood pressure, hurt bit, glucose level, etc. IDs and semi-IDs property should be well protected in each scenario. However, nonsensitive data shall be accessed by everyone for research purposes and data mining and should be protected from tampering and perturbation. To achieve these goals, Blockchain is widely used in healthcare systems. Immutability is guaranteed in Blockchain by a peer-to-peer network of nodes consisting of ordered records containing batches of transactions and the hash of the previous block. With this functionality, nonsensitive data can be stored in the Blockchain and can be accessed by everyone.

Implementation of Smart Healthcare Systems using AI, IoT, and Blockchain. DOI: https://doi.org/10.1016/B978-0-323-91916-6.00011-4

However, storage limitation in the Blockchain is a challenging issue. Thus, in many proposed solutions, healthcare data stored on cloud servers and a hash of them stored on Blockchain guarantees immutability and tampers protection of stored data. By this mechanism, transparency can be achieved.

Providing security and privacy of sensitive and semisensitive items in EMRs is critical for EMRs stored on fog, cloud, or big data storage mechanisms. As we know, security aims to protect healthcare data from harm, unauthorized use, and theft, while privacy aims to governance and use of healthcare data (Hamza, Yan, Muhammad, Bellavista, & Titouna, 2020). Blockchain can guarantee security and privacy-preserving by managing access requests and corresponding grant/deny reactions as transaction lists stored on a global ledger. When someone requests sensitive or semi-sensitive items that are encrypted and stored on appropriate infrastructure (cloud, fog, etc.), its owner replies to this request. This message is stored in Blockchain entirely on an immutable platform. Data can be accessed in its owner awareness and security and privacy are preserved at a high and acceptable level because sensitive data are encrypted and cannot be accessed without an access key. In addition, no one has full access to EMRs and only data owners grant leverage access to their data for requesters.

In emergency cases when the patient (EMR owner) cannot reply to access requests in a situation like comatose or faint, alternative access control like a break-glass key should be provided in these cases. A solution such as the Break-glass key employed for this situation is a patient pre-sets password and emergency contact person who can recover patients' medical files when they encounter an emergency.

This chapter is structured as follows: the first section describes essential preliminaries and basic acknowledgment such as security and privacy-preserving meaning and their differences, CIA definition on security, Blockchain and its structure and properties, and the second section illustrates using Blockchain for privacy-preserving on health data in data-gathering and data sharing phase separately, the third section describes need for transparency on health data and how Blockchain employed for this issue, fourth section shows and presents some designed approach on using Blockchain for privacy-preserving and transparency on healthcare data. Finally, the conclusion is drawn in the fifth section.

10.2 Preliminaries

In this section fundamental and essential information and idioms such as security, privacy-preserving and Blockchain have been described. A privacy-preserving necessity on healthcare data is explained, and its challenges are analyzed briefly. Then Blockchain structure is reviewed, and employing Blockchain on privacy-preserving is described.

10.2.1 Security and privacy-preserving on big data

These two idioms, *Security* and *Privacy*, are used interchangeably. However, they emphasize different aspects of data protection. Privacy focuses on setting up policies on data gathering, sharing, and utilizing them inappropriate ways. However, the focus of security is on protecting information against malicious attacks and accessing data for profitability. Data security is used to prevent unauthorized access to healthcare. So, technology and mechanism such as firewalls, sign-in procedures, bandwidth, content monitoring, and encryption are used to achieve security at the desired level. Nevertheless, privacy aims to ensure that the healthcare data is stored, transferred, and exposed by the data owner or on his/her awareness. This means that everyone should be aware of the disclosure of their information, and their agreement is necessary to access their data. Thus, privacy is less about protecting healthcare data from malicious use and focuses on user responsibility and preventing unauthorized users and agents from accessing the data. Security concerns can be satisfied without meeting privacy considerations, but implementing privacy cannot be achieved without implementing security issues (Fu, Wang, & Cai, 2020). This means that privacy controls granted users access and limited served data according to defined policies. In contrast, security is a mechanism and policies guarantee granted data from leakage and attacks such as man-in-the-middle, intrusion assault, etc. CIA triad analysis is a security model that helps researchers to analyze different IT security aspects of a proposed model. Confidentiality, integrity, and availability, summarized below, are three basic principles that must be ensured in systems with security concerns:

1. *Confidentiality* Ensure that information is not made available or disclosed to unauthorized users, agents, entities, or processes. Confidentiality is the most important aspect of the CIA triad and many attacks target it.
2. *Integrity* is the ability to guarantee that data is accurate and has its original information without tampering, distortion, or manipulation by anyone. A type of security attack is manipulating and intercepting stored information and altering it without the owner's knowledge.
3. *Availability* is important to ensure that the information is readily accessible to authorized users at any time and anywhere. Some security attacks, such as DoS, attempt to deny access to data for all or appropriate users.

Privacy and security on electronic-health data, like many other sensitive data, should be assured in such criteria as

1. Storing data
2. Sharing data

In storing phase, infrastructure and employed mechanism should provide confidential aspects such as:

1. *Immutability*: data integrity procedures should guarantee that unauthorized users can not alter and tamper with health data such as vital and clinical symptoms gathered by IoT devices and diagnoses and prescriptions saved by healthcare staff.
2. *Privacy-preserving*: Saved data should be anonymized, and the service provider may not have access to the private part of the user's data.
3. *Transparency*: on stored data for the authorized user. No one can repudiate the saving method on his own.

In sharing phase, critical criteria for patient and E-health system users can be categorized as

1. *Owner awareness data sharing:* the patient should be assured that his/her sensitive information is not accessible to unauthorized persons.

2. *Responsibility*: everyone should be responsible for his/her actions on the system; all requests and responses should be logged in an immutable and irrefutable manner.

Blockchain has unique features that can help E-health system providers overcome requirement properties on privacy preservation and security concerns.

10.2.2 Blockchain

Blockchain is a decentralized system of records called blocks that makes it impossible or very difficult to hack, cheat, or change the system. A Blockchain uses a distributed digital ledger that is replicated across all participants over the entire network. Each transaction is stored in a block, and newly occurred transactions are added to every ledger of participants. The replicated distributed database that is stored on multiple contributors is recognized as distributed ledger technology (DLT) (Bodkhe et al., 2020). The main property of DLT categorized as:

1. *Distributed*: all participants have distributed ledger
2. *Immutable*: changing all data on all participants is very difficult and almost impossible
3. *Time-stamped*: all transactions have a timestamp
4. *Anonymous*: participants are anonymous
5. *Consensus*: adding transactions needs to be confirmed by 51% of participants
6. *Secure*: all data are individually encrypted and signed
7. *Programmable*: a smart contract is used and can be implemented on Blockchain

Blockchain added immutable cryptographic signature of transactions to DLT. Each block contains the hash of all preblocks, so if one block changed on the chain, it would be immediately recognizable by participants and will be rejected. To add a transaction, each owner sends the transaction to the others by digitally signing a hash of the previous block and the next one's public key and adding these to the end of the chain. The receivers can verify the transaction with their private key. Adding this transaction to the chain occurs if all participants verify the hash value and accept it. This structure is shown in Fig. 10.1.

Each block consists of a header and data. The header involves three parts as follows:

1. *Mining data:* A complex computational procedure used as smart contract controls mining blocks, nonce as an incremental number used to pass difficulty, and conditions defined for calculated hash value and timestamp
2. *Hash of previous block header* to guarantee immutability of Blockchain
3. *The root of the hash tree* To minimize header data while affecting all transactions stored in the block.

The data part contains a list of transactions that will add to the Blockchain through this block. In this scenario, tampering with data on-chain became so rare and impossible. Adding a block needs an extra computation named proof-of-work (PoW), implemented by adding an incremental nonce number in the block until the block's hash satisfies required constraints such as n zero bit at the beginning of the hash. This scenario makes a competition between participants on adding a block to the chain. After block construction, it is broadcast on the network; all nodes receive it and admit the block only if all transactions are legal. Privacy can be achieved by anonymization. By

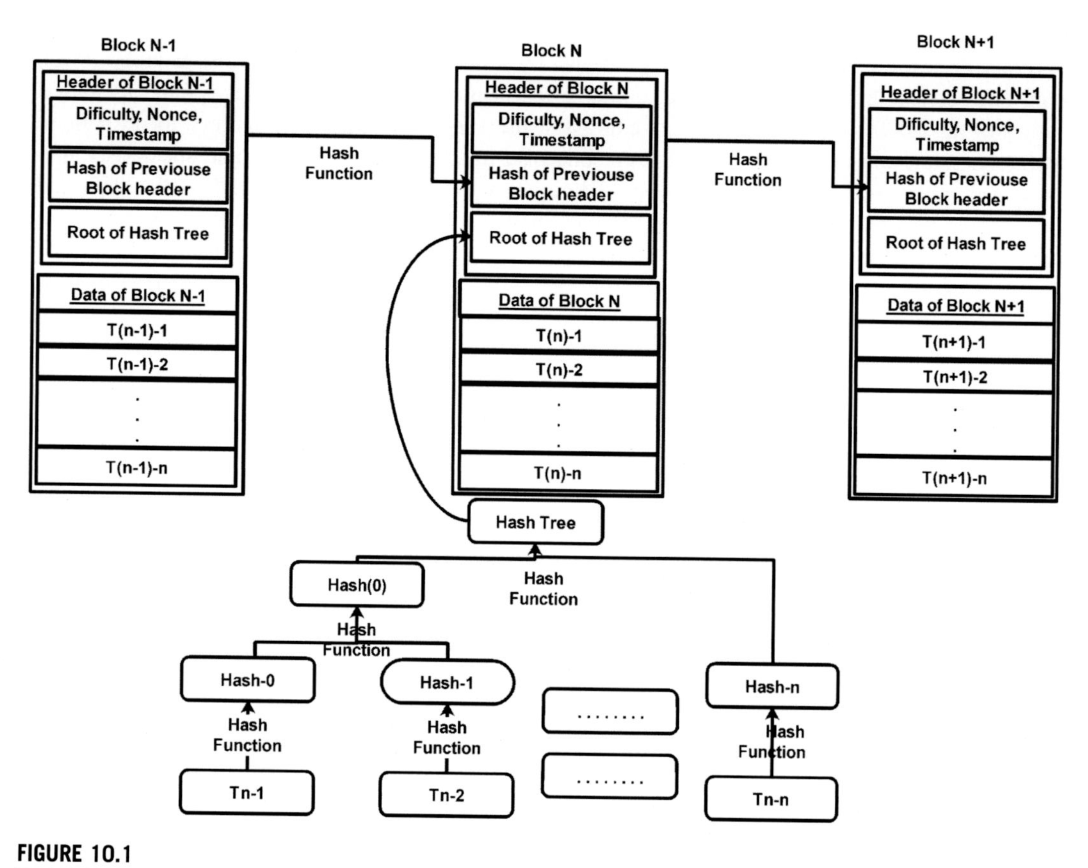

FIGURE 10.1

Blockchain structure.

the fact that public keys are anonymous, Everyone can see that some data is sent from someone to someone else. However, without sensitive information linking, that can harm the data owner or expose his/her private data (Hasselgren, Kralevska, Gligoroski, Pedersen, & Faxvaag, 2020).

10.3 Artificial intelligence for enhanced healthcare systems

Artificial intelligence (AI) has a dramatic impact on the healthcare system. Based on our studies, AI is used in medicine, health treatment, remote and robotic surgery, and predictive illness diagnosis. By saving and sharing healthcare data, several AI algorithms such as Naïve-Bayes, support vector regression, classification tree, augmented reality (AR), and virtual reality (VR) are used (Hamza et al., 2020). Using AI can improve public health conditions, reduce healthcare system costs, and increase social life standards (Kim & Huh, 2021). For example, by remote patient monitoring and applying a machine learning mechanism, analyzing prior data will be possible, and proximity of heart attack according to monitored data can be realized. Another example is using collected data to predict diseases such as chronic kidney disease by a neural network (NN) (Kim & Huh, 2021). Using AR and VR helps clinical staff to enhance the accuracy and scalability of spinal navigation (Hamza et al., 2020). Using AI in robots can assist the surgeon in high-risk surgery to do it more accurately, safer, and faster. Using AI for personalized treatment helps doctors and specialists analyze and monitor patient conditions. These techniques will eliminate hazards and outlier data and model(s) can be trained with real data. This will increase efficiency and patients can get lifestyle advice. Clinical suggestions to doctors and treatment suggestions to patients are results of accurate and real data processing by using AI, trained model(s) can predict an outbreak as a result of processing reports continuously added on blockchain.

For all these advantages, some consideration needs to be taken care such as:

1. The size of healthcare data and its growth is a big challenge
2. Append only attribute on blockchain architecture may lead to incorrect data entry. This has to be taken care of.
3. Data traceability, immutability, and integrity can assist AI algorithms in identifying the volume and type of data, but the quality of data may not be handled efficiently.
4. Execution latency and different types of interests expected by blockchain participants may mitigate the throughput of the system significantly
5. Because of the privacy and zero access mechanism on private blockchain networks, the AI mechanism is usable and more effective on public blockchain architecture, not private and consortium blockchains.

Although using AI and its related subject are preliminary from an interdisciplinary perspective, it can be employed for compressing the data as well as minimizing the redundant data (Shafay et al., 2021).

10.4 Blockchain for privacy-preserving on healthcare data

Nowadays, the internet of things (IoT) became a dominant scenario for connected devices by the Internet. According to a Statista report, more than 75 billion devices would be connected to the Internet by 2025 (Statista, 2018). One of IoT's main applications is e-health, where vital signs such as heart bits, body temperature, blood glucose, blood pressure, etc., are measured by different types of sensors. Then they are formatted on EMRs and saved on e-healthcare systems for patients' illness monitoring and medical treatment. These data are sensitive to patients and should be protected from malicious users or applications. Healthcare data is collected and stored on cloud infrastructure and shared between doctors, nurses, and medical staff. Using cloud infrastructure for e-health data is popular because, for example, when a patient wants to receive medical treatment, it is necessary to collect the history of his/her diseases, diagnoses, and treatments from his/her usual doctor and send or provide them to his new doctor or physician. It is complex, hard, and even maybe impossible. However, if all his medical data were stored in the cloud, anyone could access it anywhere. In this case, critical concerns regarding the confidentiality and privacy of patients' data will be raised. Several security and privacy breaches such as tampering with data, injecting wrong values, or even stealing information from the servers may occur. The old-style privacy-preserving methods are not efficient in healthcare data. These methods, such as k-anonymity, l-diversity, etc., are based on summarizing, suppressing, or perturbing data to anonymize them. However, in healthcare data, patients' original data are essential for treatments. New privacy-preserving systems are required to address these two aspects:

1. providing anonymity to preserve privacy
2. providing accurate data that is usable for medical staff.

Using methods such as k-anonymity, and l-diversity will address only the anonymization that is used in sharing data with the research community to investigate sensitive datasets for discovering new drugs, listing new symptoms of diseases, and other beneficial forms, but In addition to anonymity, another aspect will remain unsolved or at least unexplained. Based on the requirement of a secure healthcare system, other aspects such as privacy, security (confidentiality, integrity, availability), auditability, accountability, authenticity, and anonymity should be addressed. To satisfy these subjects, methods that describe privacy-preserving on (1) Data storage, (2) Data sharing, (3) Data auditing, and (4) Identity management can be helpful and usable.

Many studies have used Blockchain to address this issue. Using Blockchain, data can be shared on distributed ledger and accessed by all network participants accurately (without any perturbation and summarization), and high anonymity level exposes any sensitive data (Kuo, Kim, & Gabriel, 2020). However, using Blockchain on health data has several challenges:

1. *Network overhead* because of sending all transactions to all participants and PoW scenario
2. *Low throughput*, because of the complexity and amount of data that should send to the network

Under these conditions, using Blockchain without any modification is not suitable for privacy preservation in the healthcare system. Nowadays, many Blockchain approaches have been adopted to manage healthcare data, but this mechanism is still under consideration for more complete solutions to improve data privacy and transparency in healthcare systems.

As mentioned before, in healthcare systems, a huge volume of data is gathered, processed, and transferred. These properties and circumstances dramatically affect Blockchain throughput as a control mechanism for data storage and sharing processes. Thus, the hash of original healthcare data is stored as a data part in blocks and original data is saved on the cloud or such structures as IPFS. This idea plays a main role in almost all proposals and research presented in this era. These proposals can be categorized into two major aspects: saving and sharing healthcare data.

10.5 Consensus algorithms on Blockchain for privacy-preserving on healthcare data

As we know, there is no third party to authorize transactions between parties on blockchain-based architecture. Different protocols have been designed to implement consensus among all distributed nods as follows:

1. *PoW* (proof of work): The miners should perform huge computation (hashing) to achieve a nonce value that satisfies defined conditions for the target hashed value. This mechanism is used on Bitcoin. The overhead of this model is high.

2. *PoS* (proof of stake): This model is based on each node stake. Nodes with a high amount of stakes will not tamper network, but they may push the system into a centralized model.
3. *PBFT* (partial byzantine fault tolerance): This method uses 2/3 nodes to validate a new block.
4. *DPoS* (delegated proof of stake): This is similar to PoS, the major difference between PoS and DPoS is the election process on DPoS to identify Stakeholders. All nodes will participate to elect the stakeholders.
5. *PoC* (proof of capacity): Nodes with higher capacity will do the miner role and approve new blocks.
6. *PoA* (proof of authority): Only authorized nodes can approve new blocks. Although This mechanism is more efficient than many other methods, it may result in a centralized architecture.

In healthcare systems, participants are classified into two categories:

1. Personal such as patients, doctors, and nurses
2. Organizations such as hospitals, laboratories, radiography and radiology centers, and insurance institutes.

It is clear that organizations and institutes are more equipped to approve new blocks on the healthcare blockchain, but they may collude on some issues and decentralization may violate, thus DPoS is more common in this era. Although integrating several consensus algorithms instead of relying on one, used in different applications (Shi et al., 2020).

10.6 Using Blockchain for privacy-preserving in data storage phase

In healthcare systems, streamlined data should be gathered and saved for monitoring patient vital signs. These collected data sensed by IoT devices should be sent to medical health-keepers for diagnosis and healthcare treatments. These IoT devices have limited storage and computing capacity and are not suitable for processing, formatting, encrypting, and sending data. Most proposed architectures implement collecting data in two phases:

1. Sensors sense data and send it to PDA or smartphone (when sensors are wearable) or desktop or processors unit (when a patient is hospitalized in a healthcare institute or hospital)
2. PDA, smartphone, or desktop computer format data on EMR record and save it on private, hybrid, or public cloud or IPFS based architecture.

Cloud infrastructure improves efficiency and reduces the cost of healthcare systems. However, it has its drawbacks, such as a single point of failure and security and privacy concerns. The single point of failure is out of the scope of this chapter. Nevertheless, Blockchain is suitable for addressing privacy-preserving healthcare data when saved on a shared platform that everyone may access, read, or even update. In traditional healthcare systems, health data is stored in medical centers, in these scenarios, patients should register at hospitals and healthcare centers, and medical data would be under healthcare center control. Patients have access to their data, but they have no control over granting and denying access to their data. Because of this disadvantage, most of the proposed approaches are designed to provide the infrastructure where data owners have full access control.

Thus, after gathering data and formatting them on EMR records, these data are saved on accessible infrastructure such as clouds. Saving health data on the cloud may expose them to intruders and unauthorized users. To preserve privacy, data should save anonymously, the sensitive property should extract from them, or all data should be saved in an encrypted format (Hamza et al., 2020). Encrypting EMRs is used widely to assure security and privacy-preserving of sensitive data. This scenario developed as follows:

1. Vital signs such as hurt bit, body temperature, and blood glucose are gathered and sent to the formatting unit.
2. PDA or desktop computer format received data on EMR Records and encrypt them by data owner's public-key and saved them on the cloud. These data are anonymized and identified only by an associated public key that does not contains or presents any private data.

To reduce saved data on the Blockchain, EMRs are encrypted, and their hash of them is computed. Encrypted data is stored in the cloud, and the hash of them is sent to the Blockchain, as shown in Fig. 10.2. An asymmetric cryptographic method such as RSA or ECC is used to encrypt data. As we know, these encryption algorithms use a pair of keys, including a public key and a private key, to encrypt and decrypt data and also to protect against unauthorized access. The Diffie-Hellman algorithm is used as a common solution for exchanging public-key, but another innovative algorithm based on artificial neural network (AAN) is published. Some of these methods use double

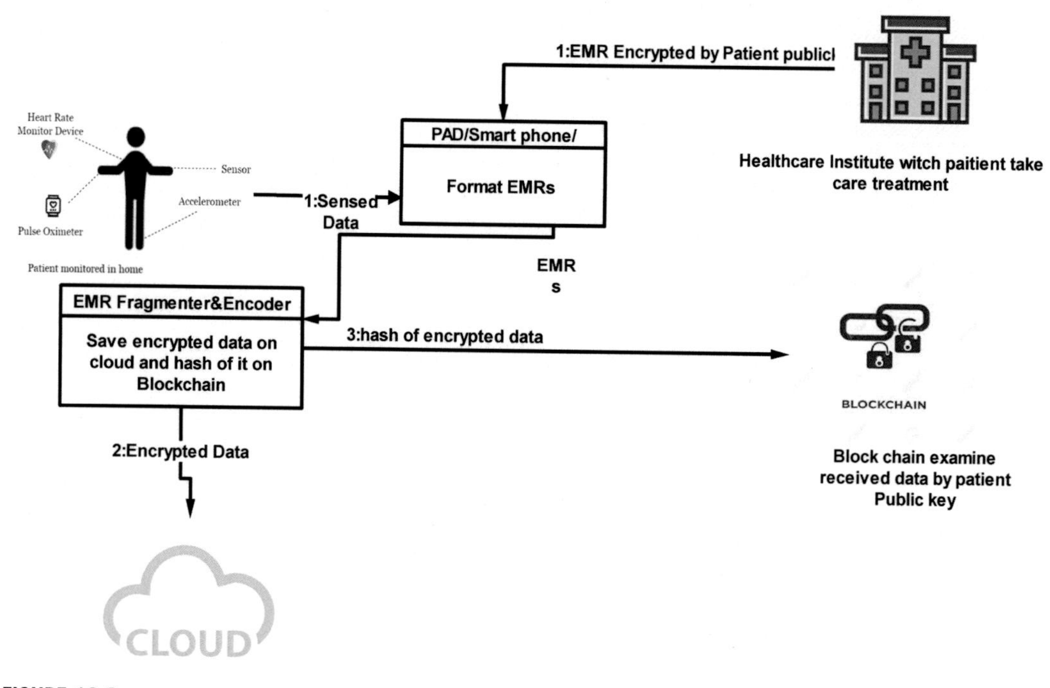

FIGURE 10.2

Storing data steps.

layer tree parity machine (DLTPM) to share ANN output by generating common input. These methods use *meta*-heuristic algorithms such as Whale Optimization as a PSO technique to generate common input (Sarkar et al., 2021). Whale Optimization, compared to other PSO methods, achieves more convergence speed due to its exclusive property. In this algorithm, all explorations and exploitations are achieved. Thus, convergence shall be done faster.

By using Blockchain in this manner, privacy-preserving and data security would be achieved. No one can tamper and distort data because it will reject by distributed ledger in a consensus checking mechanism, and no one can access patient private data because all data are anonymized.

CIA analyzing: analyzing the CIA security triad (Confidentiality, Integrity, and Availability) shows this mechanism can guarantee these aspects of security on an acceptable level.

1. *Confidentiality:* Sending and receiving data between modules are encrypted by the receivers' public key. This mechanism guarantees that transferred data is protected against sniffing.
2. *Integrity:* As mentioned before, Blockchain is resistant to modification of data. Immutability is one of the intrinsic properties of Blockchain, and data updating is very difficult or almost impossible on the Blockchain.
3. *Availability:* The availability of healthcare data is due to the cloud infrastructure and distributed feature of Blockchain. These two mechanisms guarantee availability.

10.7 Using Blockchain for privacy-preserving on data sharing

Studies have shown that the quality of healthcare treatment would be at risk when patients change their usual doctor or healthcare organization due to vendor-lock on gathered data. Healthcare organizations understand the critical worthiness of clinical and health data to their business; thus, in many cases, they are reluctant to health data sharing. Nevertheless, this data belongs to patients and should be under their control. However, data sharing is essential to provide better services and attracts more attention in recent years; the most challenging issue in sharing data is privacy-preserving and security assurance (Jaber, Fakhereldine, Dhaini, & Haraty, 2021). Sharing health data is used for:

1. Patients diagnosing processes encourage them to cooperate on recommended healthcare treatments (especially on long-term chronic diseases)
2. Performing data-mining activities on gathered data.

In both cases, someone needs to access saved data, and the data owner should grant appropriate and leveraged access to them if he recognizes them or deny their request. So, data sharing mainly consists of two major parts:

1. *Data requesting:* data requests can be issued from doctors, nurses from healthcare institutes, or any authorized healthcare staff to monitor, process, and analyze patient conditions and recommend appropriate care treatment. In this case patient request diagnosis from a doctor. However, when sharing data between a research institute and a patient, a data request is issued from a research institute.

2. *Data permission:* the best-desired solution is that the data owner can control access to his/her data and grant or denies access to them. In many depicted approaches in this era, health data is saved on cloud-based infrastructure in an encrypted format, and patients, as data owners, grant accessing their data by sending a session key to the applicator.

   An overall scenario is shown in Fig. 10.3. Sharing data between patients and doctors focused on saving the diagnoses process in a secure and immutable manner that can be done as follows:

   1. Patient sends diagnosis request to a doctor
   2. The doctor can accept or reject the patient's request
   3. All these steps are saved on the Blockchain for traceability and immutability
   4. The patient sends the session access key to the doctor and grants him access to the data
   5. Doctor retrieve shared data
   6. And save diagnosed data on shared media
   7. Hash of diagnosed data saved on the Blockchain
   8. The doctor informed the patient that diagnosed data is saved
   9. The patient gets access to his updated data

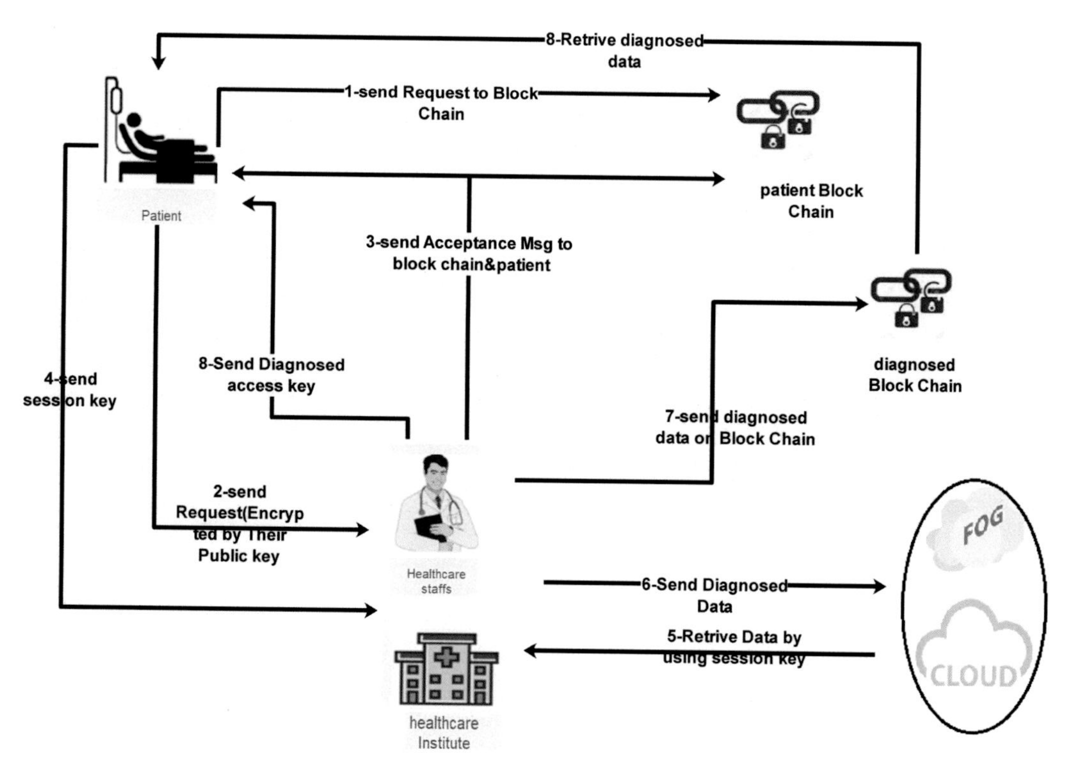

FIGURE 10.3

Basic architecture for data sharing on healthcare data.

Healthcare data sharing between patients and research institutes may differ in some aspects, such as to request initiator, response mechanism, and data sharing goals and conditions. In this case, the patient (data owner) received the applicator's request, analyzed it, and sent data on the desired accuracy level. Data perturbation is performed only on semi-identifiers, and healthcare data such as test results are sent accurately. These requests and responses are saved on the Blockchain for traceability. The security and privacy-preserving achieved on the desired level, and data owners would be assured that no one has access to their data or can tamper and perturb them. Also, no one can deny his responsibility for accessing data and manipulating them.

10.8 Blockchain for transparency in healthcare data

Applying transparency in healthcare data helps improve reports' quality and clarity. Patients would trust the healthcare system if its workflows, reports, and procedures were clear, transparent, and reliable. Transparency affects both ethical and financial issues and makes the system trust from the patients' point of view. So it would induce them to take part in long-term medical care that improves social life quality. By transparency, doctors are advised to share treatment, medication, and the disease progresses with their patients. In this model, medical interaction between patients and their health-keeper changed from a one-way manner to a cooperative model that may lead to a co-creative style and treatment. The co-creation process would lead to providing better care services to improve the patients' quality of services. So, high-level care quality is assured by frequent interactions between doctors and patients. Transparency can help better patient choices and decisions by increasing their knowledge about public health issues and professionals.

Since privacy-preserving and security is the main challenge, any proposed approach should guarantee data protection Healthcare data contain sensitive and private information, they should protect from unauthorized users or sharing by all humans. By transparency, quality and risk assessment of healthcare services present in hospitals and the medical institute became possible. If the services' results are accessible to all users, they can evaluate them by sharing them with a professional consultant. To address this problem, healthcare data is divided into two separate parts: private data and public data. Private data contains patients' identifiers and their data, but the public part contains healthcare data such as illnesses, medical tests, vital signs, etc. these data would share in an anonymized manner without a direct link to their owner. By this, transparency and privacy would achieve at an appropriate level. Public data would share in a clear text manner. Using Blockchain can play a role in immutability, and no one can repudiate his responsibility for saved data.

Another aspect of the healthcare system dramatically affected by transparency is clinical trials (CTs), which means participating in testing, validating, and verifying new drugs' safety and efficacy on patients. These processes may take a long time period, which may have extensive costs and even regulatory challenges. However, from a healthcare data management point of view, transparency, traceability, and integrity are the most important issues and would have a big effect on CTs. Blockchain would enhance CT by sharing all approved transactions on the PoW mechanism and stored as a chain of blocks. These data on CTs are stored and controlled by research organizations in traditional systems that prevent transparency and truthfully sharing of data. Using

Blockchain, transactions and events were monitored and logged by systems, and members were notified when a procedure was executed. Patients are anonymous, and the privacy of their private information is guaranteed.

The Healthcare supply chain is another part of the healthcare system that can be improved by transparency. A supply chain in healthcare systems has complex mechanisms, structures, and procedures and has spread widely. To eliminate the distribution of counterfeit drugs and fake products, supply chain monitoring is necessary. Most of the existing trace supply chains are almost centralized and suffer from intrinsic drawbacks, such as a single point of failure, lack of transparency and data distort probability by the server administrator. By using Blockchain and decentralized solutions, transparency will be improved. By using Blockchain, the interaction between manufacturer/producer, distributor, and pharmacy/consummator can be done as follows (Musamih et al., 2021; Shelke & Chakraborty, 2020):

1. The manufacturer initiates the supply process and sends it to all participants on Blockchain. Product-related information such as its image, leaflets, and other data saved on distributed infrastructure such as IPFS and its hash send to Blockchain.
2. Drug Lot will be delivered by the distributor and will pack. The result of this step will save on IPFS, and the hash of it will send to the smart contract and stored on Blockchain.
3. The last step is purchasing drugs from patients. When drugs were delivered to the patient, their image is uploaded on IPFS and the hash of it send to Blockchain.

By this procedure, the supply chain is traceable at any time, and all parties are ensured of original drug delivery, so fake and counterfeit drug distribution will mitigate.

10.9 Some exposed models using Blockchain for privacy-preserving and transparency on healthcare data

Fig. 10.4 shows a taxonomy on using Blockchain for privacy-preserving and transparency. The next three sub-sections illustrate (1) using Blockchain as an overlay network for managing privacy, (2) using multiple Blockchain for privacy-preserving for enhancing throughout, and (3) using Blockchain to aggregate data mining results. Each part is described and analyzed briefly.

10.10 Blockchain as an overlay network

Some architectures use Blockchain as an overlay network (Dorri, Kanhere, Jurdak, & Gauravaram, 2019). In these scenarios, healthcare data are stored on cloud storage infrastructure, and cloud servers compute the hash value of stored data and send it to an overlay network. The overlay network is implemented as a peer-to-peer network based on decentralized architecture. This model is shown in Fig. 10.5, consists of four parts: Patient equipped with healthcare sensors, Smart contracts, Cloud storage, and Overlay network powered by Blockchain technology.

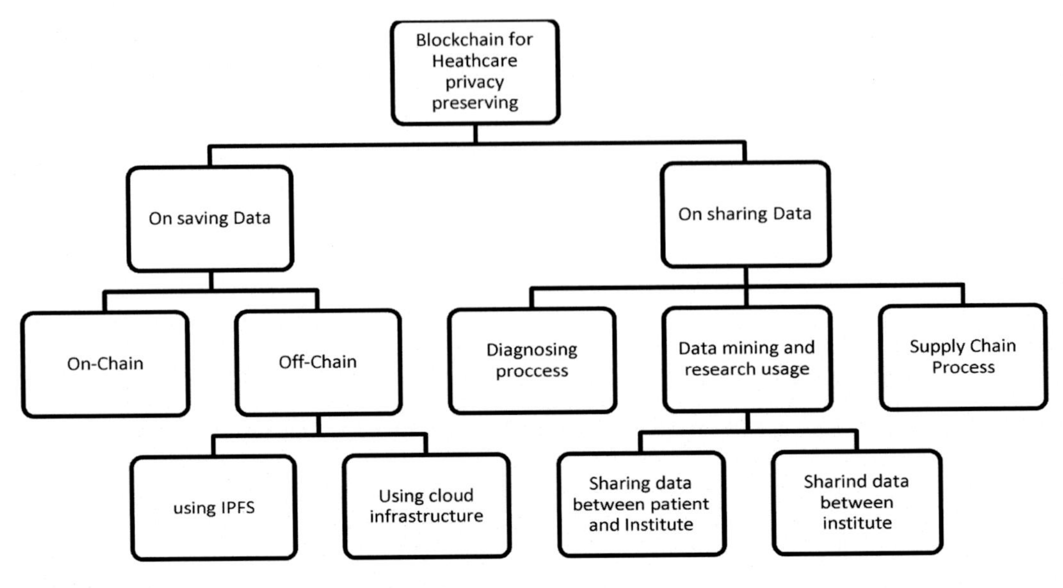

FIGURE 10.4

Taxonomy of using Blockchain on saving and sharing healthcare data.

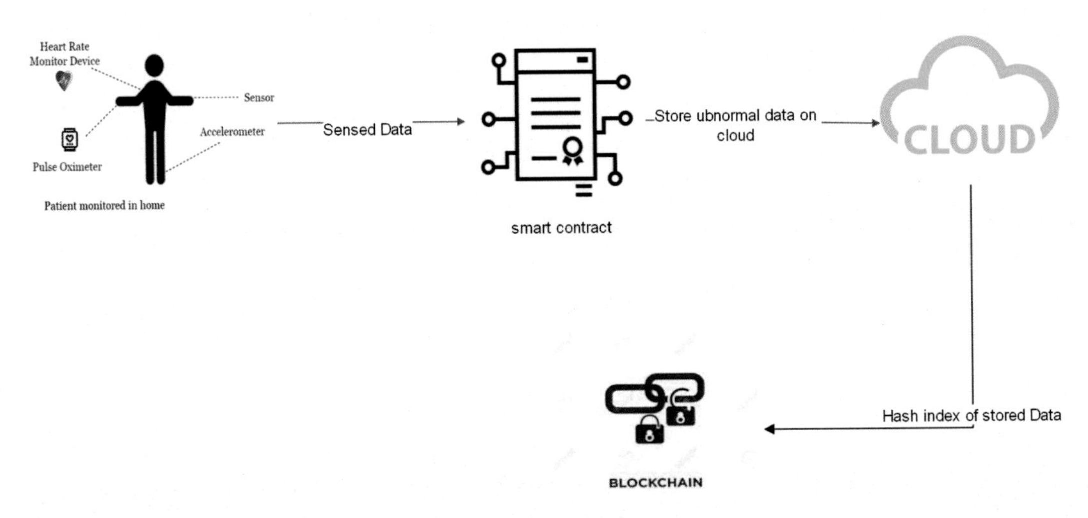

FIGURE 10.5

Using Blockchain as an overlay network to provide privacy-preserving.

The function of each part is as follows:

1. *Patient:* the patient has connected to essential sensors. These sensors collect all health data such as heartbeats, sleeping conditions, blood pressure, blood glucose, etc.

2. *Smart contracts:* Smart contracts allow defining conditions for data gathering, such as the highest and lowest patient vital signs. For example, the highest and lowest acceptable blood pressure is defined in the smart contract. Once sensed and gathered data from wearable sensors exceeds these boundaries, the smart contract will send these abnormal data to the cloud. Smart contracts run on smartphones, PDAs, or desktops that are connected to sensors. These devices send data on a secure channel to cloud storage.
3. *Cloud Storage:* healthcare data need massive storage capacity and reliable saving media, so cloud storage is used to manage them.
4. *Overlay network:* overlay network is a peer-to-peer network that is based on a distributed architecture consisting of computers, smartphones, tablets, or any other devices.

When health information is gathered from wearable sensors and analyzed as abnormal data by smart contracts, it will be formatted as healthcare records (EMRs). These records are sent to the cloud. Cloud servers calculate their hash index of them and save the hash index on the overlay network. In this way, tampering and manipulation of saved data would be impossible. Considering the case when a patient decided to share his data with someone, he will create a request, sign it, and send it to the network. The receiver's public key defines the destination of this transaction. Overlay network verifies patient's signs and then broadcasts patient's request to all nodes. Each node will receive the transaction, if it belongs to him, he will process it, and then he retrieves patient data, analyses it, creates a reply, and sends it to the patient. All these transactions are saved on the Blockchain.

As mentioned before, three main security requirements need to be addressed by models: Confidentiality, Integrity, and Availability. Confidentiality ensures authorized users can access data. Integrity guarantees data has no changes between sender and receiver when it is transferred, and availability means data is available for granted to users anytime and anywhere. In this model, only authorized users can join the network. Thus, data are exposed to registered users if it is shared with them by the data owner. So, this mechanism will guarantee confidentiality. Blockchain properties, integrity archived by Blockchain properties, and hash index of all data are saved on Blockchain. If data changes, it should be approved and accepted by PoW on Blockchain which guarantees the integration of data. Using cloud infrastructure and distributed ledger mechanism on Blockchain guarantees availability on this model. As a drawback of this depicted proposal, the mining algorithm can become a bottleneck for the system. Some proposal divides mining node into separate clusters and implements an election algorithm to elect cluster head. By this, they reduce network traffic and increase the throughput of the model.

We use JSON structure to save patient properties and diagnosed data. Most of the researchers have used XML format to manage gathered and diagnosed data. Our proposed data structure has shown in Fig. 10.6.

10.11 Using multiple blockchain for privacy-preserving on healthcare systems

Some e-healthcare systems focus on using multi-Blockchain as a solution. In these proposals, data gathered from IoT devices and diagnosis results are managed by separated Blockchains. Therefore, the patient's health data is stored at the desired time and managed by the first Blockchain named

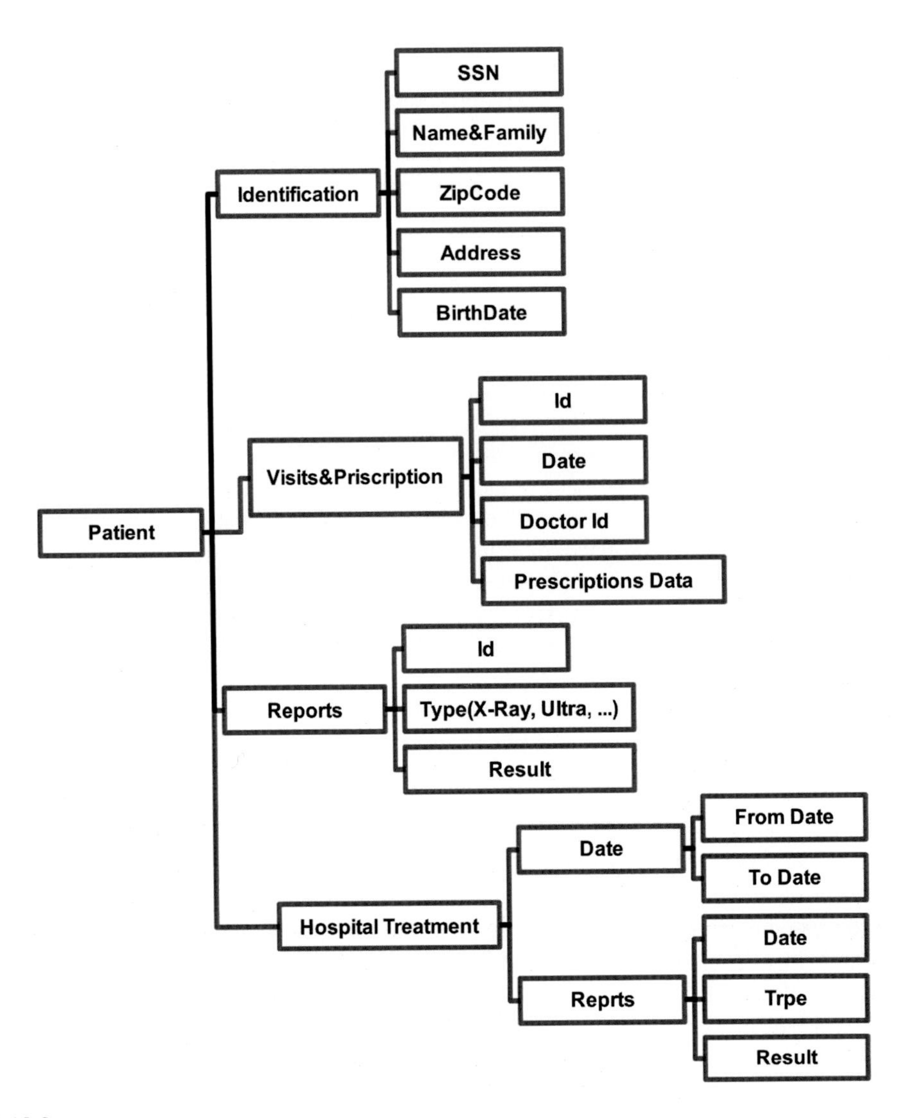

FIGURE 10.6

Electronic medical record structure.

user-data-chain. Similarly, the doctor's diagnosis is saved in time and managed by the second Blockchain named diagnose-chain. Authorized users can read each data. This architecture is shown in Fig. 10.7.

The function of each part is as follows:

1. *User nodes:* Each user manages one or more IoT devices, aggregates their data, encrypts them, and sends them to the cloud or any other storage. Some of these users play a minor role. Meanwhile, most of them create and send transactions.

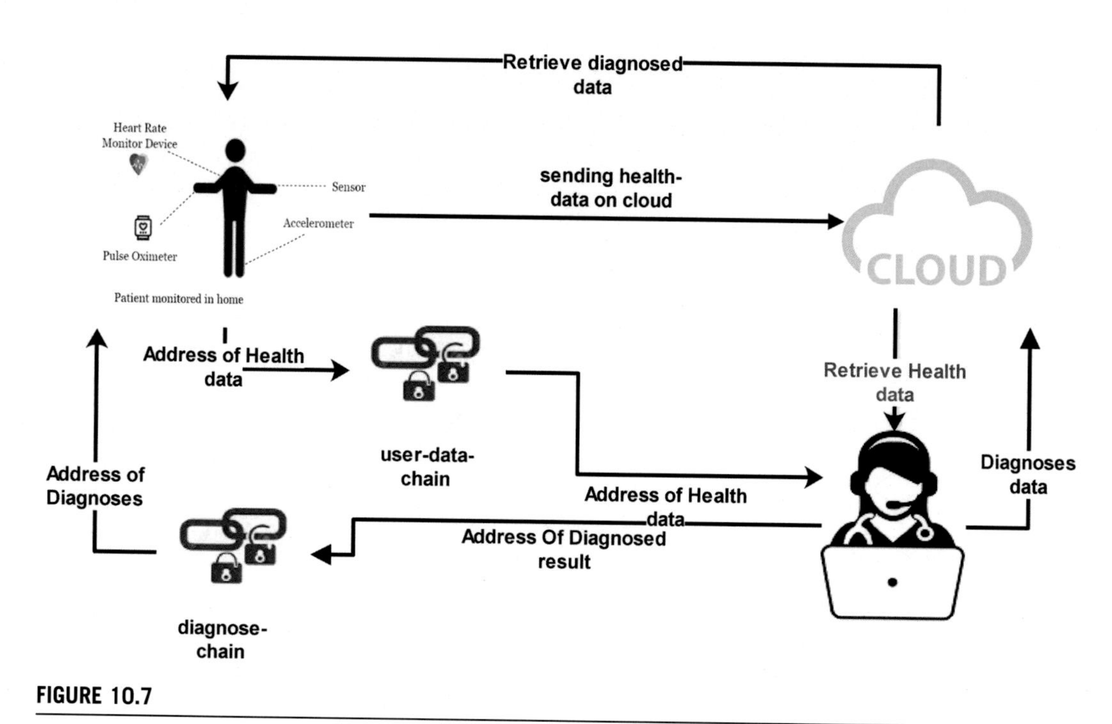

FIGURE 10.7

Using multiple-Blockchains to provide privacy-preserving on healthcare systems.

2. *Doctor nodes:* Each authorized doctor can read the health information on the user-data-chain and save his/her diagnoses and prescription on it and address it on the diagnose-chain. User nodes (patients) cannot add a node to the diagnose-chain, and doctors can not add nodes to the user-data chain.
3. *Cloud Storage:* Encrypted users' data and doctors' diagnoses are stored on cloud infrastructure.
4. *A user-data-chain* is a public Blockchain, used to publish users' data. Patients can join it to send transactions or participate in the mining process.
5. *Diagnose-chain:* is used to publish doctors' diagnoses. Authorized doctors can save diagnosis transactions, which can be added to the diagnose-chain. Patients or other authorized doctors can read these diagnoses.

Healthcare data, neither patient data gathered by IoT devices nor doctor's diagnosis, are very sensitive and need to be protected from unauthorized users. In this approach, the user-data-chain and diagnose-chain contain the hash of encrypted data that is encrypted with the user's public key or the doctor's public key. Thus adversaries can only get access to the hash of encrypted data, and saved data are immune; thus privacy-preserving has been provided on an expectable level. Security is assured by transferring data on a secure channel such as SSLs or TLSs. Accountability means that users should be in authority over their data to avoid medical disputes. Since all patient data and diagnoses are recorded on the user-data-chain and diagnose-chain, they cannot be modified according to the threat models.

10.12 Using Blockchain to manage sharing data mining result

Using data mining on health data and analyzing them would provide new medicine and treatment or increase the effectiveness of current methods and drugs. Nevertheless, big data uses technology such as the cloud to overcome its difficulties, so the challenge of privacy and security concerns will arise. The issue is to balance privacy-preserving on health data and access to them for researchers. Some traditional methods focus on data anonymization, generalization, and suppression of identifiers and semi-identifiers properties. It is difficult to create anonymous data on e-health data by traditional methods like pseudonymization to protect it against re-identification. Assuming this problem is solved with a mechanism like Blockchain, healthcare data's sensitivity hinders a centralized approach to data mining. Thus distributed data mining mechanisms can resolve these difficulties. By employing a consensus mechanism on mined data generated by each researcher, Privacy concerns will be met (Wang et al., 2019). This architecture is shown in Fig. 10.8.

Cooperation is done in two phases that guarantee privacy-preserving; in the first phase, only metadata and parameters such as mining algorithm, related parameters, data pre-processing method, etc., will be shared by the research manager as a coordinator. In the second phase, the result of mining achieved by each institute will send to the management chain. Each participant sends the hash of the training dataset with the mined results that guarantee immutability. If the results on the two institutes were different and it affects the global data mining process, institutions' truthiness can be assessed. Privacy-preserving is not compromised because the hash of private data will send

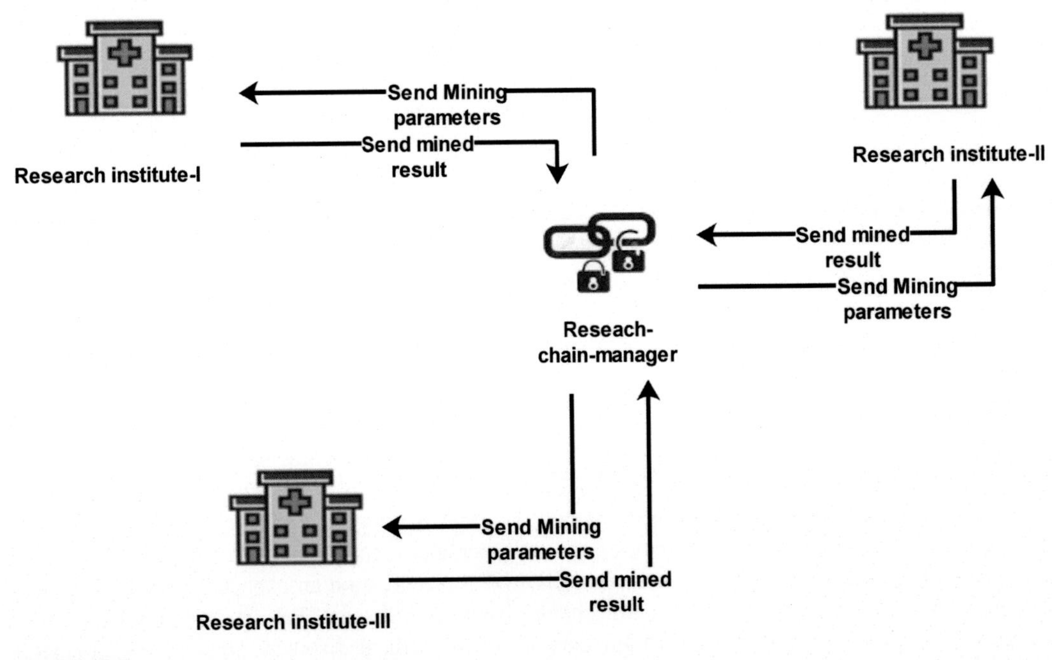

FIGURE 10.8

Using Blockchain to share mining data between researchers.

to the Blockchain. So, tracking back the patient's private data such as SSN, name, zip code, date-of-birth is impossible to protect from unauthorized access. This private data is used only by patients in charge of healthcare institutes under defined regulatory laws. In this model, researchers retrieve shared results from Blockchain, execute agreed algorithms iteratively, and analyze the result using his train dataset. Then send the hash of the training set and mining results to the research chain center. Through this sharing mechanism, there is sufficient data that assured meaningful work would be done. Advantages of this model are:

1. *Privacy-preserving:* as mentioned above, patients' private data will protect from unauthorized users because only their hash value will be shared with others.
2. *Cooperation between institutes:* healthcare institutes are reluctant to share their raw data with others. In this scenario, they only send their mining results, not their raw data.
3. *Incentive mechanism for sharing data:* research institutes are willing to have more data for their research; this model will access a massive amount of valuable data.

However, it has some disadvantages as follows:

1. *Repeated work:* All steps on data mining such as data acquisition, pre-processing, preparation, and validation need to be repeated for each participating institute on his data instead of running once on all data.
2. *Local results:* Results achieved in each local dataset will affect its local dataset behavior and may vary from other results that may evaluate as outlier data.

This model is used in different data mining and learning methods such as deep learning such as (Shafay et al., 2021) and (Chang et al., 2018). The central data mining and learning model is considered as a base model, then all training results and heuristic models on each institute are calculated separately, and the result is shared by another institute for assessment benefit of collaboration among different institutes. Several approaches such as single weight transfer and cyclical weight transfer on the deep learning model can be used to coordinate the institute in the preparation phase. This approach uses the basic idea that a model can solve new problems by using knowledge learned from solving previous problems. In practice, implementation of this architecture runs training phases on each institute's dataset and fine-training the model and its result on another domain by a different dataset. Thus the result has higher accuracy than each of the institutes separately. The result of this model has been tested and performance of distributed learning model is comparable to centralized model. Testing model accuracy shows that this model has about 2% or 3% less accurate than the centralized model. Therefore, the training model can be done by sharing mined or learned results without sharing raw healthcare data that can violate privacy issues.

10.13 Anonymity contact tracing model using Blockchain-based mechanism

Contact tracing can play an important role in mitigating contagious diseases, especially in epidemic cases like Coronavirus disease 2019 (COVID-19). Statistical models have shown that contact tracing is useful if it is done at the beginning of the outbreak (Garg, Chukwu, Nasser, Chakraborty, &

Garg, 2020). Traditional contact tracing is done by interviewing the infected person and his/her contact persons. However, this method of generating infected areas and paths is so slow and may be out-of-date and useless. By using IoT technology and mobile networks, automatic digital contact tracing can be done more efficiently and accurately. This can be done by

1. *Service provider mobile application* that tracks infected cases by service provider application and uses healthcare application form for data gathering.
2. *Call detail analysis* that uses the triangulation method to find the subscriber's location and his/her contact person.
3. *Citizen mobile application* that uses Bluetooth or GPS to monitor nearby devices. If someone infects by a disease, all his/her contact person in a specified period (e.g., 14 days for COVID-19) will be informed.
4. *Hardware solutions* such as wristbands and thermal cameras are used in some countries such as Hong Kong and China to monitor infected persons.

In some of these methods, anonymity is distorted; for example, using Bluetooth infrastructure, nearby devices should be visible to each other and not acceptable for subscribers. Using RFID tags and saving gathered data on blockchain infrastructure, anonymity and privacy-preserving will improve (Garg et al., 2020). If an identified RFID is detected as an infected case, all RFID whom be in contact with him/her will be informed if they participate on Blockchain and register their RFID on the network. This confirmation will generate by processing contact tracing data stored on the Blockchain. These data contain DFID tags, timestamps, and the distance between RFID and receivers. By this data and using the triangulation method, the distance between two RFID on a specific timestamp will calculate, and an appropriate decision will be made. So, automated contact tracing is well done, and anonymity and privacy will be preserved.

10.14 Using adaptive blockchain-based mechanism to preserve privacy in emergency situations

As mentioned before, patients with new symptoms may be referred to several medicare institutes and received different medicine or healthcare treatment. Sharing health data is necessary to facilitate patient's treatment, such as test results produced by the different institutes, prescriptions, and drugs used by them, and will help health staff diagnose and realize patient health conditions and advise more precise health treatment. For privacy-preserving, the patient defines the access policy and defines the access level to his EMRs. Each medical staff can access healthcare data if the data owner shares data and sends a session key for him to decrypt his data. This mechanism can protect data from unauthorized users, but if an emergency occurs, all EMRs of the patients are encrypted, and no one can decrypt them and used them to save a patient's life. Healthcare history is so valuable and vital for the patient, especially in emergency conditions. In this situation, using the mentioned mechanism in previous sections may delay the patients' emergency rescue. Thus, it is vital to adapt designed models to overcome this problem. A break-glass access method can be used to address this problem. By break-glass access, healthcare staff can access encrypted data without a secure session key. However, using this mechanism should not affect privacy-preserving and accountability and should prevent malicious data access.

Each patient sets a password-based break-glass key and shared it with his trusted contact persons such as his family members, his usual doctor or health care staff, or maybe his friends when emergency conditions such as car accident or heart attacks occur, and the patient was fainted or went into a coma, healthcare staffs can use patient's contact list and request break-glass key. It is necessary to log all these steps to prevent malicious usage of health data.

10.15 Comparing proposed model with similar research

Extensive research has been done on using Blockchain to improve privacy-preserving in healthcare systems. We have reviewed most of these research and articles and summarized the result in Table 10.1. The advantage and disadvantages of each proposed model are reviewed, and evaluation parameters are verified. Evaluation parameters in this study are:

1. Information anonymity: indicate that the proposed model has concern for data anonymization
2. Sensitive and nonsensitive data separation: shows that the proposed model separates identifiers, semi-identifiers, and nonidentifiers and has different treatment methods.
3. Using multi Blockchain: shows that privacy-preserving is done by one chain or more.
4. Providing full access control for data owner: shows whether data sharing is confirmed by the data owner or not.
5. Data sharing ability for research institute: shows model concentration on data sharing between research institute and aggregating results mechanism.

In our proposed model, multi Blockchain is used to save patients' data, diagnose data, and mine results separately. Sensitive and nonsensitive data are separate from each other, and sharing data is under data owners' full control.

This model can evaluate and simulate by simulation environments such as SimBlock and JaamSim. SimBlock is an open-source blockchain network simulator, developed by Distributed Systems Group, Tokyo Institute of Technology. SimBlock is event-driven and is suitable for use in blockchain network research. SimBlock also has a visualization tool, by which you can see the transition of block propagation. JaamSim is free and open-source discrete-event simulation software that includes a drag-and-drop user interface, interactive 3D graphics, input and output processing, and model development tools and editors.

Block rates and their size should consider from a different aspect of view; patient vital data that is gathered from wearable sensors and saved on the network, diagnosis data that send to a network by doctors, physicians, and health staff, and data mining and learning results generated by research institutes. Simulation parameters should be set in such a way that properly indicates the model, thus we assume the size of blocks increases from patient vital data to research data because sensed data for each patient is much less than those data shared and accessed by the research institute. Although block rate (blocks added to blockchain on each slice time) decrease from patient-blockchain to research-blockchain. The number of physician nodes compared to patients

Table 10.1 Comparing reviewed research and articles on privacy-preserving using Blockchain.

Used technique	advantage	disadvantage	Anonymization	Sensitive/ nonsensitive separation	Using multi Blockchain	Data ownership full access control	Data sharing ability
Using overlay network and node clustering (Dwivedi, Srivastava, Dhar, & Singh, 2019)	Using node clustering	Lake of sharing diagnosed data for the data owner	No	No	Yes	Yes	No
Machine learning on federated databases in health care (Chen, Wang, & Yang, 2019)	Mining result aggregation	Lake of data owner access control	Yes	No	No	No	Yes
Using node clustering and Blockchain (Chen, Xie, Lv, Wei, & Hu, 2019)	Load balancing between clusters	Lake request/ response mechanism on diagnosing process	Yes	No	No	No	No
Proposed GuardHealth to store and share Health data (Wang, Luo, & Zhou, 2020)	Using multi chain	Lake of sharing mechanism for data mining on a research institute	No	No	Yes	Yes	No
Using IPFS for Health data saving (Xu et al., 2019)	Separate vital data and diagnosed data	Lake of anonymity	No	No	Yes	Yes	No
Proposed MdiBChain	Using ECC as a lightweight asynchronous encryption method	Lake of data separation	No	No	Yes	Yes	No
Proposed adaptive model based on Break-Glass key	Cover emergency	Lake of data separation	No	No	No	Yes	No

(Physicians Per Capita) is assumed to be 3 to 1000 and the number of research institutes compared to physicians is assumed to be 2 to 10,000.

10.16 Conclusions

Since society became elder day by day and some new diseases are emerged, e-health systems became more important. Healthcare data are classified as big data because of their properties such as volume, variety, velocity, and saved on cloud infrastructures as a known solution for these problems to provide accessing data anytime and anywhere. Despite the advantages of using fog and cloud in this era, privacy-preserving, safety, and confidentiality of healthcare data on shared platforms is a major issue and became a challenge. Using Blockchain for privacy-preserving and transparency is increasing day by day. Nevertheless, Blockchain is slow and became a bottleneck in an environment such as healthcare systems. Many proposals have been presented to mitigate this challenge such as minimizing transaction volume by using hash technology, separating sensitive data from nonsensitive data, etc.

According to our studies, using multi lightweight Blockchain has not been studied in depth. We proposed this architecture, but we did not have any implementation. We are simulating and collecting results to compare it with other architecture in the near future. Nevertheless, preliminary studies and comparisons of the structure with other architectures presented are very promising. This improvement is due to different occurrences of events in healthcare systems. Indeed, the number of events that gather and store the patient's vital signs is more than when a physician examines them for diagnosing or monitoring him/her. Although, sharing data between research institutes occurs less than in other events. So, we improve overall throughput by separate management on different parts. On the other hand, separating sensitive and nonsensitive data and storing nonsensitive parts in plain text without encryption helped improve system performance.

In future work, we study the use of mechanisms such as clustering nodes and miners to increase efficiency. Also, the use of new asymmetric encryption methods can help improve system performance. Using Blockchain for privacy-preserving is an open new era, and more studies are necessary to propose a more sophisticated architecture for e-health data saving and sharing.

References

Ali, F., El-Sappagh, S., Islam, S. M. R., Ali, A., Attique, M., Imran, M., et al. (2021). An intelligent healthcare monitoring framework using wearable sensors and social networking data. *Future Generation Computer Systems*, *114*, 23–43.

Bodkhe, U., Tanwar, S., Parekh, K., Khanpara, P., Tyagi, S., Kumar, N., et al. (2020). Blockchain for industry 4.0: A comprehensive review. *IEEE Access*, *8*, 79764–79800.

Chen, X., Wang, X., Yang, K. (2019). Asynchronous blockchain-based privacy-preserving training framework for disease diagnosis. In *Proceedings of the international Conference on Big Data (Big Data)*, pp. 5469–5473. IEEE.

Chang, K., Balachandar, N., Lam, C., Brown, J., Yi, D., Beers, A., et al. (2018). Distributed deep learning networks among institutions for medical imaging. *Journal of the American Medical Informatics Association: JAMIA*, *25*(8), 945–954. Available from https://doi.org/10.1093/jamia/ocy017.

Chen, Y., Xie, H., Lv, K., Wei, S., & Hu, C. (2019). DEPLEST: A blockchain-based privacy-preserving distributed database toward user behaviors in social networks. *Information Sciences*, *501*, 100–117.

Dorri, A., Kanhere, S. S., Jurdak, R., & Gauravaram, P. (2019). LSB: A lightweight scalable blockchain for IoT security and anonymity. *Journal of Parallel and Distributed Computing*, *134*, 180–197.

Dwivedi, A. D., Srivastava, G., Dhar, S., & Singh, R. (2019). A decentralized privacy-preserving healthcare blockchain for IoT. *Sensors*, *19*(2), 326.

Fu, J., Wang, N., & Cai, Y. (2020). Privacy-preserving in healthcare blockchain systems based on lightweight message sharing. *Sensors*, *20*(7), 1898.

Garg, L., Chukwu, E., Nasser, N., Chakraborty, C., & Garg, G. (2020). Anonymity preserving IoT-based COVID-19 and other infectious disease contact tracing model. *IEEE Access*, *8*, 159402–159414.

Hamza, R., Yan, Z., Muhammad, K., Bellavista, P., & Titouna, F. (2020). A privacy-preserving cryptosystem for IoT E-healthcare. *Information Sciences*, *527*, 493–510.

Hasselgren, A., Kralevska, K., Gligoroski, D., Pedersen, S. A., & Faxvaag, A. (2020). Blockchain in healthcare and health sciences—A scoping review. *International Journal of Medical Informatics*, *134*, 104040.

Jaber, M., Fakhereldine, A., Dhaini, M., & Haraty, R. A. (2021). *A novel privacy-preserving healthcare information sharing platform using blockchain. Security and privacy issues in IoT devices and sensor networks* (pp. 245–261). Elsevier.

Jain, R., Gupta, M., Nayyar, A., & Sharma, N. (2021). *Adoption of fog computing in healthcare 4.0. Fog computing for healthcare 4.0 environments* (pp. 3–36). Springer.

Kim, S. K., & Huh, J. H. (2021). Artificial intelligence based electronic healthcare solution. In J. J. Park, S. J. Fong, Y. Pan, & Y. Sung (Eds.), *Advances in computer science and ubiquitous computing. Lecture notes in electrical engineering* (715). Singapore: Springer. Available from https://doi.org/10.1007/978-981-15-9343-7_81.

Kuo, T.-T., Kim, J., & Gabriel, R. A. (2020). Privacy-preserving model learning on a blockchain network-of-networks. *Journal of the American Medical Informatics Association*, *27*(3), 343–354.

Musamih, A., Salah, K., Jayaraman, R., Arshad, J., Debe, M., Al-Hammadi, Y., et al. (2021). A blockchain-based approach for drug traceability in healthcare supply chain. *IEEE Access*, *9*, 9728–9743.

Sarkar, A., Khan, M. Z., Singh, M. M., Noorwali, A., Chakraborty, C., & Pani, S. K. (2021). Artificial neural synchronization using nature inspired whale optimization. *IEEE Access*, *9*, 16435–16447.

Shafay, M.; Ahmad, R.W.; Salah, K.; Yaqoob, I.; Jayaraman; Omar, M. (2021). Blockchain for deep learning: Review and open challenges. TechRxiv. <https://doi.org/10.36227/techrxiv.16823140.v1>.

Shelke, Y., & Chakraborty, C. (2020). Augmented reality and virtual reality transforming spinal imaging landscape: A feasibility study. *IEEE Computer Graphics and Applications*.

Shi, S., He, D., Li, L., Kumar, N., Khurram Khan, M., Raymond Choo, K. K., et al. (2020). Applications of blockchain in ensuring the security and privacy of electronic health record systems: A survey. *Computers & Security*, 101966.

I. Statista (2018). Internet of things (IoT) connected devices installed base worldwide from 2015 to 2025 (in billions), ed.

Wang, W., Hoang, D. T., Hu, P., Xiong, Z., Niyato, D., Wang, P., et al. (2019). A survey on consensus mechanisms and mining strategy management in blockchain networks. *IEEE Access*, *7*, 22328–22370.

Wang, Z., Luo, N., & Zhou, P. (2020). GuardHealth: Blockchain empowered secure data management and graph convolutional network enabled anomaly detection in smart healthcare. *Journal of Parallel and Distributed Computing*, *142*, 1–12.

Xu, J., Xue, K., Li, S., Tian, H., Hong, J., Hong, P., et al. (2019). Healthchain: A blockchain-based privacy preserving scheme for large-scale health data. *IEEE Internet of Things Journal*, *6*(5), 8770–8781.

Yaqoob, I., Salah, K., Jayaraman, R., & Al-Hammadi, Y. (2021). Blockchain for healthcare data management: Opportunities, challenges, and future recommendations. *Neural Computing and Applications*, 1–16.

CHAPTER

Security and privacy concerns in smart healthcare system 11

Muyideen AbdulRaheem[1], Joseph Bamdele Awotunde[1], Chinmay Chakraborty[2], Emmanuel Abidemi Adeniyi[3], Idowu Dauda Oladipo[1] and Akash Kumar Bhoi[4,5]

[1]Department of Computer Sciences, University of Ilorin, Ilorin, Kwara State, Nigeria [2]Department of Electronics and Communication Engineering, Birla Institute of Technology, Mesra, Jharkhand, India [3]Department of Computer Sciences, Landmark University, Omu-Aran, Kwara State, Nigeria [4]KIET Group of Institutions, Delhi-NCR, Ghaziabad, Uttar Pradesh, India [5]Directorate of Research, Sikkim Manipal University, Gangtok, Sikkim, India

11.1 Introduction

In recent times, the internet of things (IoT), the concept first suggested by Kevin Ashton in 1999 has attracted considerable attention. Due to various rapid advances in mobile connectivity, WSNs, RFIDs, and computing in the cloud, the connection between mobile networks is now more simple than ever (Dey, Ashour, Shi, Fong, & Sherratt, 2017). Mobile devices and smartphones are both examples of personal digital assistants (PDAs), notebooks, iPads, and other hand-held embedded systems are among the IoT devices that can share data in the IoT network. These devices use wireless communication and sensor networks to exchange messages with one another and communicate useful data to the centralized system (Manogaran et al., 2018). The data obtained by IoT devices is preserved in a consolidated facility before being sent to the intended recipients. It enables people and devices to connect at any moment, from any place, with any computer, in ideal conditions, over any network, and with any service (Rathee, Sharma, Kumar, & Iqbal, 2019).

Hundreds of millions of computers, users, and utilities are now linked and exchanging data (Razzaq, Gill, Qureshi, & Ullah, 2017). Smart Healthcare System (SHS) is a beneficiary of these technologies that permit professionals in the health sector to communicate and interact with colleagues, devices, and patients in the healthcare delivery setup. Latest innovations in new and revolutionary technology, such as machine learning, robotics, and Artificial Intelligence (AI), have dramatically improved the method healthcare services are provided to health seekers in the SHS epoch (Bongomin et al., 2020). Real-time online patient control is now possible to become more versatile and flexible owing to improvements in medical monitoring services and smart technologies. Cloud networking and cellular technology have aided in no small significant way to advance SHS provision, and the role of wireless sensor networks in mobile technologies for smart healthcare is equally essential (Oueida, Aloqaily, & Ionescu, 2019).

Devices such as a mobile phones will directly obtain the data detected by the network sensors or patient-doctor, nurse-doctor, or pharmacy-doctor PDA who make critical decisions or take steps based on the data gathered by those sensors (Aceto, Persico, & Pescapé, 2018). Unauthorized

Implementation of Smart Healthcare Systems using AI, IoT, and Blockchain. DOI: https://doi.org/10.1016/B978-0-323-91916-6.00002-3

access to this vital information and medical records must be avoided otherwise, it can endanger the patient's life and, in some cases, result in death. If a patient's prescription for medical treatments and medication dosage falls into the wrong hands, it may be mishandled, with potential repercussions (Gnjidic, Husband, & Todd, 2018). As a result, for SHS, high and strict security mechanisms are required, including safe management, confidentiality, privacy, integrity, authorization, and authentication. SHS applications offer and bring several advantages and obstacles in the healthcare industry. Some of these advantages offer a flexible atmosphere for tracking patients' everyday lives and medical situations at any moment, from any place, and without respect for boundaries. Simultaneously, amongst the most serious threats to the healthcare provided by these new technologies is a lack of protection and privacy, which often exposes patient confidentiality (Abouelmehdi, Beni-Hessane, & Khaloufi, 2018).

The physiological clinical symptoms of a patient are extremely sensitive, particularly if suffering from an awkward illness (Shang et al., 2020). If a patient's data condition or poor state of care is carelessly shared, the patient might face ridicule, at the very least, and even psychological distress. Furthermore, disease knowledge can result in an individual losing their job in some cases. The patient's ability to obtain insurance protection may be harmed as an outcome of the mutual information (Gordon & Catalini, 2018).

The chapter proposes a framework for SHS privacy and security, and a practical application was conducted on the energy consumption and avalance effect. The remaining chapter is organized as follows. Section two is related work, section three discusses SHS while section four examines the security and privacy threat in SHS. Security and privacy antidote in existence are examined in section five. Section six is the conclusion and future directions.

11.2 Smart healthcare system

Smart Healthcare is a patient-based healthcare system making use of SHS devices for remote monitoring of the patient. Networks are established Sensors, actuators, and personal medical devices (PMDs) that help to bridge the gap between the modern world and electronic records. These medical tracking systems will continually track a patient's health status and relay them to an expert or authorized healthcare provider.

The entire structure of SHS is generally categorized into three separate phases or stages. The first stage consists of sensors that track a patient's vital signs and are light, low-power, and high-efficiency. These can be worn on, put into, or wrapped around the body. On-body sensors and stationary medical equipment are the two types of first-stage devices used to introduce smart healthcare. Biosensors that are connected to the human body for physiological monitoring are known as on-body sensors. In-vitro and in-vivo sensors are two different types of sensors. In-vitro electrodes are attached to the patient's psyche manually, removing the need for clinical or medical services in healthcare. In-vivo sensors are implantable devices that are positioned inside the body following surgery meeting all sterilization requirements.

Pacemakers, motion detectors, and artificial retinas are all examples of first-stage devices. All of these sensors submit health data to a PMD. These PMDs binds to the Internet or a cloud server in the second level. In SHS, PMDs for patient monitoring and assessment are either implanted in the patient's body or externally attached to the patient's body to track their medical condition. These devices connect with a base station using a wireless interface, which is then employed to read system status, and diagnostic reports, adjust device parameters and update the status accordingly (De Santis & Cacciotti, 2020).

The data is analyzed in the third level. The third tier of analysis assists healthcare professionals in assessing whether or not an intervention is needed to be based on sensed data SMS sensors and other autonomous instruments can classify data based on the patient and the examiner. A physician, for instance, needs historical knowledge about a patient's symptoms, while a chemist simply needs the patient's latest prescription log. For instance, an ambulance staff can need more information than a nurse or expert. Since sensors and apps interact wirelessly, SHS prioritizes guaranteeing that each health professional receives the information they require while maintaining the confidentiality and privacy of all collected data. Fig. 11.1 shows the general framework for SHS, the most vulnerable part for security and privacy is the transmission of data from a smart health device to a cloud storage server. The attackers can modify or change the data during transmission. In the transmission of processed data, attackers can still alter the information before getting to the organization that needs the information.

Patient protection may be jeopardized by deliberate data tampering or unintended data access by the wrong healthcare practitioner. Prototypes of SHSs are at the forefront of new security approaches, and they are poised to lead the way in the coming years. The bulk of existing IoT literature focuses on general knowledge about IoT systems rather than individual domain applications. Encryption methods are widely used to resolve security and privacy concerns. (Haus et al., 2017) claims that using elliptic curve cryptography (ECC) and Quantum cryptography is used to protect messages shared across computers and base stations participating in the IoT device scheme, able to maintain secrecy, integrity, compatibility, and non-repudiation through IoT based networks.

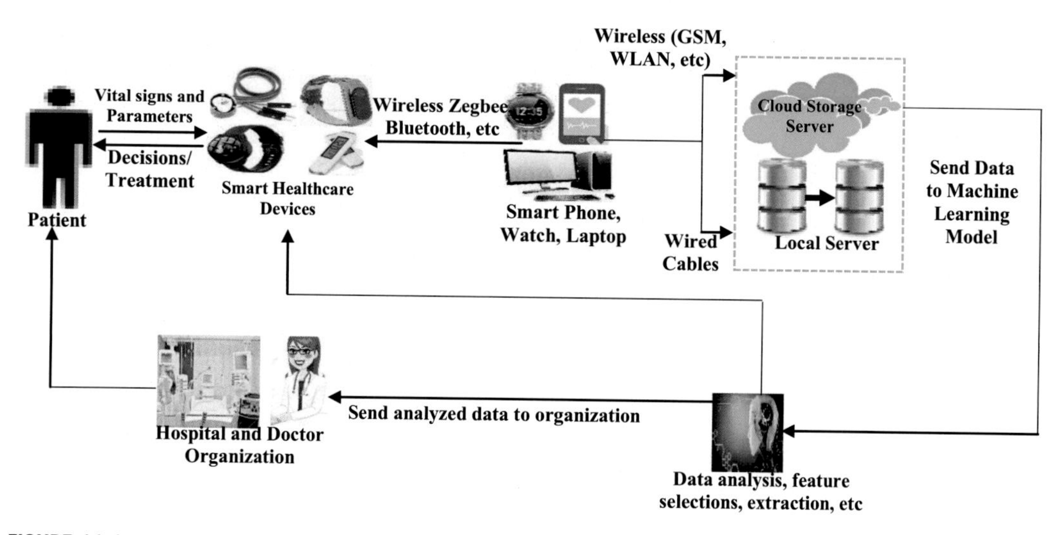

FIGURE 11.1

The general framework concepts for smart healthcare system.

The combination of ECC and identity-based encryption (IBE) systems is recommended as a solution by (Gandotra, Jha, & Jain, 2017) during the discovery and transmitting processes of device-to-device (D2D) communications, to prevent non-repudiation of data. A primary derivation process has also been introduced, which is directed at community communication.

Navarro-Ortiz, Sendra, Ameigeiras, and Lopez-Soler (2018) proposes integrating LoRaWAN, a free and standardized LowPower Wide Area Networks (LPWAN) technology, to enable portable linkage workers to utilize their existing setups. Since safety is so important for IoT systems, it was built into the LoRaWAN standard from the beginning. Mutual authentication, reputation protection, and confidentiality are the three key properties of LoRaWAN security. By using the method focused on the exchanging of period keys and confirmation codes, data transported through the IoT system are both authenticated and reliability secured, accomplishing end-to-end protection (Deep et al., 2020). The encryption scheme used is AES, which has a 128-bit key length. What the suggested solution does not reveal is its practical viability in a large IoT system, since the design is largely made up of computers, gateways, and network servers.

A hash chain, according to (Xie et al., 2019), is the safest way to ensure data confidentiality and nonrepudiation in Smart Healthcare service instrumentation a concept likens to the blockchain. Since such a system generates current autographs, it's ideal for safeguarding and controlling data such as track logs and keeping management activities. Alternatively, connection-oriented transactions, such as the sharing of control information, may be best guarded using public-key cryptography systems, which are often used to create digital signatures (Cannady, 2019). (Jameel, Hamid, Jabeen, Zeadally, & Javed, 2018) propose two attribute-based encryption systems to achieve stable shared heterogeneous communications in IoT network slicing, ensuring integrity, anonymity, and non-repudiation.

11.2.1 Applications of smart healthcare system

The SHS has smart IoT-powered devices that are portable, versatile, and have RFID tags embedded. Patients are issued these wearables on arrival for treatment (Banerjee, Chakraborty, Kumar, & Biswas, 2020). By monitoring cholesterol, heart rate, temperature, muscle tension, and other symptoms, the doctor and nurses would be able to keep track of the patient's health. They can be tracked both in the hospital and at the doctor's homes (Dhanvijay & Patil, 2019; Gupta, Chakraborty, & Gupta, 2019). In emergency conditions such as a stroke, heart attack, or cardiac arrest, the IoT-enabled smart healthcare can be incredibly helpful. Ambulances equipped with IoT systems arrive on schedule, and patients in the ambulance can be tracked remotely from the hospital. Before even arriving at the hospital, care can begin (Qadri, Nauman, Zikria, Vasilakos, & Kim, 2020). The IoT has revolutionized patient monitoring and healthcare. Sensors with a special ID are worn on the body. They collect health-related information. Data analytics is used to examine the information. This increases the result, lowers the cost, and allows for more personalized treatment (Awotunde, Adeniyi, Ogundokun, Ajamu, & Adebayo, 2021a; Chakraborty & Rodrigues, 2020; Chakraborty et al., 2021; Habibzadeh et al., 2019).

Dang, Piran, Han, Min, and Moon (2019) suggests revolutionizing algorithms for numerical efficiency that looked at upcoming technology that would enhance smart healthcare, reduce size and expense, and improve power quality. Connection is facilitated by ubiquitous networking, structured protocols, and cloud computing availability. For the study of large amounts of data requires data analytics. Robotic Ambulance service can convey the patient with an emergency kit as part of SHS, allowing for proper patient monitoring. Before the arrival of the ambulance, doctors can give first aid to the patients and offer prompt emergency attention (Chakraborty, Gupta, & Ghosh, 2013; Goyal, Sahoo, Sharma, & Singh, 2020). For people who are differently-abled, medically or mentally challenged, the SHS is incredibly helpful (Chakraborty, Banerjee, Kolekar, Garg, & Chakraborty, 2020). It solves the challenges that they face regularly (Spender et al., 2019). IoT offers emergency care to patients who live in rural locations and rugged terrains, such as mountainous areas, that are difficult to access (Chakraborty et al., 2020; Goyal et al., 2020). The shortage of railing and highway infrastructure exacerbates the issues (Bhattacharya, Banerjee, & Chakraborty, 2019). It is impossible to deliver timely medical attention to patients in those areas.

When a catastrophe occurs as a result of a natural disaster, the predicament grows exponentially. The town becomes cut off, and road access to those areas is unlikely. In such cases, the availability of SHS comes in handy. Sensor-based nodes are distributed in the affected region. They compile information and send out warnings so that relief efforts can get going quickly (Prabhu, Balakumar, & Antony, 2017). Automating day-to-day activities, the IoT and SHS increases the quality of life where machines can be used to track and make decisions. In such a case, health monitoring equipment is fitted with devices that gather medical data from patients and send it to a specialist (Newaz, Sikder, Rahman, & Uluagac, 2019).

11.2.2 Risks of using the internet of things in smart healthcare system

Intruders and attackers threaten SHS daily. According to an assessment, 70% of IoT systems are very simple to set up and hack (Yu, Zhuge, Cao, Shi, & Jiang, 2020). The disadvantages of introducing In SHS, IoT raises possible threats to patient welfare as well as the destruction of critical health records. Human accidents, computer or third-party faults, and natural occurrences, among other factors, may all contribute to this. The attack possibility will increase exponentially as the threat landscape expands with the introduction of smart devices. Furthermore, the use of IoT in SHS introduces significant vulnerabilities. The most significant causes of these flaws are:

1. Internet-of-things devices are well-connected, and others, such as networked health devices, can also connect to other devices automatically;
2. Contact and interaction Interactions within healthcare equipment and traditional networks can improve vulnerability by granting fraudulent perpetrators unauthorized access to information or records. Unauthorized control is essential in the intelligent hospital setting since a lack of an authorization system can enable unintended users to obtain access to a specific system from an end computer. The primary aim of the SHS is to provide patients with accessible healthcare services at any time and from any place, while also ensuring the protection of the SHS network to protect patient privacy and security from malicious attacks. The attackers' main goal, on the other hand, is to attack SHS devices to steal information, to attack devices to cause a denial of service (DoS) syndrome, or to shut down certain applications that track patients' conditions. Eavesdropping, which violates the patient's privacy, integrity breaches, which tamper with the message, and availability issues, such as battery draining attacks, are other types of medical device attacks (Yaacoub et al., 2020).

Most network security risks to personal clinical data protection, anonymity, and safety are:

Patient data when transmitted over transmission medium may be attacked and altered by unauthorized parties which may result in loss of patient data and privacy. The absence of adequate secure authentication of the devices in SHS may lead to other security threats like suspicious assaults. An aggressive active intruder sometimes can conduct DoS activities. If SHS computers are connected to various networks, authentication, accessibility, secrecy, and credibility are jeopardized if a proper protection protocol is not in operation. Since certain SHS systems lack a sufficient authentication protocol for wireless communication, the information contained in the computer can be easily accessible by third parties attackers. Today, smart healthcare is in its infancy and surrounded by numerous security and privacy attacks as a result of the increased usage of smart healthcare facilities.

SHS services and devices are susceptible to a range of safety risks, necessitating the introduction of suitable safety and privacy policies in smart healthcare to guarantee privacy, verification, access management, and honesty, among other aspects (Hussien, Yasin, Udzir, Zaidan, & Zaidan, 2019). These attacks can make it more difficult to embrace and deploy. As a result, smart healthcare infrastructure must be strategically planned to integrate security and privacy technologies to prevent such attacks (Stellios, Kotzanikolaou, Psarakis, Alcaraz, & Lopez, 2018). Regrettably, the majority of SHS and software are not configured to withstand security and privacy threats, which leads to a spike in network security and privacy concerns like encryption, validation, data reliability, access regulator, and confidentiality (Razzaq et al., 2017). Medical sensors in SHS sense the patient's body symptoms and transmit messages to the specialist or medical network. However, the sensors can be targeted when transmitting these messages (Patel, Sinha, Raj, Prasad, & Nath, 2018). An adversary might, for example, intercept data from wireless networks and alter the data. The data that was manipulated could be forwarded to the doctor or the system later. This could endanger the patients' lives. As a consequence, when planning the use of technologies in the healthcare environment, confidentiality should be a top priority to minimize any threats to patient privacy (Esmaeilzadeh, 2020).

Some patients' vital information must be processed, used, and treated with care, particularly if the patient has a socially stigmatized disease. Failure to properly handle this kind of patient's well-being details may result in embarrassment, incorrect care, relationship problems, or even job loss. Bad health details may also make it difficult for a person to receive health insurance coverage. As a result, it is critical to attest that the protection and secrecy of these data are maintained and transmitted safely. Vital information and data of patients should be handled with caution when processed, and used, particularly if the patient has a publicly stigmatized illness (Rivera-Segarra, Varas-Díaz, & Santos-Figueroa, 2019). Failure to properly handle this sort of patient's health records may result in embarrassment, incorrect care, marital problems, or even job loss. Bad health records may also make it difficult for a person to access health care coverage. As a result, it is critical to ensure that the reliability and safety of these data are maintained and transmitted safely (El Zouka & Hosni, 2019).

11.3 Security and privacy threats in smart healthcare system

The safety of healthcare records is a critical threat to healthcare professionals and governments all the world over. As part of the SHS, the health information of patients is formatted into electronic form, however, there is a risk that this information may be hacked or stolen entirely. For healthcare

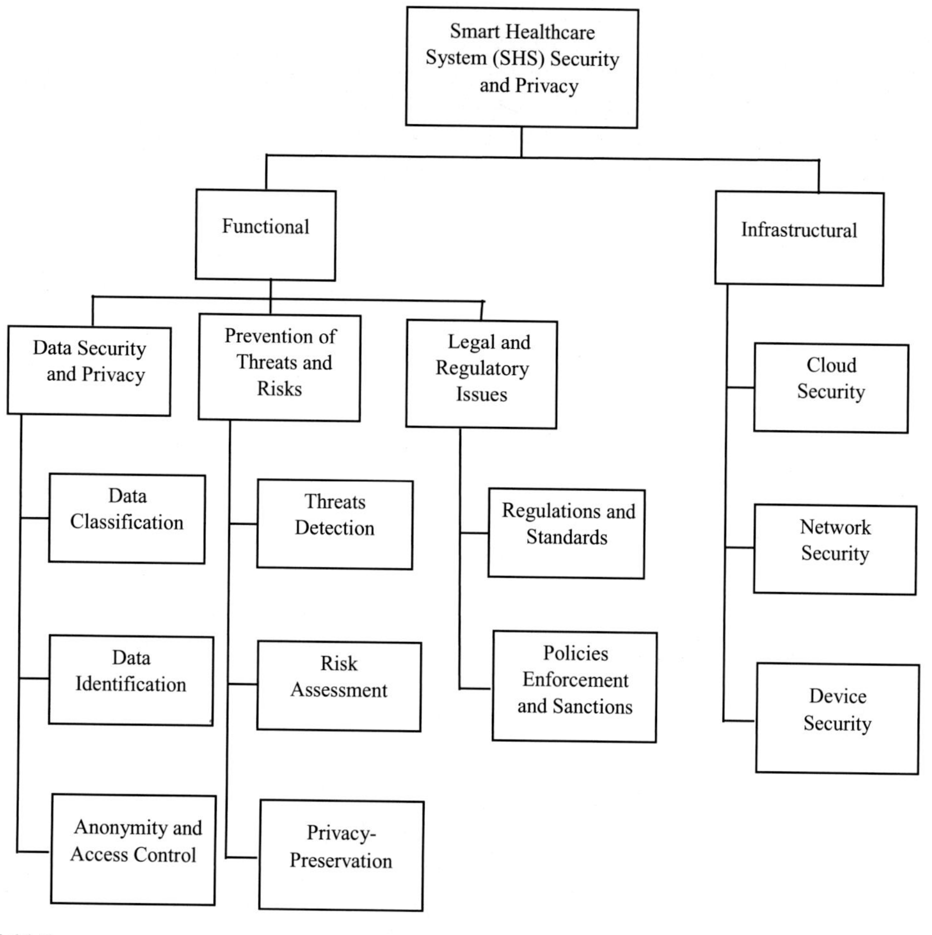

FIGURE 11.2

The security and privacy in smart healthcare system.

executives around the world, cyber security is a top priority. However, some concerns go beyond conventional cyber threats and can result in severe security breaches. Fig. 11.2 displayed the safety and confidentiality concerns in Smart Medical System.

The following are some examples of malicious threats:

11.3.1 Mode of distribution

Threats to hospital distribution can never be ignored. A data breach is possible due to the way each provider communicates with one another. Supply chains must be security reviewed at every level, from vendor transactions to pharmaceutical shipments. Other threats to healthcare security, in addition to cyber threats, can be equally serious. Employees of healthcare facilities may be involved in security breaches. If workers make errors or steal patient data on purpose, there must be a policy in place to avoid data loss. Threats to health facilities can be generally categorized based on attackers and the devices.

In various security threats to healthcare, attackers attempt to steal patient information, deny system services, or upgrade patient data for financial benefit. In the smart healthcare world, routing attacks and location-based attacks are the two major types of attacks. Attacks on routers, select and forwarding attacks, and replay attacks are some examples of routing attacks. While denial-of-service attacks, fingers and technique eavesdropping threats, and sensor-based snooping attacks on the other hand are some examples of location-based attacks. In most routing attacks, intruders aim for the data route to submit or deny data packets. In comparison, in a location-based invasion, attackers threaten the endpoint node by denying the system's resources. Managing personal information and data from individual medical providers is often a difficult challenge.

11.3.2 Mobile devices for health services

Mobile devices with safe credential storage, greater storage capacity, wireless networking interfaces, and computing resources can now be used in healthcare for not only collecting critical health parameters, as in Body Area Networks but also for healthcare management. The importance of privacy and protection in health care cannot be overstated (Avancha, Baxi, & Kotz, 2012). Healthcare facilities and clinical practices as more health and fitness programs become accessible, organizations must be aware of the risk of security vulnerabilities and healthcare information manipulation and procedures become accessible on mobile devices (Els & Cilliers, 2018). Patients and tourists, as well as doctors, nurses, and hospital personnel, use tablets, and handheld devices. This raises the risk of security breaches on both sides of the patient-care equation. An intruder will install sophisticated malware in smartphones, which will stay inactive till the person holding the malware-infected computer reaches a certain area, at which point the malware will be enabled. As a result, an adversary can use malware to weaken a hospital's reputation by triggering it whenever operators of malware-infected machines use affected systems visit the facility.

The dangers of IoT-based medical systems are growing, and any disruption or abuse might result in substantial financial loss or even life-threatening complications. Weak authentication mechanisms may enable an illegitimate attacker to obtain entry to sensitive information as well as shut down all hospital systems. As a result, ensuring the protection of patients, connected devices, and hospital networks, as well as making such an operating ecosystem immune to such attacks, is important (Abouzakhar, Jones, & Angelopoulou, 2017). EMRs are private and confidential to patients, the Health Insurance Portability and Transparency Act states (HIPAA). An intruder can use smartphones to exploit the available network to obtain illegal entry to EMRs, thus violating privacy and security.

A man-in-the-middle attack on the communication network will allow an attacker to steal a client's medical record. This enables the intruder to retrieve plain-text data from internet activity and modify a message instantaneously. The antagonist may also receive EMRs from malicious software portable apps used by patients. Publicly posting the EMR on the Network will be major damage to patients' secrecy, particularly for individuals who do not want their health challenges revealed. The method of making things out of thin air is known as reverse technology. An attacker can use malicious software on mobile devices to interact with medical equipment and feed inaccurate information using the medical devices' application layer. Erroneous results on control systems can lead to a physician making a wrong decision, which can have serious consequences for the person's condition (Imane, Tomader, & Nabil, 2018).

When it comes to medical system availability, the susceptibility of Wi-Fi/Bluetooth infrastructure to DoS attacks is a significant issue. An attacker will mount a DoS assault on Wi-Fi-enabled medical equipment if both devices are attached to the same Wi-Fi network. An intruder will install sophisticated malware in smart devices, which will stay inactive until the person holding the malware-infected computer reaches a certain area, at which point the malware will be enabled. As a result, an adversary can use malware to weaken a hospital's credibility by activating it if holders of malicious software devices use infected devices to visit the facility (Imane et al., 2018).

11.3.3 Unintentional misconduct

Healthcare protection is not always jeopardized by an unethical individual out to damage others. Unintended personnel behavior that resulted in a breach of patient data protection accounted for 12% of security incidents in the healthcare industry. These errors can range from misplacing a patient's chart to a security device that isn't working properly. They can also occur when old machines are discarded with patient data (Watson & Payne, 2020). Hackers, network intruders, former employees, or others can steal or access information, interrupt operations, or cause system harm. This is a pure technological hazard in which an intruder gains access to a healthcare system from an external network and steals patient records. As a consequence, it is an unsolved problem on the horizon (National Research Council, 1997). During an emergency, healthcare institutions encourage doctors to circumvent access authorization, a practice known as the break-the-glass (BTG) method. In BTG cases, the healthcare system's usual work cycle is disrupted. This BTG approach allows doctors or other staff members to exploit or reveal sensitive details about patients without their permission. To safeguard against malicious or unintentional information misuse, healthcare providers keep records. The new strategy is preventative and ineffective in BTG situations (National Research Council, 1997; Wickramage, Fidge, Sahama, & Wong, 2017). Patient-centered information accountability and transparency are needed. The patient should be able to identify possible vulnerabilities to their confidential data and incident responses by revoking permission on their health data by keeping a device activity log (Jayabalan & O'Daniel, 2017). If a doctor or an internal user violates a patient's confidentiality, the doctor's or internal user's privileges should be revoked (Wickramage et al., 2017). Even if his previous position satisfies the access policy, a revoked person cannot access future electronic healthcare records (Chenthara, Ahmed, Wang, & Whittaker, 2019).

11.3.3.1 Insider abuse

Insider abuse was responsible for 15% of security breaches in the healthcare sector in 2013 (Chernyshev, Zeadally, & Baig, 2019). This term applies to situations in which employees of a company steal property or data, or engage in other criminal activity. Surprisingly, the number of people who get work in the healthcare industry solely for the intent of infiltrating the system and gaining access to patient health information makes insider misuse stand out. They usually steal this information to gain access to funds or commit tax fraud. Insider danger is becoming more and more of a concern for companies. Insiders' in-depth knowledge of security procedures and monitoring policies puts companies in perilous positions if these attacks are carried out. As a result, identifying insiders is a major task that has piqued researchers' interest for more than a decade. The allowed sensor nodes' authentication could be breached, alternatively, the perpetrator might snatch a token or details from the networks and launch an assault on the whole system.

Recognizing elements in attacks (Cappelli, Moore, & Trzeciak, 2012), recognizing behavioral causes (Maasberg, Zhang, Ko, Miller, & Beebe, 2020), and detecting anomalies suggestive of unusual and malicious insider behavior have all been addressed in detail (Cotenescu & Eftimie, 2017; Glancy, Biros, Liang, & Luse, 2020). The insider attacks were categorized into four parts in an effort by the CMUCERT Insider Threat project a pioneering assess insider attacks (Cappelli et al., 2012), namely financial and data frauds, espionage, sabotage, and intellectual property (IP) theft. The research identified and described critical paths using a framework called System Dynamics that the majority of insiders stick to in a sequence of isolating suspicious behavior and MERIT (Management and Information on the dangers of security breaches) replicas. Also, the authors differentiate between insiders who unwittingly encourage an assault or reveal the SHS to needless harm, as well as stakeholders who behave maliciously by breaching protocols (to enable their everyday activities) or becoming reckless (targets of phishing assaults) (Yuan & Wu, 2021).

Insider threats arise when employees inside an SHS exploit their privileged access to compromise the protection, credibility, or availability of the SHS's systems (Nurse et al., 2014). The seriousness of the insider threat is well understood, as shown by many real-world examples and comprehensive studies (Cappelli et al., 2012; Nurse et al., 2014; Sarkar, 2010). We assume that in an era of SHS, where all is a device that can connect, preserve, and exchange sensitive business data threats would become substantially more difficult for SHSs to manage; this is a view held by many others (Nurse, Erola, Agrafiotis, Goldsmith, & Creese, 2015). In certain cases, since standard perimeters are becoming so ambiguous, it makes no sense to allow these gadgets to be "insiders" to recognize anything as having the potential for permitted entry. Because of this, it's important to understand how to handle the danger posed by insiders in SHS environments. Regrettably, there has been no thorough study of this danger to date.

11.3.4 Data integrity attack

An attacker can tamper with a patient's data, deceive the recipients more by inserting incorrect patient information, and then submit the erroneous information in a Data Integrity attack. These threats attacks may result in erroneous treatment, erroneous patient status, and erroneous emergency calls to specific people. The alteration of data has the potential to result in the death of a patient. DoS attacks are common on each layer of the network and can be conducted in a variety of ways.

Data integrity is among the most critical security issues in SHS because it influences both data storage and data transmission. Data is continuously transmitted in the SHS, with some of it being deposited and exchanged third-party vendors are used to offering utilities to consumers. Data confidentiality must be maintained throughout its lifespan. Security flaws may arise as a result of multiple service access interfaces. Attackers can amend or delete data deposited in the schemes. Malicious applications, for instance, may be inserted into the system, causing data loss. To ensure data privacy, smart city systems must reduce this risk. The data lifecycle in SHS which include several processes and data which does not follow the applicable requirements should be rejected using acceptable methods. Since SHS data is vigorous in design and large in size, data reliability is a major issue (Altulyan, Yao, Kanhere, Wang, & Huang, 2020).

11.3.5 Denial of service attack

In a DoS attack, the intruder floods the system's data exchange with unidentified traffic, making services inaccessible to others while other nodes would be unable to transmit data until the busy channel is detected (Abdelrahman et al., 2021). The attacker usually exploits the action by tempering a quota of the flags in control ledges in a DoS attack. Since nodes in the IEEE 802.11 standard do not counter-check all it is hard to trace such an attack because of the labels in control frames. Without certification or approval to view data, patient data may be obtained in a DoS attack (Butt, Jamal, Azad, Ali, & Safa, 2019). The DoS attack often keeps the data channel in the device busy, preventing any other data from reaching any other sensor in the network. DoS attacks cause data communication among networks that will be missing or unavailable. This form of attack jeopardizes the accessibility of systems or healthcare facilities, as well as network functionality and sensor obligations.

DoS attacks on IoT networks are the most frequent and easier to enforce. They can be seen in a variety of ways and are described as an intrusion that can jeopardize the network's or systems' ability to accomplish projected purposes. Since its inception, the IoT has been extensively chastised for the lack of consideration paid to safety issues in the design and deployment of its hardware, applications, and infrastructure elements. This negligent approach has led to numerous vulnerabilities that have already been successfully exploited by hackers and cybercriminals to compromise IoT elements so that they can be misused for different purposes, including the staging of DoS and DoS threats (Ayo, Folorunso, Abayomi-Alli, Adekunle, & Awotunde, 2020). DoS attacks and DoS (DDoS) sharing allow network facilities/data inaccessible to users. When DoS attacks have been compromised by various nodes are reported, it is known as a DDoS attack. Many IoT devices are designed from low-cost generic hardware parts, considering their often complex and economically appealing external nature. These processors and software typically have built-in security issues, making them impossible for owners and managers to monitor. Furthermore, the facilities and teamwork of firmware and software updates for the wireless problem are still at a primitive level. So, it's also hard to update or repair these unsecured IoT computers.

11.3.6 Fingerprint and timing-based snooping

When data is transmitted from sensor to sensor, data is encrypted, and the analyzing data is transferred from sensor to sensor or sensor to private site, wireless networks are unable to identify vulnerabilities. This physical layer intrusion requires just time slots and a packet signature. A signature is a collection of Radio Occurrence waveform features distinctive to a specific transmitter. An attacker listens quietly to all sensor data transmissions with timestamps and fingerprints in this assault. After seeing this, the intruder attaches each message to a particular transmitter using the fingerprint and uses different periods of inference for each sensor place. When an invader obtains this data, he or she can interrupt health conditions (Banda, Bommakanti, & Mohan, 2016; Reinbrecht, Susin, Bossuet, Sigl, & Sepúlveda, 2017).

Fingerprint and timing-based snooping (FATS) is a powerful attack that enables attackers to intercept radio devices produced by home automation from outside the residence while maintaining undetected, as is the essence of all passive attacks (Chow, Xu, & He, 2014). After the algorithm has been educated enough, attackers will be able to see the names of smart devices and their locations. After that, attackers will be able to see the everyday activities of the occupants of the house. The FATS assault targets remote signatures as well as actual surveillance of home automation interactions (Srinivasan, Stankovic, & Whitehouse, 2008; Xu, Zheng, Saad, & Han, 2015). Wireless biometrics is a physical characteristic of radio frequency-based interactions that allows radios to be distinguished. This is also valid if the machines are of the same type and made by the same maker (Xu et al., 2015). Moreover, the attacker algorithm compares the timestamps of each caught signal to determine similarities. Time intervals will also be used to calculate the distance across each computing phone and the RF signal sniffer used by the intruder.

11.3.7 Router attack

Since it allows for the delivery of intelligence over the internet and facilitates linkage mobility in large hospitals, data routing is critical for healthcare-based systems. Routing, however, poses a few challenges, owing to the transparent existence of wireless networks. The intruder targets the information sent between sensors in some wireless sensor nodes in this invasion. This is because the most critical requirement of a wireless health care system is the safe transmission of medical records to the intended recipient, who may be a physician or a specialist. There are very few implementations that make use of multi-trust guiding in this attack steering of basic and crucial details displaying patients' daily care ranking. Multi-trust guiding is important for expanding the system's incorporation district and, as a result, offering stability at the cost of complexity.

Routers play a crucial role in network communications by facilitating the exchange of data. Router attacks can take advantage of protocol bugs, router software anomalies, and poor authentication. Distributed DoS and brute force attacks are two forms of attacks that can occur. Attacks affect network services and business processes as they are happening. For a link request between computers and servers, the TCP protocol uses synchronization packets known as TCP/SYN packets. The source machine When an SYN flood attack happens, it sends a huge number of TCP/SYN packets with a forged URL. Since the path is inaccessible, the channel's destination node is unable to establish a connection to the root. If a router seems unable to verify a TCP message, resources will quickly become exhausted. Since the scope of the attack will consume the router's resources, this is a form of DoS.

When a hacker attempts to guess the password and gain access to a router, it is known as a brute force attack. To break the password, the intruder will use software with a dictionary of words. If the password is relatively weak, the attack could take a short time depending on the strength of the password and the combinations used to find a match. This form of attack isn't only confined to business routers; it can also happen at home if a hacker is within range of the router. A disgruntled employee with access to the network topology, router login, password information, and knowledge of the network topology can gain unauthorized access to routers and compromise the network. Passwords should be updated regularly, and strict access controls should be enforced to avoid this situation. To reduce their vulnerability to attacks, routers must have stable and up-to-date software with strong configurations.

11.3.8 Select forwarding attack

In this attack, the attacker gains entrance to one or more sensors to carry out the attack. As a consequence, this form of forwarding is referred to as community-oriented specific forwarding. When an attacker gains access to a sensor in this attack, it drops data packets and sends them to neighboring sensors to raise suspicion. This attack has a major impact on the device, particularly if the sensor is close to the base location. As the effect of the packet loss caused by the SF attack, it can be difficult to pinpoint the source of packet loss. The receiver receives incomplete data, thus, the attack is highly dangerous to any patient or smart medical health system.

A selective forwarding assault on wireless communication has a significant effect on network efficiency and consumes a significant amount of energy. Previous countermeasures presumed that the attacker's misconduct could be detected by all peers inside the communication range. Although, whereas smart networks need a minimum signal-to-noise ratio to properly acquire frames, and because node intrusion is inevitable in densely scattered wireless sensor networks, previous approaches have struggled to accurately detect misbehaviors. In a selective forwarding assault (SFA), an intruder acts as a normal node in the transmission period and dismisses traffic from adjacent nodes selectively (Zhang & Zhang, 2019). Non-critical data may be forwarded normally, but vital data, such as information caused by an adversary in a military application, may be discarded. It will do significant harm to WSN because it is unethical to lose confidential information in a monitoring application (Liu, Dong, Ota, & Long, 2015). In the field of WSN defense, detecting and isolating SFA is a significant research subject.

11.3.9 Sensor attack

Sensors often left or joined the network due to unintended sensor malfunction in the suspicious behavior on a cellular network conducted by outer attackers. The sensor may be drained and turned off in the sensor control necessitates the use of cellular network constraints. In this case, an attacker may simply replace the sensor in the network with a malicious one, and then carry out malicious activities with ease. As a result, if the patient data is not well distributed through several sensors, the hacker can change the data as much as he wants. False data can be inserted or served as a result of the absence of an authorization format as legitimate.

11.3.10 Replay attack

When a reverse attack can occur if an intruder gains unapproved access to a computer. When the sender stops transmitting data, the attacker tests the system's operations and gives a signal to the recipient. At that point, the attacker takes over as the first source. The main aim of the attacker in these attacks is to create network assurance. The attacker sends the receiver a notification that is mainly used in the validation process. A replay attack is identified as a security violation in which data is processed without permission and then rerouted to the recipient to try to lure the latter into doing something illegal, such as mistaken identification or authentication, or a duplicate operation. Any threat has an impact on the system in any way. Unlawful access, data modification, refusal of continuous surveillance, data target route adjustment, and data reduction are the most significant effects on a health monitoring system.

An unlawful user gains access to the Smart Health system, collects network traffic, and sends the message to the recipient as the original sender in this form of attack (Rughoobur & Nagowah, 2017). The attacker wants to gain the system's confidence. A replay attack is a security breach in which any data is stored without permission and then resent to the recipient. This assault can harm a SHS by gaining unauthorized access and then stealing sensitive patient information.

11.3.10.1 Security problem in radio frequency identification

Radio frequency identification's (RFID) mechanism isn't without defects. It can be used for a variety of purposes. RFID faces numerous security threats and challenges, all of which must be addressed and implemented in WHD (wearable health care devices) Maintaining confidentiality and key management:

Radio frequency identification (RFID) is one of the most commonly used tools for automatically identifying objects or individuals. RFID application is widely used in many areas, including distribution networks, engineering, and transportation control systems, and is based on a combination of tags and readers. Despite its many benefits, however, the technology poses many challenges and concerns that aren't attracting more researchers, especially security and privacy concerns. Like other electronics and networks, RFID systems are vulnerable to both physical and electronic attacks. Hackers who want to steal private information, gain access to protected places, or bring a system down for personal gain are becoming more popular as technology progresses and becomes more prevalent. Whenever an illegitimate RFID reader listens in on communications between a label and a reader and acquires classified information, this is referred to as spying. For this procedure to succeed, the hacker must also understand the basic protocols, tags, and reader information.

A hacker's brain and a cell phone are all that are required for the assault on force research. By monitoring the energy usage levels of RFID tags, power analysis assaults can be mounted on RFID devices, according to leading experts. Researchers uncovered this intrusion method when researching the power pollution levels of smart cards, namely the disparity in supply voltages between valid passwords and a wrong password. RFID tags and readers, like most products, can be reverse-engineered; however, performance will necessitate a detailed understanding of the protocols and features. To accept files from the IC, attackers will disassemble the chip to figure out how it operates.

A man-in-the-middle attack happens during signal transmission. The attacker waits for communication between a label and a user before intercepting and manipulating the data, similar to eavesdropping. The attacker intercepts the novel indication and then directs the wrong data while posing as a regular RFID component. A DoS attack refers to any malfunction of an RFID device that is linked to an attack. Physical attacks, such as using noise interference to jam the device, hindering radio indications, or even deleting or deactivating RFID labels, are common.

Cloning and hacking are essentially two distinct practices that are often carried out concurrently. Cloning is the method of copying data from an original tag and using the mutated tag to obtain access to a secured region or entity. This type of attack has been used in access control or inventory management operations since the intruder must know the details on the label to duplicate it. RFID tags may not presently have enough storage capacity to house viruses; nevertheless, viruses can pose a significant threat to an RFID system in the future. When an RFID tag is read at a plant, a virus programmed on the tag by an unknown source has the potential to bring the RFID device to a halt. The virus spreads from the sticker to the reader and then to corporate servers and applications, causing linked devices, RFID modules, and networks to go down.

11.3.11 Distributed denial of service attacks

Distributed denial of service attacks (DDoS) attacks have become more harmful since the launch of non-legacy IoT devices. Attackers may now take advantage of IoT devices' weak security implementation to takeover them and employ them to launch an assault on the anticipated system or network. It has been observed that as the cost of installing more IoT devices increases, so does the number of attacks. The primary objective of a DDoS assault is to refute legitimate users' access to channel and latency facilities, culminating in a service disruption (Bhardwaj, Mangat, Vig, Halder, & Conti, 2021; Bhati, Bouras, Qidwai, & Belhi, 2020; Huseinović, Mrdović, Bicakci, & Uludag, 2020). The intruder starts with non-legacy IoT systems, such as CCTV cameras, camcorders, baby tracking devices, and wearable devices, which have inadequate built-in security and other drawbacks, such as poor computing power and energy density.

IoT systems are not always easy to hack, but also inexpensive. Instead of spending in and maintaining expensive networks to launch strong DDoS attacks, attackers can gain possession of compromised IoT devices for free or at a fraction of the cost of running a server. Companies do not keep records of a device's security credentials until it hits the market. Several authentication bugs in the code are exploited by hackers. Security patches for these devices that correct the defective software are not released by the manufacturers. If compromised IoT devices have been taken control of, the attacker is free to change the device's security credentials. If the infected computer is ever tracked for the period of assault, the vendor or manufacturer of the device will be unable to retune the safety permits and regain regulator from the invader. The invader intends to use the system to inflict as much harm on the target as possible for as long as possible.

A DDoS outbreak is an effort to bring a targeted server down partially or entirely by flooding it with internet traffic. The primary goal of this attack is to interrupt the victim's server or network's normal traffic flow. DDoS attacks are parametric assaults wherein non-legacy IoT computers with weak security, such as video cameras, baby monitors, and printers, are exploited to form a botnet. High volumes of transmission from compromised IoT devices are redirected to servers, causing normal services to be disrupted (Salim, Rathore, & Park, 2019). DDoS assaults on e-medical servers in the IoT will jeopardize actual patient tracking as well as the overall efficiency of e-health services. This paper has analyzed strategies for DDoS attacks in IoT, as well as a secure solution for protecting servers against such assaults has been proposed (ul Sami, Asif, Bin Ahmad, & Ullah, 2018).

The security of patient data is highly significant and must be preserved to demonstrate the system's reliability (Awotunde, Chakraborty, & Adeniyi, 2021b). Servers in e-health systems are extremely important because they allow for real-time monitoring of patients. Patients' health monitoring may be jeopardized if these servers go offline for a short period of time or further. With the expansion of networks, there has been a rise in connection flaws, and also the execution of complex attacks on critical infrastructure by trained hackers. E-medical is a sensitive and difficult environment in which network connectivity is critical, and if the network is subjected to a major attack such as DDoS attacks, as well as life-saving procedures for patients, can be challenging. As a consequence, a protocol to guarantee stability and prevent/detect such assaults has been developed (Ogundokun, Awotunde, Sadiku et al., 2021).

11.4 Security and privacy solution in smart healthcare system

Certain e-health systems now have as a consequence, more efficient communications and computational capability of the growing usage of IoT technologies in the healthcare system (Awotunde, Abiodun, Adeniyi, Folorunso, & Jimoh, 2021). As a consequence, by opening more interactive channels, connected objects will jeopardize device security and personal privacy. According to Demirkan (2013), a smart health service (SHS) should provide incentives for a medical company to improve innovations with fewer hazards and enhanced historical knowledge, converging electronic medical records (EMRs), cloud services, media platforms, sophisticated equipment, and research instrument, and converging EMRs, advanced sensors, and data The SHS technologies can generate value for taxpayers, service professionals, and analysts by tracking, evaluating, and processing healthcare data everywhere, anywhere. For instance, aged people may be able to access medical facilities at home (Amrutha et al., 2017).

Registered structure patient data centers for data storage and dissemination, people can gain insight into and choose whether to exchange their clinical signals with clinicians for analysis of diseases (Prakash & Ganesh, 2019). In an emergency, smart health systems are particularly helpful due to their compact nature (Ambhati, Kota, Chaudhari, & Jain, 2017). A diabetic patient, for instance, can pass out unexpectedly at work. Ambulance staff also request his or her medical records in this environment. Because of smartphone apps that track patients' diet, workout, rest, and blood glucose levels, it has become much simple to study basic medical conditions. Both computational models raise a host of security concerns that must be addressed with caution when exchanging data online. Smart health, as a subfield of smart cities, benefits from the same set of technology though relying heavily on access to health data. As a consequence, all functionality should be considered when developing security and privacy-preserving (PP) solutions. This necessitates that all of the parties concerned be truly linked to intelligent healthcare facilities.

Society benefits from both responsible clinical research and confidentiality protection. Health research is important for the advancement of human healthcare services. For experiments to be acceptable, patients who participate in them must be shielded from harm and their needs must be upheld. The aim of protecting people's information is to protect the rights of individuals. The primary goal of collecting personally identified health records for health research, on the other side, is to aid the community. Conversely, it is important to highlight that safety is important for society because it allows complex activities, like research and public health, to be conducted in ways that protect citizens' dignity. Individuals can benefit from health research because it provides access to novel therapies, improved diagnostics, and more effective approaches to disease prevention and care, for example.

While the healthcare sector moves toward increased use of electronic medical reports, data protection will become increasingly necessary, and Congress has already introduced a variety of bills to ease and control the transition. In the future, improvements in information security would certainly make audit trails and access restrictions quicker to implement. Although the panel does not recommend a clear technical solution, at least four technological approaches to enhancing information privacy and protection that can be particularly effective in health care have been proposed by others: (1) data mining with privacy protection and statistical transparency limitation, (2) personal electronic medical record devices, (3) unbiased consent management tools, and (4) pseudonymization Each aims to minimize or eliminate the transmission of publicly identifiable information.

11.4.1 Biometrics

Identifying patients based on biologically specific characteristics (face, fingerprint, iris, voice) means that the right people get the right treatment, making global healthcare safer and more reliable. Patient misidentification has a detrimental effect on care quality, patient safety, and well-being: in the United States, the use of Social Security numbers does not solve the issue. SSNs can be entered wrongly, stolen, and fraudulently used, and their use for patient identification is not standardized under US law; as of 2018, there is no national standard for patient identification (Lindgren, 2019).

Exchanged, stolen, misplaced, forged, mistyped, and duplicated information are all possibilities. A biometric identification, such as a face or a fingerprint, is unique to each person. Biometrics can also be used in cases where patients are unable to provide any sort of physical identification, such as in hospitals or emergency rooms. Patients are, in essence, their ID. In the healthcare industry, data protection is important, particularly as high-profile hacks become more popular. Medical records are now a target, posing a danger to patients and providers alike. Biometrics assist in the security of medical records, but are they necessary in the healthcare industry? Since biometrics are one-of-a-kind, they're perfect for patient identification.

Healthcare facilities may be able to authenticate a person's identity based on a scan of their face or the sound of their speech. Not only does this make the operation more effective for all sides, but it also has co-location advantages. Patients' digital information is transferred with them as they move from one facility to another. Biometrics also helps to deter theft by making it more difficult for outsiders to access data from a device when they reach a facility. However, there are some disadvantages to this new technology. The introduction of biometrics into healthcare facilities is accompanied by the purchase of costly hardware.

11.4.2 TinySec

TinySec is a solution used in WBAN (Karlof, Sastry, & Wagner, 2004) to accomplish data cryptography and authorization at the connection layer. It is widely used in Wireless Sensor Networks and is included in the official TinyOS distribution. TinySec creates secure packets by encrypting message data and computing a MAC for the entire packets, including the header, using a cluster common key between sensor nodes. It is based on a single key that has been designed into each node before distribution, so a network-wide public key serves as a security baseline. The TinySec protocol has the downside of not being able to defend against node capture. The technique ensures.

11.4.3 ZigBee services security

The technique was established by a group of industry players to offer ultra-low-power wireless networking with innovative logic. In comparison to IEEE802.15.4, the ZigBee network layer specifies auxiliary security services such as authentication and key-exchange procedures. This allows nodes to access the network and allocate keys, with the help of a trust center, defined by the ZigBee protocol, which provides certain coordinator roles (Porambage, Okwuibe, Liyanage, Ylianttila, & Taleb, 2018). The technique enhances access to the SHS network and protects the data across the network.

11.4.4 Bluetooth protocols security

Logical link control and adaptation protocol (LLCAP), link manager protocol (LMP), and Baseband are among the protocols used in the technique. The LLCAP is in charge of higher-level multiplexing support and packet reassembly, which will assist with the communication quality of operation. LMP, on the other hand, is in charge of security matters such as encryption, authentication, and key exchange. The baseband establishes a connection between Bluetooth devices which allows data to be exchanged in packets (Lounis & Zulkernine, 2018). This protects data transmission across the SHS network by authenticating the devices transmitting the data through the channels.

11.4.5 Elliptic curve cryptography

In wireless networks, this technique is seen as a feasible alternative for public-key cryptography (Abikoye, Ojo, Awotunde, & Ogundokun, 2020). Elliptic Curve Cryptography (ECC) is mostly used for its properties of high computing, limited key size, and lightweight signatures (Lara-Nino, Diaz-Perez, & Morales-Sandoval, 2020). Apart from the fact that the energy needs remain high, the technique offers an option for high wireless device protection.

11.4.6 Encryption techniques

This technique provides the appropriate protection for SHS through the encryption of the network data with different keys with a high degree of protection across three separate mechanisms that make for successful encryption: IBE, symmetric key encryption, and conventional public-key encryption (Bhatia & Verma, 2017). This will help in the protection of SHS from the healthcare devices to the cloud storage server, only the authorized users will have the key to decrypt the data and information involved (Awotunde, Ogundokun, Jimoh, Misra, & Aro, 2021).

11.4.7 Hardware encryption

A ZigBee-ready compatible RF transceiver is used to implement hardware cryptography, rather than software-based encryption, as previously mentioned. Using 128-bit keys, the transceiver carries out IEEE 802.15.4 authentication activities encrypting data with AES. The operations make use of a decryption and encryption mode known as counter CTR (Abdulraheem, Awotunde, Jimoh, & Oladipo, 2021; Braham, Butakov, & Ruhl, 2018). The method can be used to protect the sensors and devices used in capturing SHS data.

The storage and transfer of data can be secured and safeguarded in data security to ensure the data's integrity, validity, and most crucially, authenticity (AbdulRaheem et al., 2021). It also ensures that data may only be read and edited by authorized individuals. Another important goal to consider while constructing an SHS is PP. When shared data is communicated across an open and insecure channel, it primarily accounts for the severity and sensitivity of the data. PP necessitates both content and context. Content privacy protects patient information against data leakage, but attaining patient privacy is difficult since an attacker can identify a patient's health status based on the attending doctor's identity. It's also critical to protect contextual privacy. Contextual privacy refers to the safeguarding of the communication's context. The smart gateways can be installed in any healthcare sensor, allowing for authentication and authorization while reducing sensory overload and meeting all security standards. The system, in particular, relies on the datagram transport layer security (DTLS) handshake protocol, which is widely recognized as a critical component of IoT security. The DTLS is a communications protocol that protects datagram-based applications against eavesdropping, manipulation, and message forgery by allowing them to communicate securely. When compared to the centralized delegation design, the results of the investigation proved that the proposed architecture provided higher security. The system, on the other hand, was not resistant to possible attacks.

11.5 Security and privacy requirements in smart healthcare system

To guarantee the protection and maintain the integrity of a patient's medical records at all stages, smart healthcare schemes involved finite safe-keeping actions. To execute complex safety operations that meet all of these criteria, supporting infrastructure is required (Hartmann, Hashmi, & Imran, 2019). The SHS system's two most critical features are protection as well as the confidentiality of personal records. Data must be protected from unauthorized entry as it is being transported, compiled, handled, and securely stored (Al-Janabi, Al-Shourbaji, Shojafar, & Shamshirband, 2017). Privacy, in contrast, signifies the right to track the storage and use of one's private data. For example, a patient could insist that his information not be exchanged with insurance providers, as this information may be used to deny him insurance (Awotunde et al., 2021c). More precisely, mission-critical data inside a device is highly vulnerable, because if it is exposed to unauthorized employees, it may result in the patient losing their work, an embarrassment in public, and emotional illness, among other things (Sithole, 2019).

As another example, consider an intruder gain access to records by the node physically capturing and altering the data; as a result, incorrect information is forwarded to the practitioner, potentially resulting in the death of a patient. Someone could use the patient's medical history to discover the patient's romantic rivalries. As a consequence, greater care and attention should be provided to protecting this sensitive and essential data from unauthorized access, usage, or alteration. The following are the main security and privacy standards that must be met for the SH system to be secure and widely accepted by its users:

Data confidentiality: The security of sensitive data from disclosure is referred to as data confidentiality, and it is regarded as a critical issue in SHS. Since network nodes used in medical situations are required to send personal and sensitive information about a patient's health, their details must be shielded from unwanted access, which may jeopardize the patient's existence in danger. During transmission, this vital data may be "overheard," causing harm to the patient, the provider, or the device itself. By establishing a mutual key between secured cluster heads and their coordinators on a secure communication channel, encryption will improve the security of this sensitive data.

Data integrity: Data integrity consists of the steps to protect a message's quality, as well as its authenticity and continuity. It applies to both individual messages and messages in a tunnel. Data protection, on the other side, does not safeguard data against outward tampering so information can be tampered with when it travels across an untrustworthy network to an intruder that can quickly regulate the patient details before it enters the network administrator. By mixing several bits, altering data contained inside packets, and then transmitting the message to the PS, adjustments may be made more specific. This interception and alteration can cause severe health problems, and in extreme cases, death. As a consequence, authentication mechanisms must be used to avoid information from being accessed and altered by a possible adversary.

Data freshness: Data cleanliness strategies can efficiently shield data integrity and confidentiality from an adversary capturing and replaying older data, confusing the network coordinator. It makes certain the old data isn't reprocessed, and that the frames are appropriate. There are currently two kinds of data freshness. Strong freshness guarantees both frame ordering and delay, while weak freshness only guarantees frame ordering but no delay. High freshness is used for initialization whenever a signal is sent to the base station, while network nodes with a short service time are used for weak freshness.

Availability of the network: It implies that a therapeutic physician has an easy connection to clinical data. Because such a computer contains critical, extremely sensitive, and potentially life-saving details, the network must protect it and be open to patients at all times in the event of an emergency. In the event of a network failure, it is important to transfer operations to a different network.

Data authentication: Data authentication can be needed in medical and non-medical applications. As a result, nodes in a network should be able to confirm the information is being sent from a trusted source instead of an imposter. As a result, by exchanging an undisclosed key, the network and coordinator nodes for all data measure message authentication code (MAC). The network coordinator will be assured that the message is being carried out by a reliable node if the MAC code is calculated correctly.

Secure management: The decryption and encryption operations involve safe control by the coordinator to provide key distribution to a network. During node association and disassociation, the coordinator's job is to connect and safely remove network nodes.

Dependability: The device must be dependable and trustworthy. Failure to retrieve the correct data is another big problem in the network, as it could put the patient's life in jeopardy. Error-correcting code techniques can be used to solve this problem.

Secure localization: The location of the patient must be accurately estimated in most network applications. Due to a lack of tracking methods, an intruder may send incorrect information, such as a false signal about the patient's location. The authors spoke about localization systems and how they can be hacked.

Accountability: Medical professionals need to safeguard patient privacy records in the medical sector. If a provider fails to safeguard if he or she abuses his or her duty for this knowledge, he or she should be punished and kept responsible to prevent potential abuses. The author addressed the issue of transparency and suggested a method for dealing with it.

Flexibility: The patient must be able to designate which APs monitor medical data within a network. In an emergency, for example, authority to analyze a patient's details may be issued on the spot to a particular doctor who isn't identified as having approval. In other words, if a patient switches hospitals or doctors, the access controls should be transferable.

Privacy rules and compliance requirements: The need to safeguard patient records is a global concern. To safeguard individual health, one of the most significant privacy steps is to create rules/policies on who has permission to change confidential data about the patient. The health-care provisions list a variety of legislation and actions. Currently, diverse sets of privacy regulations/policies exist everywhere in the country. The American Health Insurance Portability and Accountability Act (HIPAA) is a set of guidelines for doctors, clinics, and other healthcare services that are designed to protect an individual's health and medical history safely. HIPAA explains the steps that must be taken to secure patient information as it is used for administrative or correspondence purposes. If a contractor exchanges confidential information for financial gain, the Act introduces civil and criminal sanctions, including a $250,000 fine and/or ten years in prison.

11.6 Practical application of a secure medical data using a TEA encryption algorithm

TEA encryption is almost identical to TEA decryption with reversed feature operation since it is a Feistel structure. The encrypted communications process begins with the cipher texts of RP5 and LP5, which are now referred to as right ciphertext RC and left ciphertext LC. The original 32-bit half RC's output is left-shifted 4-bit and attached to the third K2 subkey. The outcome is stored in your memory as RC1. The initial 32-bit half RC is added to the decimal value 2654435769, which describes the Golden Ratio constant, and the result is stored as RC2. The fourth K3 subkey is then subjected to a 5-bit right-shifted 32-bit half RC, with the outcome stored as RC3. The XOR computation is then performed on RC1, RC2, and RC3, yielding RC4. Furthermore, the RC4 function is applied to the original 32-bit half LC, and the result is saved as RC5. The TEA's round one half-cycle is completed with this operation.

In the second half-loop round of TEA, the original processed value RC5 is left-shifted 4-bit, and the outcome is then substituted with the first subkey K0. The outcome will be stored in memory as LC1. The recorded outcome, RC5, is compounded by the Golden Ratio value, 2654435769, and the outcome is preserved in a memory called LC2. Before being preserved as LC3, the stored value RC5 is right-shifted the next 5-bit and connected to K1. The XOR operation is then performed on LC1, LC2, and LC3, yielding LC4. In addition, the LC4 value is added to the initial 32-bit half RC, and the result is preserved as LC5. The second half-cycle round of TEA comes to an end with this process.

This method completes the entire TEA cycle. To fulfill the requirement of a full TEA, the entire TEA cryptography procedure was repeated thirty-two times. The decoded ciphertext is then compared to the encoding technique's input plaintext to verify they seem to be the same value or document.

$$\mathrm{RC1} = \mathrm{RC} \ll 4 + \mathrm{K2} \tag{11.1}$$

$$\mathrm{RC2} = \mathrm{RC} + \mathrm{delta} \tag{11.2}$$

$$\mathrm{RC3} = \mathrm{RC} \gg 5 + \mathrm{K3} \tag{11.3}$$

$$\mathrm{RC4} = \mathrm{RC1} \oplus \mathrm{RC2} \oplus \mathrm{RC3} \tag{11.4}$$

$$\mathrm{RC5} = \mathrm{RC4LC} \tag{11.5}$$

Equations of the first-round decryption (11.1)–(11.5)

$$LC1 = RC5 \ll 4 + K0 \tag{11.6}$$

$$LC2 = RC5 + \text{delta} \tag{11.7}$$

$$LC3 = RC5 \gg 5 + K1 \tag{11.8}$$

$$LC4 = LC1 \oplus LC2 \oplus LC3 \tag{11.9}$$

$$LC5 = LC4 + RC \tag{11.10}$$

Equations of second-round decryption (11.6)–(11.10)

The raw data may be in any form, like organized, semi-structured, and unstructured data, as well as text and video, audio, or a combination of the three. Data can be sent directly to the TEA encryption algorithm to access the cipher code. Apart from text data and streams of ones and zeros, this data is translated into binaries. The binary stream is sent to the TEA encryption algorithm to access the ciphertext, and the sender and receiver both know the secret key. After getting ciphertext from the source, the TEA converts it into plaintext. The file must be converted to zeros and other streams (binary) before being translated to the source file; otherwise, the sender will provide plaintext. Algorithms 11.1 and 11.2 both show how to encrypt and decrypt the TEA algorithm.

Algorithm 11.1: TEA encryption algorithm.

Encrypt (plaintext p, key k):

1. Begin
2. Assign delta = 9E3779B9
3. Compute k0, k1, k2, and k3 from k
4. slip p into 32-bit bock rp and lp
5. Assign cycle = 0
6. compute rp1 as rp LSHIFT 4 AND k0
7. compute rp2 as rp AND delta
8. compute rp3 as rp RSHIFT 5 AND k1
9. compute rp4 as rp1 XOR rp2 XOR rp3
10. assign rp4 AND lp to rp5
11. compute lp1 as rp5 LSHIFT 4 AND k2
12. compute lp2 as rp5 AND delta
13. compute lp3 as rp5 RSHIFT 5 AND k3
14. compute lp1 XOR lp2 XOR lp3
15. assign lp4 AND rp to lp5
16. cycle = cycle + 1
17. Repeat step 6 through step 16 until cycle = 32

Algorithm 2: TEA decryption algorithm.

Decrypt (cipher c, key k):

1. Begin
2. Assign delta = 9E3779B9
3. Compute k0, k1, k2, k3 from k

4. slip c into 32-bit bock rc and lc
5. Assign cycle = 0
6. compute rc1 as rc LSHIFT 4 AND k2
7. compute rc2 as rc AND delta
8. compute rc3 as rc RSHIFT 5 AND k3
9. compute rc4 as rc1 XOR rc2 XOR rc3
10. assign rc4 AND lc to rc5
11. compute lc1 as rc5 LSHIFT 4 AND k0
12. compute lc2 as rc5 AND delta
13. compute lc3 as rc5 RSHIFT 5 AND k1
14. compute lc1 XOR lc2 XOR lc3
15. assign lc4 AND rc to lc5
16. cycle = cycle + 1
17. Repeat step 6 through step 16 until cycle = 32

The proposed TEA procedure is executed USING Xilinx. Xilinx is a device that allows you to produce and simulate the HDL (Hardware Description Language) design. The encryption results of the TEA model are presented in Table 11.1 and Fig. 11.3 respectively. In all three models, the

Table 11.1 shows the energy consumption of TEA algorithms in relation to their block sizes.

	1	2	4	8	16	32	64	128	254
Eb (pj/bit)	1487.81	894.45	607.38	508.93	513.45	612.34	956.95	1926.09	4095.64

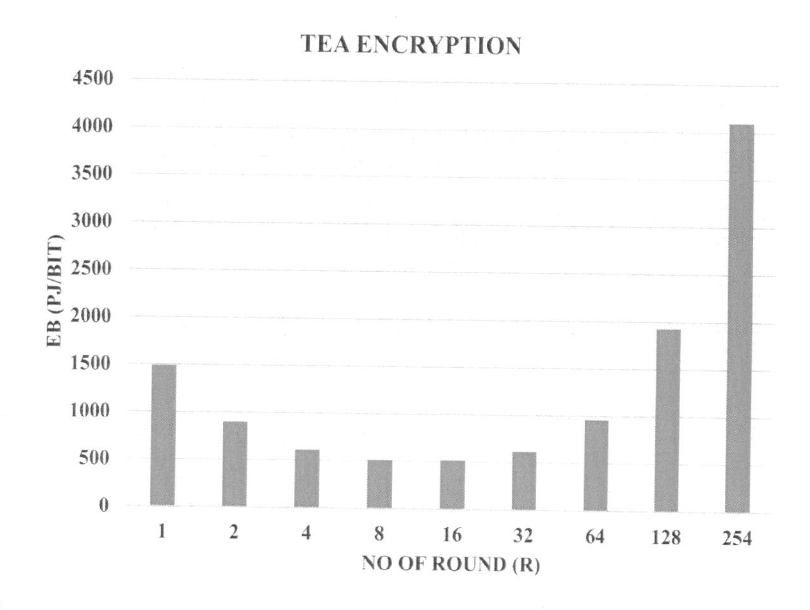

FIGURE 11.3

The energy consumption by different techniques.

minimum energy is consumed at 16 rounds, according to the graphs. Table 11.2 shows the energy consumption of different algorithms in relation to their block sizes.

The energy consumption of various existing lightweight encryption algorithms, as well as the proposed algorithm, is shown in Table 11.2. As compared to other algorithms, TEA uses the least amount of energy (Table 11.3).

The outcomes of this test demonstrate the number of bit shifts that happen in ciphertext once a bit in the key is modified. In checking, the ciphertext is various bits, but the key is the same bit (128-bit), and the key is modified to 5 bits to see the AE in the TEA. According to the test findings, TEA has a cumulative AE of on average 70.75%, and an AE of more than 50% indicates a high-security encryption value (Fig. 11.4).

Table 11.2 Comparison of energy consumed by various algorithms.

Algorithm	Block size (bits)	Energy consumption $\left(\frac{\mu J}{byte}\right)$
AES-128	128	4.31
DES	64	21.8
IDEA	64	9.87
DESXL	64	24.15
KLEIN-80	64	7.0
MCrypton-64	64	21.5
TEA	*64*	*0.138*
TEA	*128*	*2.25*

Table 11.3 The TEA algorithm result based on avalance effect.

Bit chipertext		Changed key bit amount				
		1 bit	2 bit	3 bit	4 bit	5 bit
32	CTB	36	48	64	89	86
	AE	29%	39%	51%	66%	61%
64	CTB	60	84	63	76	70
	AE	67%	92%	59%	95%	78%
128	CTB	123	138	96	81	89
	AE	88%	79%	57%	46%	52%
254	CTB	232	247	205	190	202
	AE	93%	99%	85%	73%	78%
AE Average		69.25%	77.25%	70%	70%	69.75%
(∑AE) Average		70.75%				

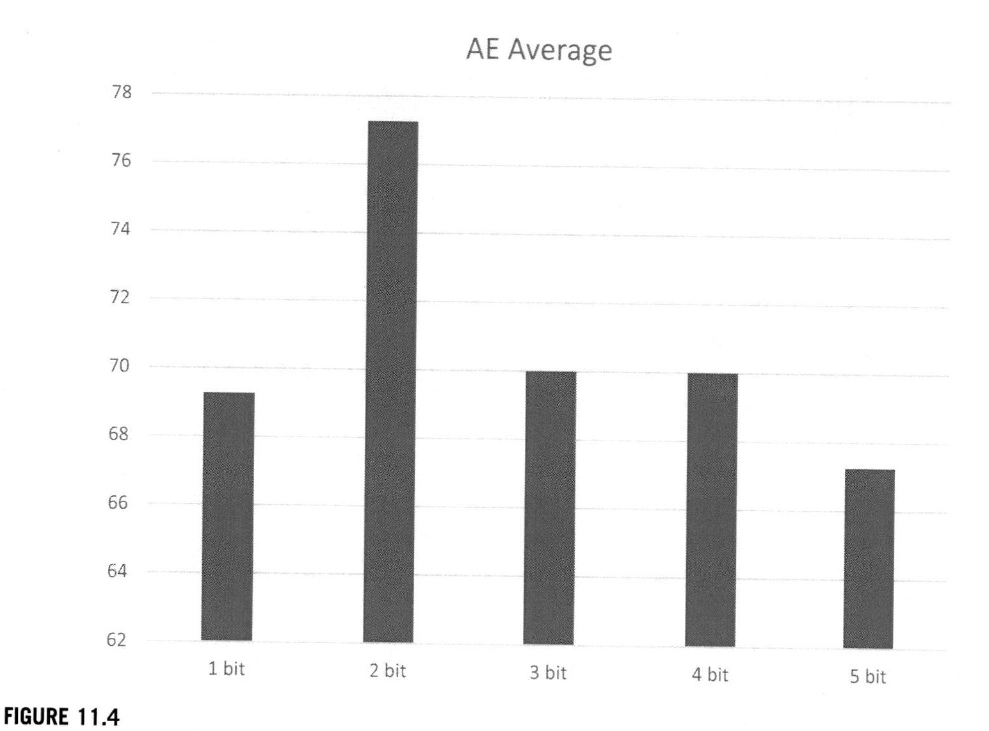

FIGURE 11.4

The TEA algorithm result based on avalance effect.

11.7 Conclusion and future directions

This chapter provides a detailed conversation of the latest technology in Smart Health Service, with a focus on what has been accomplished in the domains of confidentiality, risks, and defense mechanisms It further defines and discusses the user's fundamental protection and privacy issues about SHS specifications and problems. The demand for efficient and the need for cost-effective healthcare options have always been on the rise. There are several feasible technologies to be built for medical applicants with vast resources and increasing study into the smart healthcare domain. Acceptance of SHSs is a gradual yet steady phase. The researchers need to focus more on the security and privacy of health data and carry the healthcare professionals along to convince the health seeks to adapt to the digital era. Collaboration between researchers and healthcare professionals, can build a bridge to fill the gap issues and illnesses can be tackled in science, and smarter practices can be adopted. Even though SHSs will enhance and increase the quality of life, if security and privacy are undermined, these benefits can be quickly outdone. On both the professionals' and researchers' end, further precautions must be taken to deal with risks and secure potentially sensitive information. As a result, the vision and long-term viability of this quickly evolving sector are dependent on the cooperation of academics, healthcare providers, and the general public.

References

Abdelrahman, A. M., Rodrigues, J. J., Mahmoud, M. M., Saleem, K., Das, A. K., Korotaev, V., & Kozlov, S. A. (2021). Software-defined networking security for private data center networks and clouds: Vulnerabilities, attacks, countermeasures, and solutions. *International Journal of Communication Systems*, *34*(4), e4706.

Abdulraheem, M., Awotunde, J. B., Jimoh, R. G., & Oladipo, I. D. (2021). An efficient lightweight cryptographic algorithm for IoT security. *Communications in Computer and Information Science*, *2021*(1350), 41–53.

AbdulRaheem, M., Balogun, G. B., Abiodun, M. K., Taofeek-Ibrahim, F. A., Tomori, A. R., Oladipo, I. D., & Awotunde, J. B. (2021). An enhanced lightweight speck system for cloud-based smart healthcare, October *Communications in Computer and Information Science*, *1455*, 363–376, 2021.

Abikoye, O. C., Ojo, U. A., Awotunde, J. B., & Ogundokun, R. O. (2020). A safe and secured iris template using steganography and cryptography. *Multimedia Tools and Applications*, *79*(31), 23483–23506.

Abouelmehdi, K., Beni-Hessane, A., & Khaloufi, H. (2018). Big healthcare data: Preserving security and privacy. *Journal of Big Data*, *5*(1), 1–18.

Abouzakhar, N. S., Jones, A., & Angelopoulou, O. (2017, June). Internet of things security: A review of risks and threats to healthcare sector. In *Proceedings of the IEEE international conference on internet of things (iThings) and IEEE green computing and communications (GreenCom) and IEEE cyber, physical and social computing (CPSCom) and IEEE smart data (SmartData)* (pp. 373–378). IEEE.

Aceto, G., Persico, V., & Pescapé, A. (2018). The role of information and communication technologies in healthcare: Taxonomies, perspectives, and challenges. *Journal of Network and Computer Applications*, *107*, 125–154.

Al-Janabi, S., Al-Shourbaji, I., Shojafar, M., & Shamshirband, S. (2017). Survey of main challenges (security and privacy) in wireless body area networks for healthcare applications. *Egyptian Informatics Journal*, *18*(2), 113–122.

Altulyan, M., Yao, L., Kanhere, S. S., Wang, X., & Huang, C. (2020). A unified framework for data integrity protection in people-centric smart cities. *Multimedia Tools and Applications*, *79*(7), 4989–5002.

Ambhati, R. K., Kota, V. K., Chaudhari, S. Y., & Jain, M. (2017, March). E-IoT: Context oriented mote prioritization for emergency IoT networks. In *Proceedings of the international conference on wireless communications, signal processing and networking (WiSPNET)* (pp. 1897–1903). IEEE.

Amrutha, K. R., Haritha, S. M., Haritha, V. M., Jensy, A. J., Sasidharan, S., & Charly, J. K. (2017). IoT based medical home. *International Journal of Computer Applications*, *165*(11).

Avancha, S., Baxi, A., & Kotz, D. (2012). Privacy in mobile technology for personal healthcare. *ACM Computing Surveys (CSUR)*, *45*(1), 1–54.

Awotunde, J. B., Abiodun, K. M., Adeniyi, E. A., Folorunso, S. O., & Jimoh, R. G. (2021, November). A deep learning-based intrusion detection technique for a secured IoMT system. In *Proceedings of the international conference on informatics and intelligent applications* (pp. 50–62). Springer, Cham.

Awotunde, J. B., Adeniyi, A. E., Ogundokun, R. O., Ajamu, G. J., & Adebayo, P. O. (2021a). MIoT-based big data analytics architecture, opportunities and challenges for enhanced telemedicine systems. *Studies in Fuzziness and Soft Computing*, *410*, 199–220, 2021.

Awotunde, J. B., Chakraborty, C., & Adeniyi, A. E. (2021b). Intrusion detection in industrial internet of things network-based on deep learning model with rule-based feature selection. *Wireless Communications and Mobile Computing*, *2021*, 7154587, 2021.

Awotunde, J. B., Jimoh, R. G., Folorunso, S. O., Adeniyi, E. A., Abiodun, K. M., & Banjo, O. O. (2021c). Privacy and security concerns in IoT-based healthcare systems. *Internet of Things*, *2021*, 105–134.

Awotunde, J. B., Ogundokun, R. O., Jimoh, R. G., Misra, S., & Aro, T. O. (2021). Machine learning algorithm for cryptocurrencies price prediction. *Studies in Computational Intelligence, 972*, 421–447.

Ayo, F. E., Folorunso, S. O., Abayomi-Alli, A. A., Adekunle, A. O., & Awotunde, J. B. (2020). Network intrusion detection is based on deep learning model optimized with rule-based hybrid feature selection. *Information Security Journal: A Global Perspective*, 1–17.

Banda, G., Bommakanti, C. K., & Mohan, H. (2016). One IoT: An IoT protocol and framework for OEMs to make IoT-enabled devices forward compatible. *Journal of Reliable Intelligent Environments, 2*(3), 131–144.

Banerjee, A., Chakraborty, C., Kumar, A., & Biswas, D. (2020). *Emerging trends in IoT and big data analytics for biomedical and health care technologies. Handbook of data science approaches for biomedical engineering* (pp. 121–152). Academic Press.

Bhardwaj, A., Mangat, V., Vig, R., Halder, S., & Conti, M. (2021). Distributed denial of service attacks in cloud: State-of-the-art of scientific and commercial solutions. *Computer Science Review, 39*, 100332.

Bhati, A., Bouras, A., Qidwai, U. A., & Belhi, A. (2020, July). Deep learning based identification of DDoS attacks in industrial application. In *2020 Fourth world conference on smart trends in systems, security and sustainability (WorldS4)* (pp. 190–196). IEEE.

Bhatia, T., & Verma, A. K. (2017). Data security in mobile cloud computing paradigm: A survey, taxonomy and open research issues. *The Journal of Supercomputing, 73*(6), 2558–2631.

Bhattacharya, S., Banerjee, S., & Chakraborty, C. (2019). Iot-based smart transportation system under real-time environment. *Big Data-Enabled Internet Things, 16*, 353–372.

Bongomin, O., Yemane, A., Kembabazi, B., Malanda, C., Mwape, M. C., Mpofu, N. S., & Tigalana, D. (2020). The hype and disruptive technologies of industry 4.0 in major industrial sectors: A state of the art.

Braham, T. G., Butakov, S., & Ruhl, R. (2018, January). Reference security architecture for body area networks in healthcare applications. In *Proceedings of the international conference on platform technology and service (PlatCon)* (pp. 1–6). IEEE.

Butt, S. A., Jamal, T., Azad, M. A., Ali, A., & Safa, N. S. (2019). A multivariant secure framework for smart mobile health application. *Transactions on Emerging Telecommunications Technologies*, e3684.

Cannady, S. (2019). A study on the efficiency of encryption algorithms for wireless sensor networks (Doctoral dissertation), Pace University.

Cappelli, D. M., Moore, A. P., & Trzeciak, R. F. (2012). *The CERT guide to insider threats: How to prevent, detect, and respond to information technology crimes (Theft, Sabotage, Fraud)*. Addison-Wesley.

Chakraborty, C., & Rodrigues, J. J. (2020). A comprehensive review on device-to-device communication paradigm: Trends, challenges and applications. *Wireless Personal Communications, 114*(1), 185–207.

Chakraborty, C., Banerjee, A., Kolekar, M. H., Garg, L., & Chakraborty, B. (2020). (Eds.) *Internet of things for healthcare technologies*. Springer.

Chakraborty, C., Gupta, B., & Ghosh, S. K. (2013). A review on telemedicine-based WBAN framework for patient monitoring. *Telemedicine and e-Health, 19*(8), 619–626.

Chakraborty, C., Roy, S., Sharma, S., Tran, T., Dwivedi, P., & Singha, M. (2021). IoT based wearable healthcare system: Post COVID-19. The impact of the COVID-19 pandemic on green societiesenvironmental sustainability (pp. 305–321).

Chenthara, S., Ahmed, K., Wang, H., & Whittaker, F. (2019). Security and privacy-preserving challenges of e-health solutions in cloud computing. *IEEE Access, 7*, 74361–74382.

Chernyshev, M., Zeadally, S., & Baig, Z. (2019). Healthcare data breaches: Implications for digital forensic readiness. *Journal of Medical Systems, 43*(1), 1–12.

Chow, C. Y., Xu, W., & He, T. (2014). *Privacy enhancing technologies for wireless sensor networks. The art of wireless sensor networks* (pp. 609–641). Berlin, Heidelberg: Springer.

Cotenescu, V., & Eftimie, S. (2017). Insider threat detection and mitigation techniques. *Scientific Bulletin "Mircea Cel Batran" Naval Academy*, *20*(1), 552.

Dang, L. M., Piran, M., Han, D., Min, K., & Moon, H. (2019). A survey on internet of things and cloud computing for healthcare. *Electronics*, *8*(7), 768.

De Santis, M., & Cacciotti, I. (2020). Wireless implantable and biodegradable sensors for postsurgery monitoring: Current status and future perspectives. *Nanotechnology*, *31*(25), 252001.

Deep, S., Zheng, X., Jolfaei, A., Yu, D., Ostovari, P., & Kashif Bashir, A. (2020). A survey of security and privacy issues in the Internet of Things from the layered context. *Transactions on Emerging Telecommunications Technologies*, e3935.

Demirkan, H. (2013). A smart healthcare systems framework. *IT Professional*, *15*(5), 38–45.

Dey, N., Ashour, A. S., Shi, F., Fong, S. J., & Sherratt, R. S. (2017). Developing residential wireless sensor networks for ECG healthcare monitoring. *IEEE Transactions on Consumer Electronics*, *63*(4), 442–449.

Dhanvijay, M. M., & Patil, S. C. (2019). Internet of things: A survey of enabling technologies in healthcare and its applications. *Computer Networks*, *153*, 113–131.

El Zouka, H. A., & Hosni, M. M. (2019). Secure IoT communications for smart healthcare monitoring system. *Internet of Things*, 100036.

Els, F., & Cilliers, L. (2018). A privacy management framework for personal electronic health records. *African Journal of Science, Technology, Innovation and Development*, *10*(6), 725–734.

Esmaeilzadeh, P. (2020). Use of AI-based tools for healthcare purposes: A survey study from consumers' perspectives. *BMC Medical Informatics and Decision Making*, *20*(1), 1–19.

Gandotra, P., Jha, R. K., & Jain, S. (2017). A survey on device-to-device (D2D) communication: Architecture and security issues. *Journal of Network and Computer Applications*, *78*, 9–29.

Glancy, F., Biros, D. P., Liang, N., & Luse, A. (2020). Classification of malicious insiders and the association of the forms of attacks. *Journal of Criminal Psychology*.

Gnjidic, D., Husband, A., & Todd, A. (2018). Challenges and innovations of delivering medicines to older adults. *Advanced Drug Delivery Reviews*, *135*, 97–105.

Gordon, W. J., & Catalini, C. (2018). Blockchain technology for healthcare: Facilitating the transition to patient-driven interoperability. *Computational and Structural Biotechnology Journal*, *16*, 224–230.

Goyal, P., Sahoo, A. K., Sharma, T. K., & Singh, P. K. (2020). Internet of things: Applications, security and privacy: A survey. Materials today: Proceedings.

Gupta, A. K., Chakraborty, C., & Gupta, B. (2019). Monitoring of epileptical patients using cloud-enabled health-IoT system. *Traitement du Signal*, *36*(5), 425–431.

Habibzadeh, H., Dinesh, K., Shishvan, O. R., Boggio-Dandry, A., Sharma, G., & Soyata, T. (2019). A survey of healthcare internet of things (HIoT): A clinical perspective. *IEEE Internet of Things Journal*, *7*(1), 53–71.

Hartmann, M., Hashmi, U. S., & Imran, A. (2019). Edge computing in smart health care systems: Review, challenges, and research directions. *Transactions on Emerging Telecommunications Technologies*, e3710.

Haus, M., Waqas, M., Ding, A. Y., Li, Y., Tarkoma, S., & Ott, J. (2017). Security and privacy in device-to-device (D2D) communication: A review. *IEEE Communications Surveys & Tutorials*, *19*(2), 1054–1079.

Huseinović, A., Mrdović, S., Bicakci, K., & Uludag, S. (2020). A survey of denial-of-service attacks and solutions in the smart grid. *IEEE Access*, *8*, 177447–177470.

Hussien, H. M., Yasin, S. M., Udzir, S. N. I., Zaidan, A. A., & Zaidan, B. B. (2019). A systematic review for enabling of develop a blockchain technology in healthcare application: Taxonomy, substantially analysis, motivations, challenges, recommendations and future direction. *Journal of Medical Systems*, *43*(10), 1–35.

Imane, S., Tomader, M., & Nabil, H. (2018, November). Comparison between CoAP and MQTT in smart healthcare and some threats. In *Proceedings of the international symposium on advanced electrical and communication technologies (ISAECT)* (pp. 1–4). IEEE.

Jameel, F., Hamid, Z., Jabeen, F., Zeadally, S., & Javed, M. A. (2018). A survey of device-to-device communications: Research issues and challenges. *IEEE Communications Surveys & Tutorials*, *20*(3), 2133–2168.

Jayabalan, M., & O'Daniel, T. (2017, November). Continuous and transparent access control framework for electronic health records: A preliminary study. In *Proceedings of the second international conferences on information technology, information systems and electrical engineering (ICITISEE)* (pp. 165–170). IEEE.

Karlof, C., Sastry, N., & Wagner, D. (2004, November). TinySec: A link layer security architecture for wireless sensor networks. In *Proceedings of the second international conference on embedded networked sensor systems* (pp. 162–175).

Lara-Nino, C. A., Diaz-Perez, A., & Morales-Sandoval, M. (2020). Lightweight elliptic curve cryptography accelerator for internet of things applications. *Ad Hoc Networks*, *103*, 102159.

Lindgren, S. (2019). Identities in critical condition: The urgent need to reevaluate the investigation and resolution of claims of medical identity theft. *Mitchell Hamline L. Rev.*, *45*, 42.

Liu, A., Dong, M., Ota, K., & Long, J. (2015). PHACK: An efficient scheme for selective forwarding attack detection in WSNs. *Sensors*, *15*(12), 30942–30963.

Lounis, K., & Zulkernine, M. (2018, October). Connection dumping vulnerability affecting Bluetooth availability. In *Proceedings of the international conference on risks and security of internet and systems* (pp. 188–204). Springer, Cham.

Maasberg, M., Zhang, X., Ko, M., Miller, S. R., & Beebe, N. L. (2020). An analysis of motive and observable behavioral indicators associated with insider cyber-sabotage and other attacks. *IEEE Engineering Management Review*, *48*(2), 151–165.

Manogaran, G., Varatharajan, R., Lopez, D., Kumar, P. M., Sundarasekar, R., & Thota, C. (2018). A new architecture of internet of things and big data ecosystem for secured smart healthcare monitoring and alerting system. *Future Generation Computer Systems*, *82*, 375–387.

National Research Council. (1997). For the record: Protecting electronic health information.

Navarro-Ortiz, J., Sendra, S., Ameigeiras, P., & Lopez-Soler, J. M. (2018). Integration of LoRaWAN and 4G/5G for the Industrial Internet of Things. *IEEE Communications Magazine*, *56*(2), 60–67.

Newaz, A. I., Sikder, A. K., Rahman, M. A., & Uluagac, A. S. (2019, October). Healthguard: A machine learning-based security framework for smart healthcare systems. In *Proceedings of the sixth international conference on social networks analysis, management and security (SNAMS)* (pp. 389–396). IEEE.

Nurse, J. R., Buckley, O., Legg, P. A., Goldsmith, M., Creese, S., Wright, G. R., & Whitty, M. (2014, May). Understanding insider threat: A framework for characterising attacks. In *Proceedings of the security and privacy workshops* (pp. 214–228). IEEE.

Nurse, J. R., Erola, A., Agrafiotis, I., Goldsmith, M., & Creese, S. (2015, September). Smart insiders: Exploring the threat from insiders using the internet-of-things. In *Proceedings of the international workshop on secure internet of things (SIoT)* (pp. 5–14). IEEE.

Ogundokun, R. O., Awotunde, J. B., Sadiku, P., Adeniyi, E. A., Abiodun, M., & Dauda, O. I. (2021). An enhanced intrusion detection system using particle swarm optimization feature extraction technique. *Procedia Computer Science*, *193*, 504–512.

Oueida, S., Aloqaily, M., & Ionescu, S. (2019). A smart healthcare reward model for resource allocation in smart city. *Multimedia Tools and Applications*, *78*(17), 24573–24594.

Patel, R., Sinha, N., Raj, K., Prasad, D., & Nath, V. (2018, November). Smart healthcare system using IoT. In *Proceedings of the international conference on nanoelectronics, circuits and communication systems* (pp. 149–156). Springer, Singapore.

Porambage, P., Okwuibe, J., Liyanage, M., Ylianttila, M., & Taleb, T. (2018). Survey on multi-access edge computing for internet of things realization. *IEEE Communications Surveys & Tutorials*, *20*(4), 2961–2991.

Prabhu, B., Balakumar, N., & Antony, A. (2017). Wireless sensor network based smart environment applications. Wireless sensor network based smart environment applications (January 31, 2017). *IJIRT*, *3*(8).

Prakash, R., & Ganesh, A. B. (2019). Internet of things (IoT) enabled wireless sensor network for physiological data acquisition. In *Proceedings of the international conference on intelligent computing and applications* (pp. 163–170). Springer, Singapore.

Qadri, Y. A., Nauman, A., Zikria, Y. B., Vasilakos, A. V., & Kim, S. W. (2020). The future of healthcare internet of things: A survey of emerging technologies. *IEEE Communications Surveys & Tutorials*, *22*(2), 1121–1167.

Rathee, G., Sharma, A., Kumar, R., & Iqbal, R. (2019). A secure communicating things network framework for industrial IoT using blockchain technology. *Ad Hoc Networks*, *94*, 101933.

Razzaq, M. A., Gill, S. H., Qureshi, M. A., & Ullah, S. (2017). Security issues in the Internet of Things (IoT): A comprehensive study. *International Journal of Advanced Computer Science and Applications*, *8*(6), 383.

Reinbrecht, C., Susin, A., Bossuet, L., Sigl, G., & Sepúlveda, J. (2017). Timing attack on NoC-based systems: Prime + Probe attack and NoC-based protection. *Microprocessors and Microsystems*, *52*, 556–565.

Rivera-Segarra, E., Varas-Díaz, N., & Santos-Figueroa, A. (2019). "That's all Fake": Health professionals stigma and physical healthcare of people living with Serious Mental Illness. *PLoS One*, *14*(12), e0226401.

Rughoobur, P., & Nagowah, L. (2017, December). A lightweight replay attack detection framework for battery depended IoT devices designed for healthcare. In *Proceedings of the international conference on infocom technologies and unmanned systems (trends and future directions) (ICTUS)* (pp. 811–817). IEEE.

Salim, M. M., Rathore, S., & Park, J. H. (2019). Distributed denial of service attacks and its defenses in IoT: A survey. *The Journal of Supercomputing*, 1–44.

Sarkar, K. R. (2010). Assessing insider threats to information security using technical, behavioural and organisational measures. *Information Security Technical Report*, *15*(3), 112–133.

Shang, Y., Pan, C., Yang, X., Zhong, M., Shang, X., Wu, Z., ... Sang, L. (2020). Management of critically ill patients with COVID-19 in ICU: Statement from front-line intensive care experts in Wuhan, China. *Annals of Intensive Care*, *10*(1), 1–24.

Sithole, T. G. (2019). Assessing resilience of public sector information systems against cyber threats and attacks. A South African perspective (Doctoral dissertation), University of Pretoria.

Spender, A., Bullen, C., Altmann-Richer, L., Cripps, J., Duffy, R., Falkous, C., ... Yeap, W. (2019). Wearables and the internet of things: Considerations for the life and health insurance industry. *British Actuarial Journal*, *24*.

Srinivasan, V., Stankovic, J., & Whitehouse, K. (2008, September). Protecting your daily in-home activity information from a wireless snooping attack. In *Proceedings of the tenth international conference on Ubiquitous computing* (pp. 202–211).

Stellios, I., Kotzanikolaou, P., Psarakis, M., Alcaraz, C., & Lopez, J. (2018). A survey of IoT-enabled cyberattacks: Assessing attack paths to critical infrastructures and services. *IEEE Communications Surveys & Tutorials*, *20*(4), 3453–3495.

ul Sami, I., Asif, M., Bin Ahmad, M., & Ullah, R. (2018). DoS/DDoS detection for e-healthcare in internet of things. *International Journal of Advanced Computer Science and Applications*, *9*(1), 297–300.

Watson, K., & Payne, D. M. (2020). Ethical practice in sharing and mining medical data. *Journal of Information, Communication and Ethics in Society*.

Wickramage, C., Fidge, C., Sahama, T., & Wong, R. (2017, December). Challenges for log based detection of privacy violations during healthcare emergencies. In *GLOBECOM 2017-2017 IEEE global communications conference* (pp. 1–6). IEEE.

Xie, J., Tang, H., Huang, T., Yu, F. R., Xie, R., Liu, J., & Liu, Y. (2019). A survey of blockchain technology applied to smart cities: Research issues and challenges. *IEEE Communications Surveys & Tutorials*, *21*(3), 2794–2830.

Xu, Q., Zheng, R., Saad, W., & Han, Z. (2015). Device fingerprinting in wireless networks: Challenges and opportunities. *IEEE Communications Surveys & Tutorials*, *18*(1), 94–104.

Yaacoub, J. P. A., Noura, M., Noura, H. N., Salman, O., Yaacoub, E., Couturier, R., & Chehab, A. (2020). Securing internet of medical things systems: limitations, issues and recommendations. *Future Generation Computer Systems*, *105*, 581–606.

Yu, M., Zhuge, J., Cao, M., Shi, Z., & Jiang, L. (2020). A survey of security vulnerability analysis, discovery, detection, and mitigation on IoT devices. *Future Internet*, *12*(2), 27.

Yuan, S., & Wu, X. (2021). Deep learning for insider threat detection: Review, challenges and opportunities. *Computers & Security*, 102221.

Zhang, Q., & Zhang, W. (2019). Accurate detection of selective forwarding attack in wireless sensor networks. *International Journal of Distributed Sensor Networks*, *15*(1), 1550147718824008.

Index

Note: Page numbers followed by "*f*" and "*t*" refer to figures and tables, respectively.

T

U

V

Printed in the United States
by Baker & Taylor Publisher Services